THE METEOROLOGY
OF POSIDONIUS

This volume describes the meteorology of the Stoic philosopher Posidonius from the existing fragments, and discusses his relation to earlier thinkers on this subject, as well as the methods he used to obtain information about and to find explanations of meteorological phenomena.

The book examines ancient meteorology, an aspect of ancient thought largely neglected by scholars. Hall produces a detailed account of how Posidonius and other ancient thinkers approached and attempted to explain meteorological phenomena – phenomena familiar to everyone. which could not be ignored in attempts to understand the natural world, but were difficult to explain satisfactorily and convincingly despite the efforts of important ancient thinkers. The volume explores particular classes of phenomena, including climatic events and geological processes, providing a comprehensive overview of Posidonius' ideas on these topics. Concluding chapters allow for an assessment of Posidonius' particular contribution to the field and his influence on later writers working on this subject.

The Meteorology of Posidonius provides an important resource for students and scholars working on ancient philosophy and ancient science, particularly ancient meteorology.

J.J. Hall read Classics at Trinity College, Cambridge, and then did research, gaining a Ph.D. for a dissertation on ancient theories of wind. His career was spent on the staff of Cambridge University Library. Now retired, and still in Cambridge, U.K., he has written this book.

Issues in Ancient Philosophy
Series editor: George Boys-Stones, University of Toronto, Canada

Routledge's *Issues in Ancient Philosophy* exists to bring fresh light to the central themes of ancient philosophy through original studies which focus especially on texts and authors which lie outside the central 'canon'. Contributions to the series are characterised by rigorous scholarship presented in an accessible manner; they are designed to be essential and invigorating reading for all advanced students in the field of ancient philosophy.

The Stoic Doctrine of Providence
A Study of Its Development and of Some of Its Major Issues
Bernard Collette

Investigating the Relationship Between Aristotle's *Eudemian* and *Nicomachean Ethics*
Giulio Di Basilio

Thales the Measurer
Livio Rossetti

Gorgias's Thought
An Epistemological Reading
Erminia Di Iulio

The Theology of the *Epinomis*
Vera Calchi

Cosmos and Perception in Plato's Timaeus
In the Eye of the Cognitive Storm
Mark Eli Kalderon

The Meteorology of Posidonius
J.J. Hall

For more information about this series, please visit: https://www.routledge.com/Issues-in-Ancient-Philosophy/book-series/ANCIENTPHIL

The Meteorology of Posidonius

J.J. Hall

Routledge
Taylor & Francis Group

LONDON AND NEW YORK

First published 2024
by Routledge
4 Park Square, Milton Park, Abingdon, Oxon OX14 4RN

and by Routledge
605 Third Avenue, New York, NY 10158

Routledge is an imprint of the Taylor & Francis Group, an informa business

British Library Cataloguing-in-Publication Data
A catalogue record for this book is available from the British Library

ISBN: 978-0-367-02372-0 (hbk)
ISBN: 978-1-032-53030-7 (pbk)
ISBN: 978-0-429-39993-0 (ebk)

DOI: 10.4324/9780429399930

Typeset in Times New Roman
by SPi Technologies India Pvt Ltd (Straive)

In memoriam magistrorum meorum

F.G. Turner
A.N.W. Saunders
F.H. Sandbach

Contents

List of figures   *ix*
Preface   *x*
Acknowledgements   *xii*
Abbreviations and methods of citation   *xiii*

1 The definition of "meteorology" used in this book   1

2 The biography and later reputation of Posidonius   4

3 Sources for the study of Posidonius' meteorology   10

4 The history of Greek meteorology before Posidonius   20

5 Earlier authors on meteorology used by Posidonius   35

6 The region in which meteorological phenomena occur   41

7 Climatic zones   54

8 Thunder and lightning   65

9 Lights in the sky: comets, the Milky Way and
   other phenomena   75

10 Exhalations   89

11 Winds   96

12 Earthquakes and volcanoes   113

13 The sea and its tides   128

14 Rain, snow, hail and cloud   142

15 Rivers: the Nile floods   152

16 Rainbows, haloes and mock-suns   160

17 Weather prediction and divination   171

18 Meteorology and providence   181

19 Epicurean meteorology compared with that of the Stoics   184

20 The place of meteorology among the different branches of
    knowledge   191

21 Sources and methods in Posidonius' meteorology   196

22 Assessment of the meteorology of Posidonius
    and his successors   212

    Bibliography   *221*
    Index   *229*

Figures

11.1 The wind-rose, in Posidonius and related authors   104
13.1 Suggested interpretation of Posidonius' explanation of
      the daily cycle of the tides, as described by Priscianus
      (Posidonius F219 lines 105–9 EK)   135

Preface

I have been working in this project for nearly a decade. It began after I had attended a conference on Posidonius held in Cambridge in 2013, at which very little was said about his meteorology. Ancient meteorology had been the subject of my Ph.D. dissertation years before, and the omission of the subject at that conference encouraged me to look into Posidonius' meteorology for myself. Since then this project has gradually expanded from the article I originally intended into this quite lengthy book. I must thank the publishers for their patience as I have repeatedly missed promised dates of completion. I have decided that I must now regard the book as finished, although I am aware of more work which I ideally should have done. One regret I would like to express is that the new edition of the *Placita* of Aëtius by Mansfeld and Runia (2020) was brought to my attention too late for me to make full use of it, so that in places I have continued to rely on the old edition of Diels (1879).

I most warmly acknowledge the encouragement and advice I have received from Liba Taub during the development of this project. I thank David Sedley for his helpful criticisms of the first draft of what has become Chapters 20 and 21 of this book (such faults as those chapters now contain are of course my responsibility), and George Boys-Stones as editor of the series *Issues in ancient philosophy* for his comments on the text of the book as submitted to him in the spring of 2022. I must also thank Max Leventhal for his editorial assistance given some years ago at an earlier stage of the project, and Charlie Pemberton for similar assistance given recently. I thank Cambridge University Press for permission to use many quotations from I.G. Kidd's translation of the fragments of Posidonius and I thank the University Presses of Harvard and Oxford for permission to quote extensively from translations of ancient texts of which they hold the copyright (more formal acknowledgement of Harvard's permission is given below). Finally, I thank my wife for her patience with the amount of time I have spent on this project.

Translations of ancient texts not attributed to any author are my own.

I should like to add a further point here. As Posidonius' own works are lost, there is nearly always room for at least a measure of doubt whether he really said or did the things which our sources say he said or did. I hope I may be permitted to say this here, as something which applies throughout this book, and may be excused when I sometimes say in what follows that Posidonius said or did this or that, and do not say on every occasion that the evidence indicates that he probably said it or did it.

J.J. Hall
Cambridge, December 2022

Acknowledgements

I append formal statements of permission to quote material owned by Harvard University Press:

ARISTOTLE, VOL. III, translated by E. S. Forster and D. J. Furley, Loeb Classical Library, Volume 400, Cambridge, Mass.: Harvard University Press, first published 1955. Loeb Classical Library ® is a registered trademark of the President and Fellows of Harvard College. Used by permission. All rights reserved.

ARISTOTLE, VOL. VII, translated by H. D. P. Lee, Loeb Classical Library Volume 397, Cambridge, Mass.: Harvard University Press, first published 1952. Loeb Classical Library ® is a registered trademark of the President and Fellows of Harvard College. Used by permission. All rights reserved.

Abbreviations and methods of citation

I use the following abbreviations and unusual citation forms:

De mundo =   Pseudo-Aristotle, *De mundo* (usually cited without author's name).

DK =   H. Diels and W. Kranz, *Die Fragmente der Vorsokratiker. Sechste Auflage*. 3 vols. Berlin: Weidmann, 1951–2 (and later editions with the same pagination).

EK =   L. Edelstein, and I.G. Kidd, eds. *Posidonius I: The fragments*. 2nd ed. Cambridge: Cambridge University Press, 1989.

LSJ =   H.G. Liddell, R. Scott and H.S. Jones. *A Greek-English lexicon. New edition*. Oxford: Oxford University Press, 1925–40.

Mete. =   *Meteorologica* (of Aristotle; sometimes cited without author's name).

NQ =   *Naturales quaestiones* (of Seneca).

SVF =   H.F.A. von Arnim, *Stoicorum veterum fragmenta*. 4 vols. Leipzig: Teubner, 1903-24.

With the above exceptions I cite modern works by the name(s) of author(s) or editor(s) followed by the date of publication, e.g. Lee (1952). If no author or editor is named, or if the work is a reference work better known by its title than by the name of the editor, then I cite the work by the title followed by the date of publication. Details of these works will be found in the bibliography.

I normally cite ancient works by author and title, either giving the full title in Latin or English (whichever is more commonly used), or the abbreviation of the title used by LSJ or by P.G.W. Glare, ed., *Oxford Latin dictionary*, Oxford: Oxford University Press, 1982.

Ancient authors of only one surviving work (e.g. Diogenes Laertius, Lucretius, Strabo) I normally cite by author's name only, followed by the number of the book and chapter (or whatever numerical designation is commonly used in referring to a particular passage), e.g., Strabo II.2.2.

I cite simply as "Aëtius" the collection of *Placita philosophorum* published by Diels (1879) as the *Placita* of Aëtius and by Mansfeld and Runia (2020) in *Aëtiana V*, and (following LSJ) I cite the *Refutatio omnium haeresium* as "Hippolytus, *Haer.*" This is purely for convenience and implies no view about the actual authorship of these works.

1 The definition of "meteorology" used in this book

Meteorology, "The study of, or the science that treats of, the motions and phenomena of the atmosphere", as the *Oxford English Dictionary*[1] defines it, deals with phenomena that were familiar and important in classical antiquity, as in every other period of human history. Some of these phenomena were essential to the ancients' way of life: rain to water their crops, wind to propel their ships. Some were a danger, like lightning strikes. Some, if not obviously dangerous, yet were weird, like rainbows and haloes round the sun. Well known as they were, few of them had any obvious cause. They lacked the regularity of the movements of sun, moon, planets and stars, which ancient astronomers studied with considerable success, and which suggested the conclusion that the atmospheric phenomena had a different nature and cause from the astronomical. No living being appeared to be controlling atmospheric phenomena; certainly human beings had no control of them.

It was natural to wonder about the causes of these phenomena, and many Greek thinkers did. Some ancient physical theorising – for instance, the theory of the four elements, earth, air, fire and water – offered a general explanation of physical change but it could not explain the details of actual changes. Students of meteorology were seeking the causes of observed phenomena; whatever hypothesis a meteorological thinker adopted, it had to explain a phenomenon which was known with relative precision. Of course this is not the only study pursued in antiquity of which this is true, but it is a study which goes back to the beginnings of Greek philosophy, and was pursued by some of the greatest thinkers and writers – by Anaximander, Anaxagoras and Democritus (to name just three of many pre-Aristotelians active in this field); by Aristotle and Theophrastus; by Epicurus; in later antiquity, by Lucretius and Seneca among Latin writers; and, among the Greeks, by Posidonius. It must be worthwhile to study how he and other ancient thinkers dealt with meteorological phenomena.

In this study I shall I shall take "meteorology" as including a number of phenomena which do not come within the modern definition, but which share the features which atmospheric phenomena had in antiquity, of being precisely known phenomena which as a rule occurred irregularly and with no obvious cause: I use the word in the sense which Aristotle claims that his predecessors

DOI: 10.4324/9780429399930-1

had given to its Greek original, μετεωρολογία,[2] and which he himself uses in his *Meteorologica*, Books I–III; that is, I take it as covering the nature of the sublunary region down to the earth's surface; "exhalations" from water and earth; shooting stars, the aurora borealis, comets, and the Milky Way; rain and other sorts of precipitation; wind; rivers; the sea; earthquakes; thunder, lightning and related phenomena; and optical phenomena such as rainbows. I shall also say something on related subjects which Aristotle in *Mete.* I–III mentions only in passing: climatic "zones"; the seasons; and weather-signs. (I do not include minerals, which Aristotle discusses briefly at the end of *Mete.* III[3]; also, some subjects which come within Aristotle's definition, such as the aurora borealis, will be hardly mentioned in this work, because we lack evidence of Posidonius' views.)

It is not part of my purpose to discuss who in antiquity, other than Aristotle, regarded this range of topics as constituting μετεωρολογία; but it is clear that there were later ancient writers who treated very nearly this range of topics as a discrete subject, to be dealt with together in one work or one section of a work. Seneca's *Naturales quaestiones* covers just this range of phenomena, except that he mentions the sea only very briefly and omits the Milky Way; the same is true of Chapter 4 of the pseudo-Aristotelian *De mundo*, though that chapter omits the Milky Way and hardly mentions rivers; in the account of Stoic doctrine in Diogenes Laertius VII there is one section, chapters 151–4, which deals with nearly the same range of phenomena: it deals with atmospheric happenings (wind, rainbows, precipitation, and thunder, lightning and related phenomena) together with comets and phenomena like them, and earthquakes. (Definitions of the four seasons are also included but the aurora, the Milky Way, rivers and the sea are not mentioned – however, it is natural, in such a brief summary, that some relevant topics should escape mention.) Book III of the *Placita* of Aëtius covers very nearly the same range of subjects as *Mete.* I–III.

None of these works uses μετεωρολογία to define this range of topics. Nearest is Aëtius, who uses τὰ μετάρσια and defines this as "things from the circle of the moon reaching to the position of the earth"[4]; Seneca uses "sublimia", defining this as "things happening between sky and earth"[5]; the *De mundo* says "the most notable phenomena in and about the inhabited world",[6] Diogenes "things happening in air".[7] But they all treat this range of topics as a discrete subject.

Whether Posidonius did the same we cannot be certain: our knowledge of his meteorology depends on reports of his views on individual phenomena made by a variety of ancient authors. But it seems very likely: as we shall see, Posidonius made extensive use of Aristotle's *Meteorologica*; there is clearly a relationship between Posidonius' meteorology and that of the *De mundo*, though there is room for doubt which is the earlier[8]; Seneca in *NQ* often mentions Posidonius, and Diogenes Laertius VII.151–4 cites him three times. (There is no obvious close connection between Posidonius and Aëtius.) Posidonius clearly did not follow Aristotle in calling this range of topics μετεωρολογία, since he used μετέωρος in a wider sense: we have citations of works of his which include some form of μετέωρος in the title and which deal, not only with rainbows, but also with the definition of *cosmos* and of *surface* (ἐπιφάνεια), with the sun, and with the difference between a philosophical and a mathematical astronomy.[9] But it is still likely that, like Seneca, the *De mundo* and Diogenes Laertius VII.151–4, he regarded the topics of Aristotle's μετεωρολογία as constituting a discrete subject. Even if he did not, it still makes sense, for the reasons given above, to deal in a single study with his treatment of all the phenomena which come within Aristotle's definition of μετεωρολογία.

Notes

1 *Oxford English Dictionary* (1989).
2 *Mete*. 338a26ff.
3 *Mete*. 378a17ff. Minerals and the like are presumably included in Aristotle's description of the subject in *Mete*. 338a26ff, when he mentions γῆς ὅσα μέρη καὶ εἴδη, "the various kinds and parts of earth" (*Mete*. 338b25, with the translation of Lee [1952]), but the subject is not seriously discussed in *Mete*. I–III.
4 Book III proœmium, τὰ ἀπὸ τοῦ κύκλου τῆς σελήνης καθήκοντα μέχρι πρὸς τῆν θέσιν τῆς γῆς.
5 *NQ* II.1.1–2, "inter caelum terramque versantia". II.1.3 explains that earthquakes are included because they are due to wind.
6 *De mundo* 394a7, τῶν ἀξιολογωτάτων ἐν αὐτῇ [sc. οἰκουμένῃ] καὶ περὶ αὐτὴν παθῶν. I use Furley's translation (in Forster and Furley [1955]).
7 Diogenes Laertius VII.151, τῶν ἐν ἀέρι γινομένων.
8 See below, Chapter 3, p. 13–14.
9 Posidonius F14–18 EK.

2 The biography and later reputation of Posidonius

Posidonius was born, almost certainly, between 140 and 130 B.C.: definite information is lacking, but we know he was a pupil of Panaetius, who is believed to have died about 110 B.C.,[1] so he can hardly have been born after 130. We know he was still alive in 60 B.C., because Cicero tells Atticus, in a letter of that year, that Posidonius has declined a request that he should write a book on Cicero's consulship.[2] The precise dates sometimes given for his lifetime, 135–51 B.C., depend, as Kidd shows, on unreliable and conjecturally emended sources,[3] but may well be right. He was born at Apamea in Syria.[4] Apart from this, nothing, not even his father's name, is recorded about his origins. However, Apamea was originally established by Macedonian soldiers and had been an important Seleucid military base,[5] so its Greek population was probably substantial, and Posidonius was presumably born a member of it, since he had a Greek name, and must have had a Greek education to have become, as the dates suggest, a pupil of Panaetius at an early age. Nothing in what we know of his biography, therefore, makes it likely that he was seriously influenced by the ideas of non-Greek Syrians,[6] especially considering that he must have left Apamea by the time he became the pupil of Panaetius, and Cicero includes him in a list of philosophers who, after leaving their original home, never returned to it.[7]

Panaetius was head of the Stoic school in Athens from, probably, 129 B.C. until his death in about 110 B.C.,[8] and Cicero tells us that Posidonius was his pupil, presumably in Athens.[9] At a date and for reasons which are not known, Posidonius left Athens, moved to Rhodes and began to teach philosophy there.[10] He became a citizen of Rhodes and took part in its public life.[11] In 87–86 B.C. he went on an embassy from Rhodes to Rome, about which we know only that he talked with Caius Marius during the latter's last illness[12]: it was not unprecedented for a distinguished philosopher or other teacher to be an ambassador from a Greek city to Rome.[13] Posidonius also, at an unknown date, held the office of *prytanis*,[14] the highest Rhodian magistracy[15]; some *prytaneis*, we know, were politically important at Rhodes,[16] but as there were five or six of them at any one time, and the term of office was only six months,[17] it is unlikely that they all were. Of Posidonius' tenure of this office we know only

DOI: 10.4324/9780429399930-2

what Strabo records Posidonius himself as saying, that at the time when he held it a bituminous earth was found in Rhodes, which destroyed a vermin that attacked vines, provided it was mixed with a great quantity of oil.[18] Presumably his attention was especially drawn to the problem because he was *prytanis*, and perhaps he was in some way responsible for providing the cure.

Besides travelling to Italy, he also made at least one journey into more westerly parts of Europe, as we know from Strabo, who records that Posidonius said he had been in, or seen things in, a number of places there. In this way we know that he had been in Liguria[19]; he had observed the customs of Celts in Transalpine Gaul who had not adopted a Greco-Roman life style[20] (archaeology suggests that this did not require him to travel far into Gaul)[21]; he had spent 30 days in Gadeira (Cadiz)[22]; he had seen gold and silver mines in southern Spain[23]; and he had returned from Spain to Italy by sea, a journey long delayed by contrary winds, in the course of which he had seen the coast of North Africa, the Balearic Islands and Sardinia.[24] The date of this journey is most likely to have been in the nineties B.C.[25] In the course of his travels he recorded observations, meteorological, astronomical, zoological, ethnographic and others: those that are meteorological – in the broad sense in which I am using the term – will be discussed in several chapters of this book. He may well have made journeys to other places besides those just mentioned: the fragments of his works include information about many other places around the Mediterranean; but evidence of autopsy is lacking.

It was as a Stoic philosopher that Posidonius made his name: when his professional character is mentioned, it is as a philosopher, or as a Stoic, or as both, that ancient authors – such as Cicero, Pliny the Elder, Plutarch, Athenaeus – almost always characterise him.[26] Most of his teaching must have been of his own brand of Stoic philosophy. Of his pupils, by far the most famous was Cicero, who, though not a Stoic himself, always speaks of Posidonius with respect.[27] Another famous Roman who came to hear Posidonius was Pompey, who came to him in 66 B.C., when in command of the Roman campaign against piracy, and in 62 B.C. after defeating Mithridates; he treated Posidonius (and was treated by him) with great respect, but obviously was not a regular pupil.[28] Not much is recorded about his other pupils and associates, but we do know of two who shared his interest in meteorology (in the wide sense). Seneca mentions Asclepiodotus as a pupil of Posidonius interested in volcanoes and earthquakes.[29] Cicero writes to Atticus (after Posidonius' death) that he has asked Athenodorus Calvus to send him a summary (τὰ κεφάλαια) of a work on ethics by Posidonius[30]; this Athenodorus presumably had known Posidonius well, and is probably the same Athenodorus whose name Strabo couples with that of Posidonius as an expert on tides.[31] About Posidonius' school after his death, and about his family life, we have only a notice in the *Suda* which tells that he had a daughter (name not recorded) who married one Menecrates of Nysa and bore him a son, Jason, described as a philosopher and an author (evidently a minor one) who was a pupil of Posidonius and succeeded him as head of his school.[32]

Posidonius' reputation was as a Stoic philosopher, and many of his works were on matters central to Stoic philosophy. Thus, he wrote on epistemology[33] and on logic;[34] he evidently wrote much on the principles of Stoic physics (on the nature of the cosmos and its periodical destruction and rebirth, on its government by Providence, and so on[35]), and much also on ethics and human psychology.[36] But he was a man of extremely wide interests, who also wrote on subjects which most of his contemporaries would have regarded as not proper work, or not necessary work, for a philosopher. Thus, he wrote a major work on the history of the Greeks, the Romans and the neighbouring peoples, starting where Polybius had left off (146 B.C.) and going on at least until the mid-eighties B.C.[37] – a work which included much that we might class as anthropological, on the social customs of the peoples he wrote about.[38] He wrote on divination (at considerable length)[39]; on geography[40]; on geometry[41]; on astronomy, including mathematical calculations about sizes and distances which in theory he regarded as outside a philosopher's province[42]; and, as the following chapters of this book will show, he wrote on meteorology – a subject discussed by many philosophers down to the time of Aristotle, Theophrastus and Epicurus, but which in the two following centuries philosophers had largely ignored.

His public reputation in his lifetime was high, as is shown by the Rhodians' choice of him, a non-native, as an ambassador to Rome, and by the desire to hear him shown by Pompey when a victorious Roman general. His influence after his death was great, but perhaps less than his lifetime reputation might have suggested. Apart from the non-Stoic Cicero, no significant philosopher or writer on philosophy is recorded among his pupils, and his school did not remain important. But many writers thought him important enough to be worth quoting, though often not without criticism, and often on subjects not central to philosophy. Many of Cicero's references to Posidonius are to his views on divination, about which Cicero disagreed with him.[43] A generation later, Strabo in his *Geographica* often quotes and often criticises Posidonius, as we shall see. In the 1st century A.D. Seneca has many references to him, on various matters, often ethical, in the *Epistulae*, and on meteorology in the *Naturales quaestiones*.[44] In the following two centuries Galen quotes with approval his views on human psychology and on ethics,[45] and he is an important source for the astronomy of Cleomedes.[46] Diogenes Laertius cites him frequently in his account of Stoic philosophy in his Book VII.39–160, especially on Stoic physics.[47] None of these men regarded Posidonius as (say) Lucretius regarded Epicurus, as the fount of all wisdom. Even the professed Stoics – Strabo, Seneca and Cleomedes – were ready to criticise and disagree with him, as the following chapters will show. But to all of them he was a man whose views were to be taken seriously.

His reputation lasted into late antiquity. Later in this work I shall discuss important information about him found in 6th-century authors: Simplicius, who gives details of the distinction he drew between natural philosophy as regards heavenly bodies and a mathematical astronomy, and Priscianus Lydus, who gives the most detailed surviving account of his theory of tides. Simplicius states that his source is not Posidonius' own work,[43] and Kidd writes of Priscianus, "there is no clear indication that he had read Posidonius himself".[49] Posidonius' own works may already have been lost by the sixth century, and certainly none of them survives today.

In some ways Posidonius was a rather conservative thinker, defending aspects of traditional Stoic teaching which one might have thought were particularly open to question, and which Panaetius, Posidonius' teacher, had in fact doubted or rejected: the reality of divination, and the doctrine of the periodical destruction of the cosmos by fire, followed by its rebirth.[50] In another important respect he moved with the times. For much of the 3rd and 2nd centuries B.C., Plato and Aristotle had received relatively little attention from philosophers: the heads of Plato's Academy, starting with Arcesilaus in about 265 B.C., were teachers of Scepticism rather than Platonism, and Aristotle's Peripatos was in decline after the death of Strato in about 270.[51] From the late-2nd century onwards there came a change, and philosophers, even if not committed Platonists or Peripatetics, began to take Plato and Aristotle seriously.[52] In this, Panaetius was a pioneer: Cicero says of him "semper ... habuit in ore Platonem, Aristotelem, Xenocratem, Theophrastum, Dicaearchum, ut ipsius scripta declarant", "he was always talking about Plato, Aristotle, Xenocrates [a pupil of Plato], Theophrastus, Dicaearchus [two early Peripatetics], as his own writings tell."[53] The *Index Stoicorum Herculanensis* calls Panaetius ἰσχυρῶς φιλοπλάτων καὶ φιλοαριστοτέλης "strongly a lover of Plato and Aristotle".[54] Posidonius followed Panaetius in this. Galen writes of his approval of Plato, θαυμάζων τὸν ἄνδρα καὶ θεῖον ἀποκαλῶν, "admiring the man and calling him divine",[55] and Strabo says of Posidonius πολὺ ... ἐστι τὸ αἰτιολογικὸν παρὰ αὐτῷ καὶ τὸ Ἀριστοτελίζον, "there is much enquiry into causes in him, that is, 'Aristotelising'".[56] Several other passages of Galen speak of Posidonius' approval of Plato's views about human emotions and psychology,[57] and Aristotle is coupled with Plato in two of them.[58] Plato, who had minimal interest in meteorology, will not often be mentioned in this book, but Aristotle will be mentioned frequently as the source of opinions of Posidonius on meteorological matters.

In making a serious study of meteorology Posidonius was, among the thinkers of his time, striking out on a line of his own; but before we examine his views on the subject it will be well to say first something about our sources of information about his meteorology, and about ancient meteorology generally.

Notes

1 Kidd (1988) 8. For the probable date of Panaetius' death see van Straaten (1946) 23.
2 Cicero, *Att.* II.1.2 = Posidonius T34 EK; Kidd (1988) 26.
3 Kidd (1988) 8–9.
4 See, e.g., Strabo XIV.2.13; Athenaeus VI.252E (= Posidonius T2a and 2b EK).
5 Strabo XVI.2.10; Jones (1971) 243.
6 Kidd (1988) 7.
7 Cicero, *Tusc.* V.107 (= Posidonius T3 EK).

8   Van Straaten (1946) 19–23.
9   Cicero, *Off.* III.8, and *Div.* I.6 (= Posidonius T9 and T10 EK); Kidd (1988) 12–13.
10 Posidonius teaching philosophy at Rhodes: Plutarch, *Cicero*, 4.5 (= Posidonius T29 EK); Kidd (1988) 7–8.
11 Strabo XIV.2.13 (Posidonius T2a EK) ἐπολιτεύσατο ... ἐν Ῥόδῳ. Kidd (1999) translates "was a citizen of Rhodes". The phrase could mean "took part in politics at Rhodes", but that implies citizenship.
12 Plutarch, *Marius*, 45.7 (Posidonius T28 EK). On this embassy see Kidd (1988) 22–3. The connection with Marius' death fixes the date of the embassy.
13 A famous occasion was in 155 B.C., when the philosophers Carneades, Diogenes of Babylon, and Critolaus came as ambassadors from Athens to Rome (Cicero, *Tusc.* IV.3.5; Plutarch, *Cato Maior* 22; Long [1974] 94.) It was probably a few years after 86 B.C. that Apollonius Molon, a famous teacher of rhetoric, was an ambassador from Rhodes to Rome (Cicero, *Brutus* 312; Kidd [1988] 23).
14 Strabo VII.5.8 (Posidonius T27 and F235 EK).
15 Livy XLII.45.4 "summo magistratu". Compare Plutarch, *Praecepta rei publicae gerendae* 813D. On this office and Posidonius' tenure of it see Kidd (1988) 21–2.
16 See, e.g., Polybius XXVII.3.3 and Livy XLII.45.4, on Hegesilochus.
17 Gschnitzer (1973) 766–9.
18 Strabo VII.5.8 (Posidonius F235 EK). See Kidd (1988) 828.
19 Strabo III.4.17 (Posidonius T23 EK).
20 Strabo IV.4.5 (Posidonius T19 EK).
21 Kidd (1988) 17–18 and 937.
22 Strabo III.1.5 (Posidonius T15 EK).
23 Strabo III.2.9 (Posidonius T20 EK). That Posidonius had seen the mines is implied by his saying "in general anyone who saw the area would have said" (Kidd's [1999] translation of καθόλου ἂν εἶπε ... ἰδών τις τοὺς τόπους).
24 Strabo XVII.3.4 and III.2.5 (Posidonius T21 and T22 EK).
25 Kidd (1988) 16–17.
26 Looking just at the Testimonia section in EK, he is "philosophus" to Cicero at *Tusc* II.61 and V.107 (T38 and T3 EK), "maximum Stoicorum" at *Hortensius* fr. 50 Grilli (T33 EK); at Pliny *Nat.* VII.112 (T36 EK) he is "sapientiae professione clari" (which Kidd [1999] translates as "the famous professor of philosophy"); Plutarch calls him φιλόσοφος at *Marius* 45.7 and *Cicero* 4.5 (T28 and T29 EK); Galen in *De sequela* contrasts him with other Stoics, thus implying that he was one (T58 EK); to Athenaeus he is φιλόσοφος at IX.401A, Στωϊκός at XII.549D–E, τὸν ἀπὸ τῆς Στοᾶς φιλόσοφον ("the philosopher from the Stoa", i.e., "of the Stoic school") at XIV.657E (T24, T7 and T8 EK). There are variations: Plutarch, *Pompeius* 42.5 (T39 EK) includes him among "sophists" (σοφιστῶν), apparently using the word to cover both philosophers and rhetoricians (see Kidd [1988] 29–30); Pseudo-Lucian, *Longaevi* (T4 EK) calls him "philosopher and historian" (φιλόσοφός τε ἅμα καὶ ἱστορίας συγγραφεύς); in discussions of astrology, Posidonius is "astrologus idemque philosophus" to Saint Augustine, *De civitate Dei* V.5, and a "mathematicus" in Boethius, *De diis et praesensionibus* 20.77 (T69 and T70 EK).
27 See Plutarch, *Cicero*, 4.5 (Posidonius T29 EK) and passages of Cicero printed as T30–T34 EK.
28 See passages of Strabo, Pliny the Elder, Cicero and Plutarch printed as Posidonius T35–T36 and T38–T39 EK.
29 Seneca *NQ* II.26.6 and VI.17.3 (Posidonius T41 EK).
30 Cicero, *Att.* XVI.11.4 and XVI.14.4 (Posidonius T44 and F41a–b EK).
31 Kidd (1988) 36.
32 *Suda*, s.v. Ἰάσων 52 (Posidonius T40 EK).
33 His Περὶ κριτηρίου (*On the criterion*, sc. *of truth*): Posidonius F42 EK.

34 See Posidonius F19 EK, on relational syllogisms.
35 E.g., Posidonius F4, F8, F13–14, F20–21 EK.
36 Posidonius F29–35, F139–187 EK.
37 Posidonius F51–78 and F252–283 EK. For his starting point see T1a EK and Kidd (1988) 5; for his end date see Kidd (1988) 277–8.
38 E.g., Posidonius F67 and F274–275 EK, on customs of the Celts.
39 Cicero, *Div.* I.6, mentions a work on the subject by Posidonius in five books (Posidonius F26 EK).
40 Notably Περὶ ὠκεανοῦ, *On the ocean*: Posidonius F49 EK.
41 Posidonius F46–47 and F195–8 EK.
42 For calculations of sizes and distances see Posidonius F115 EK (diameter of the sun) and F202 EK (circumference of the earth). For such calculations not being part of a philosopher's work see F18 EK.
43 See Posidonius F106–110 EK (passages from Cicero's *De divinatione*).
44 See EK, Index of sources, pp. 262–3.
45 E.g., Posidonius F150b, 151, 156, 159, 160 EK.
46 See Posidonius T57 and F19 EK = Cleomedes II.7.126 (II.7 lines 11–14 Todd) and I.11.65 (I.8 lines 158–162 Todd).
47 Of the collection of fragments of Posidonius in EK, 23 are from Diogenes' account of Stoic physics (VII.132–160) but only 15 from his account of other parts of their philosophy (VII.39–131), See EK's Index of sources, p. 259.
48 The Simplicius passage is *In Ph.* II.2 (p. 291.21–292.31 Diels; Posidonius F18 EK; see [e.g.] Chapter 21 below, p. 202), for which Simplicius cites as his source Alexander of Aphrodisias, who made λέξιν τινὰ τοῦ Γεμίνου … ἐκ τῆς ἐπιτομῆς τῶν Ποσειδωνίου Μετεωρολογικῶν, "a quotation from Geminus … from his epitome of Posidonius' *Meteorology*" (translation from Kidd [1999]). Simplicius also cites Posidonius more briefly, about element theory, at *In Cael.* IV.3 (Posidonius F93a EK; see below, Chapter 5 note 12 [p. 39]); here he might be using the same source (Kidd [1988] 375).
49 Kidd (1988) 782. On Priscianus Lydus and Posidonius' tidal theory see below, Chapter 13 (p. 133–6).
50 See Diogenes Laertius VII.149 on divination, and VII.142 on the cosmos (=Posidonius F7 and F13 EK, omitting Diogenes' mentions of Panaetius' dissent, which are Panaetius frs. 66 and 73 van Straaten).
51 See, e.g., Long (1974) 9–10.
52 For a brief account see Frede (1999) 771–8; for the first century B.C., Falcon (2012) 17–21.
53 Cicero, *Fin.* IV.28.79 (Panaetius fr. 55 van Straaten). Cicero also speaks of Panaetius' high praise of Plato at *Tusc.* I.32.79 (Panaetius fr. 56 van Straaten).
54 *Index Stoicorum Herculanensis*, col. LXI (Panaetius fr. 57 van Straaten).
55 Galen, *De placitis Hippocratis et Platonis*, IV,421 = p. 284 33f De Lacy (Posidonius T97 EK).
56 Strabo II.3.8 (Posidonius T85 EK, with translation from Kidd [1999]).
57 See Posidonius T91 and T95–99 EK.
58 Posidonius T95 and T96 EK.

3 Sources for the study of Posidonius' meteorology

Sources for the views of Posidonius

As Posidonius' own works are lost, our knowledge of his views depends on "fragments", on quotations and reports of his opinions found in the texts of other authors; and, however probable it may be that some ancient authors repeat his views without naming him, clearly any study must begin from ancient sources in which he is named. As Edelstein says, "the attested fragments ... have to be the basis of any ... interpretation of Posidonius"[1]: These fragments are second-hand testimony; but the later writers who report Posidonius' views had, as a rule, no reason to report them without at least approximate correctness – he was a respected author for centuries after his death, he had written much, his views must have been widely known, and any total misrepresentation of them would have been obvious. We may accept the reports as at least roughly correct unless there is definite reason to suppose otherwise. However, though we talk about "fragments", it may be doubted, at least with meteorology, whether any "fragment" preserves Posidonius' actual words; and those who report his views were not always – perhaps not often – concerned to give a full and fair account of them. They may quote selectively to suit their own views or interests, or misrepresent his views to make it easier to argue against them; they may be careless, or required to summarise; or, when they mention Posidonius, they may omit to make clear how much of what they are saying actually relates to him.

The earliest writer to consider is his pupil Cicero. Cicero shows minimal interest in meteorological theory, but he did write about the possibility of predicting the future, a part of which is prediction of weather and climate. It seems unlikely that he would misrepresent his respected teacher's views, even though he disagrees with them, but what he says still presents a problem: at *De divinatione* I.129–31 he refers to Posidonius believing that there are in nature signs of future happenings, and cites two examples of such signs mentioned by earlier authors: did he find these examples in Posidonius?[2]

A more important writer on Posidonius' meteorology is Strabo. Born in Posidonius' lifetime, Strabo clearly had access to Posidonius' works and made extensive use of them. He is an important source; but in several places we shall

DOI: 10.4324/9780429399930-3

find that what he said (or what our manuscripts say he said) seems so unlikely,[3] or so self-contradictory,[4] that we may suspect an error in the manuscripts; elsewhere, Strabo shows a desire to convict Posidonius of error or self-contradiction, and sometimes we may suspect that this has led Strabo to a tendentious misinterpretation of Posidonius.[5] Also, Strabo remarks on Posidonius' un-Stoic keenness to find the causes of phenomena,[6] and tells us little about the causal explanations that Posidonius postulated. Strabo is not to be blamed for this: no doubt he could see that Posidonius' explanations were speculative, and we know that they were mostly wrong; but Strabo has deprived us of much information about Posidonius' thought which he was well placed to provide.[7]

Chronologically, the next important source is Seneca's *Naturales quaestiones*. As a Stoic, one would expect Seneca to show great respect for the one important earlier Stoic who had dealt seriously with meteorology, and (unlike Strabo) he shows no reluctance to attempt explanations of the causes of meteorological phenomena. But he is not always respectful to Posidonius: he treats his theory of hail with marked irony.[8] More generally, it is clear, especially from his treatment of Aristotle's meteorology, where we can compare what he says with Aristotle's text, that he tends, after giving a reasonably accurate summary of an earlier theory, to diverge from it when he appears to be giving details.[9] He by no means always agrees with Posidonius: on comets he explicitly dissents from him and the Stoic school, maintaining, contrary to their view, that comets are permanent bodies ("aeterna opera").[10] He describes Posidonius' theory of thunder and lightning as based on a theory of dry and wet exhalations,[11] but when he tells us his own theory of thunder and lightning, exhalations are mentioned only as an afterthought.[12] On rainbows he says that he agrees with Posidonius[13]; but his explanation of haloes[14] is quite different, if Alexander is even roughly right to say that Posidonius here agreed with Aristotle.[15] We must be extremely cautious in attributing to Posidonius anything in Seneca which is not attributed to Posidonius by name.

A third important source is the account of Stoic doctrine in Diogenes Laertius VII, especially the account of Stoic meteorology in VII.151–4; Posidonius is cited three times in this one short passage. Diogenes, or his source, had no apparent reason to distort Posidonius' views, but his account is only a summary, giving few details; sometimes it leaves it doubtful where a citation of Posidonius begins or ends,[16] and sometimes we may suspect that an important part of Posidonius' theory has been omitted in the process of summarising.[17]

Another summariser, Aëtius, who provides some information about Posidonius' meteorology, can give cause for a different kind of doubt. His work deals, in summary form, with a series of subjects about which ancient thinkers differed (e.g., Fate; Tides), giving the opinion about each subject of a number of different thinkers; here, it is not difficult to see where the account of each thinker's opinion begins and ends, but there is – besides the chance that something important has been omitted – a danger that, in the course of repeated editing and summarising, an opinion has been attributed to the wrong thinker. This possibility applies to Aëtius' account of Posidonius' theory of tides.[18]

The reliability of other sources will have to be considered when I quote them; what I have said is enough to show that there is a possibility of doubt when we consider almost any report of Posidonius' meteorology. But there are factors we can look for which, if found, strengthen the probability that a report is true:

Is the report confirmed by another ancient author?

Is it consistent with other things which Posidonius is reported to have said?

Does it seem probable in the light of what was said on the same subject by other ancient authors (especially those likely to have influenced, or to have been influenced by, Posidonius)?

If the report is about something observable, is it likely, in view of what is reported about Posidonius' biography, that he had an opportunity to observe it, or to speak with people who had observed it?

If the report is about something observable, and likely to have remained unchanged until today (the direction of a prevailing wind, say), is the report true?

The more of these questions that can be answered in the affirmative, the surer we can be that a report about Posidonius is true. It is my belief, and I shall try to show, that there are enough affirmative answers to enable us to construct a reasonably accurate, if incomplete, account of Posidonius' meteorology.

Apart from the attested fragments, there are a number of ancient texts which clearly have a relation to Posidonius' meteorology and have sometimes been regarded as evidence for his meteorological views even in places where he is not named. I have already discussed Seneca's *Naturales quaestiones*; it will be best to mention here some other texts.

One of these texts is the pseudo-Aristotelian *De mundo*.[19] That there is a connection between Posidonius and the section on meteorology, Chapter 4, of this treatise is certainly true: *De mundo* gives an account of the rainbow which agrees almost word for word with one reported by Diogenes Laertius to be that of Posidonius.[20] But this agreement is not consistent. *De mundo* says, if our manuscripts are correct,[21] that hail is formed νιφετοῦ συστραφέντος, "when a snow-storm is solidified"[22] – a theory quite different from the theory of hail which Seneca attributes to Posidonius. Diogenes Laertius describes as Stoic, probably Posidonian, what seems to be a classification of earthquakes into four named types[23]; *De mundo* 396a1–12 has what is clearly a related classification, but with seven named types of earthquake; and where these can be identified with types named by Diogenes, the names are different: what Diogenes calls χασματίαι, κλιματίαι (probably), and βρασματίαι are in *De mundo* named as ῥῆκται, ἐπικλίνται and βράσται.[24] There is some relationship here between Posidonius and the *De mundo*; but clearly the author of *De mundo* (or some intermediate source) did not just copy Posidonius.

In particular, the author of *De mundo* is an Aristotelian. He accepts Aristotle's theory, rejected by Posidonius and other Stoics, that the heavens are composed of a fifth element, different from the sublunary elements earth, water, air and fire;[25] unlike Posidonius and most Stoics, he regards the cosmos as eternal;[26] his deity, though not Aristotle's Unmoved Mover, yet resembles him in dwelling at the boundary of the cosmos,[27] exerting power but himself undertaking no toil[28] – a view "most probably intended as a criticism of the Stoic view that the god is causally operative in the world by permeating it".[29] An author who takes so much from Aristotle, even though the Stoics rejected it, would naturally accept also Aristotle's view that the theory of dry and wet exhalations provides the basis for the explanation of almost all meteorological phenomena; and his doing so is no evidence that Posidonius, or any other Stoic, held that view.

A further question arises: what is the date of the *De mundo*? It purports to be by Aristotle, but almost all scholars agree that this is impossible. As meteorological proofs of this may be mentioned that the author of *De mundo* had heard of tides and their relation to the moon's movements,[30] something of which Aristotle shows no knowledge[31]; also, as just mentioned, *De mundo* has an elaborate classification of different types of earthquake, quite different, but perhaps ultimately derived, from the simple division into two types which Aristotle makes in his *Meteorologica*.[32] There is much other evidence: apparent anachronisms, such as the mention in *De mundo* of the island "Taprobane" (Sri Lanka)[33]; inconsistency with Aristotle's known views; differences of style; apparent doubts of Aristotelian authorship in some ancient writers.[34] However, though *De mundo* cannot be by Aristotle, it may have been written soon after his death. Knowledge of the tides may have reached old Greece by 300 B.C.[35]; knowledge of Sri Lanka as an island apparently resulted from the eastern campaigns of Alexander the Great.[36] It may be that the *De mundo* antedates Posidonius.

We may doubt whether Posidonius, who clearly made a detailed study of meteorological topics, would have made much use of a brief summary such as that in *De mundo*. If *De mundo* does antedate Posidonius, then the likelihood is that, where they agree, they depend on a common source; and in meteorology, where that source is not Aristotle, the obvious source is Theophrastus. Both *De mundo* and, in his meteorology, Posidonius (as we shall see) frequently follow Aristotle, so are likely to have used Aristotle's successor also; neither is likely to have used Epicurus; and we lack knowledge of any other serious student of meteorology between the lifetimes of Theophrastus and Posidonius. Strohm argues that Theophrastus was an important source for the *De mundo*[37]; how much use Posidonius made of Theophrastus I will discuss later. Setting this aside for the time being, there is one point in *De mundo* Chapter 4 which suggests that it is later than Posidonius. Posidonius appears either to have elaborated Aristotle's twofold division of earthquakes into a fourfold division, or to have supplemented Aristotle's twofold division with an alternative twofold division of his own.[38] (About Theophrastus' position I don't think we can be certain.) The *De mundo* gives us a more elaborate classification of earthquakes obviously related to that of Posidonius but considerably different from it. It is possible that this is an independent development from Aristotle and Theophrastus, but it is surely rather likelier that it is developed from Posidonius' classification, and that *De mundo* is therefore later than Posidonius; which agrees with the tentative conclusion reached on other grounds in the recent review by J.C. Thom.[39]

Somewhat similar to the meteorological chapter of *De mundo*, but without the author's commitment to Aristotelianism, are the meteorological fragments of Arrian.[40] Here, too, a relationship to Posidonius clearly exists. The fragments were once thought to be from a text which antedated Posidonius and was used by him,[41] until Wilamowitz showed in a brief note that this improbably early dating is based on a misunderstanding.[42] Since then it has been generally accepted that the author of these fragments is the well-known Arrian who, in the 2nd century A.D., recorded the discourses of Epictetus, wrote the life of Alexander and governed Cappadocia.[43] Some have supposed that Arrian followed Posidonius closely in meteorology, and that the fragments can be used as a source for Posidonius.[44] This is not easy to assess, because a great part of the fragments concerns matters for which we have no attested evidence for Posidonius, but it is clear that the relationship is not such a simple one: Arrian agrees with Posidonius in his theory of comets[45]; but he had a different view about the height up to which clouds and rain occur.[46] His agreement with Posidonius is not consistent.

Cleomedes, in his elementary treatise on astronomy (his one surviving work, probably 2nd century A.D.), is another author sometimes assumed to be repeating Posidonius' views, even in passages where Posidonius is not named. Cleomedes does say, at the end of his work, τὰ πολλὰ δὲ τῶν εἰρημένων ἐκ τῶν Ποσειδωνίου εἴληπται, "the greater part of what has been said has been taken from the works of Posidonius".[47] But Cleomedes' previous sentence says that the work has been compiled ἐκ συγγραμμάτων τινῶν ... καὶ παλαιῶν καὶ νεωτέρων, "from works both ancient and more modern".[48] At the end of Book I, writing about the size of the sun, he refers as his authorities to τινων περὶ μόνου τούτου συντάγματα πεποιηκότων, ὧν ἐστι καὶ Ποσειδώνιος, "some who have written monographs about this, among whom there is even Posidonius" – here Posidonius is perhaps implied to be the most important, but is only one of several authors used.[49] Elsewhere Cleomedes recounts Posidonius' view that land at the equator is habitable, but goes on to quote criticisms of this view, concluding that it is probably incorrect.[50] Clearly, Cleomedes did not simply copy Posidonius. He had evidently read Posidonius, and where he names Posidonius as holding a particular view there is no reason to doubt him; but when he propounds a view without naming Posidonius it is prima facie unreasonable to regard this as an authority for Posidonius' view equal to or greater than an ancient source which does name Posidonius. On a subject about which no source names Posidonius as holding a particular view, it is reasonable to regard Cleomedes' view as an indication of what Posidonius is likely to have thought.

On meteorological topics about which no opinion is attributed to Posidonius by name, another source worth examining may be found in those testimonia which describe a view as "Stoic" without naming any individual. If they cannot be shown to preserve Posidonius' own view, they (like Cleomedes) at least provide an indication of what he is likely to have thought; for Posidonius was, as already mentioned, in some respects a rather conservative thinker, likely to have adopted a view held by previous Stoics unless he saw good reason to the contrary. In Diogenes Laertius' account of Stoicism, his name frequently appears in lists of Stoic authors who endorsed the standard Stoic view of some topic[51]; passages which record him as holding a view different from that of most other Stoic authors are much fewer.[52] In more respects than not, he was an orthodox Stoic. Also, he was much the best-known Stoic to have dealt seriously with meteorology: it may well have happened that an author wishing to find the Stoic view of a meteorological subject consulted Posidonius' work and recorded his view as that of "the Stoics". Ideally, all reports of "Stoic" views on meteorology should be evaluated for their likely relationship to Posidonius.

One can very rarely be confident that Posidonius held a view on a meteorological topic which is expounded by Seneca, the *De mundo*, Arrian or Cleomedes, or attributed to unnamed "Stoics", but which no ancient author attributes to Posidonius by name. My aim is to base this study on the attested fragments, and I believe that, though there will be inevitable gaps and uncertainties, there is enough evidence to give a reasonably accurate picture of Posidonius' meteorological work. I shall from time to time cite passages from the authors and works just mentioned where Posidonius is not named, but it seems likely that what they say fills a gap in what is attested about Posidonius, or elucidates a theory held by him, or illustrates how a successor developed his ideas, but the results are generally less than certain.

Sources for other relevant authors

By "relevant" I mean authors whose works Posidonius may have used in constructing his meteorology, or which, even if not used by him, are worth comparing with his. Some such works survive complete: Aristotle's *Meteorologica*, Theophrastus' *De ventis*, Pseudo-Aristotle's *Problemata XXVI* (on winds), and (in Latin translation) Aristotle's (?) *De inundatione Nili*.[53] A possible but unlikely source for Posidonius, but worth comparing with his meteorology, is Epicurus' *Letter to Pythocles*; probably just later than Posidonius' work, but also worth comparison, is Book VI of Lucretius' *De rerum natura*.

There were other ancient writers on meteorology who may have influenced Posidonius but whose works are lost. Many pre-Socratics were concerned with meteorology, and they are important, not because they are the likeliest authors for Posidonius to have used, but because so many ancient meteorological theories evidently originated with them. Verbatim fragments are few[54]; there is a good deal of information in Aristotle's *Meteorologica*, notably on comets, the Milky Way, earthquakes, and thunder and lightning[55]; but his aim is not an objective history, but to refute older theories so as to make way for his own. Much of our information comes from much later authors, such as Aëtius and Diogenes Laertius, already discussed in relation to Posidonius, with the dangers of errors introduced during several stages of transmission and of summarising.

With the pre-Socratics there is a further chance of distortion, from changes in the Greek language, for instance the change in the meaning of ἀήρ from "mist" to "air".

Another important group of thinkers for whom no complete texts survive are the early Stoics. They were not particularly interested in meteorology, but it is likely that Posidonius, as a fellow Stoic, would have paid particular attention to what they did have to say on the subject. Two "fragments", one from Diogenes Laertius and one from Seneca, attribute explanations of lightning and comets by name to Zeno,[56] and three extracts preserved by Stobaeus attribute by name to Chrysippus summary accounts of the seasons of the year, of mist, rain and related phenomena, and of lightning and related phenomena. The last of these three is attributed to Aëtius,[57] but the first two are attributed to Arius Didymus,[58] who is arguably an earlier and presumably better authority, though his date is in fact doubtful.[59] About the authority of Aëtius and Diogenes, and of Seneca, I have written above. Other passages printed in *Stoicorum veterum fragmenta* report Stoic accounts of meteorological phenomena without naming any individual Stoic; these are mostly from Diogenes and Aëtius but do include one passage of Cicero. They present the additional problem, already mentioned, that we must try to infer which Stoics are meant in each passage.

There is one other text which must be considered here: the text, partly preserved in Syriac and in an almost complete form in an Arabic translation from the Syriac, entitled, in Daiber's English translation, *Theophrastus' treatise on meteorological phenomena*.[60] If this title is accurate, it could have been an important source for Posidonius; following the practice of F.A. Bakker,[61] I call it in what follows "the Syriac Meteorology". It contains explanations of thunder, lightning, thunderbolts, clouds, rain, snow, hail, dew, hoar-frost, winds, lunar haloes and earthquakes. Daiber argues that it is a translation of a complete work by Theophrastus, though not a finished work, but one which he composed as a draft for lectures.[62] It is, however, a very strange text. Its most arresting feature is that it offers multiple causes for most of the phenomena discussed, including seven of thunder, four of lightning and four of earthquakes, usually supported by an analogy with a familiar phenomenon which has similar effects (e.g., thunder is compared with the noise produced by red-hot iron plunged into water). Theophrastus, in works surviving in Greek, usually gives just one cause for each phenomenon, and though he does sometimes propose multiple causes, he does it, as Bakker has shown, in a different way from the Syriac Meteorology.[63] The regular proposal of multiple causes in meteorology is rather the practice of the Epicureans; the frequent mention of familiar analogies is characteristic of Lucretius Book 6.[64]

On the other hand, there are features of the Syriac Meteorology which link it with Theophrastus and the Peripatetics rather than the Epicureans.[65] The idea, found in the Syriac text of the Syriac Meteorology but not in the Arabic, that the horizontal movement of wind is the resultant of natural movements in opposite directions, vertically upwards and vertically downwards,[66] is known from other evidence to have been the view of Theophrastus[67] and seems to have been peculiar to him (though Aristotle did use a similar idea in other contexts[68]). The idea that some matter has a natural upward motion to its natural place[69] is Peripatetic, not Epicurean, and so is the idea that there are two exhalations (though the Syriac Meteorology differs from Aristotle in calling them, in Daiber's translation, "fine" and "thick" or "light" and "heavy", rather than "dry" and "wet").[70] The Syriac Meteorology (section 14(14)–(16)) says: "It is not correct to say that God should be the cause of disorder in the world; nay, he is the cause of its arrangement and order". This is most certainly not Epicurean, and M. van Raalte has shown it to be doubtful whether it is Theophrastus' view either.[71] There are such doubts attached to this text that one must clearly be extremely cautious in regarding it, or any part of it, as a source from which Posidonius may have derived ideas.

Notes

1 See EK p. xviii.
2 See below, Chapter 17 (p. 176–7).
3 As with the circumference of the earth: see below, Chapter 7 (p. 56).
4 As with the inundation of the lands of the Cimbri, in the opinion of many scholars: see below, Chapter 12 note 46 (p. 125).
5 As with the existence of mountains on or near the equator: see below, Chapter 7 note 42 (p. 64).
6 Strabo II.3.8, quoted below, Chapter 4 (p. 27).
7 Posidonius' account of tides is a good example of this: see below, Chapter 13 (p. 133).
8 On Posidonius' theory of hail: see below, Chapter 14 (p. 144–5).
9 See Hall (1977).
10 *NQ* VII.22.1. For the views of Posidonius and other Stoics see below, Chapter 9 (p. 79–80).
11 See below, Chapter 8 (p. 69).
12 *NQ* II.21 ff and II.57; exhalation is mentioned only in II.57.3.
13 See below, Chapter 16 (p. 161).
14 *NQ* I.2.1–2.
15 See below, Chapter 16 (p. 164–5).
16 As with the theory of earthquakes: see below, Chapter 12 (pp. 117, 119).
17 As with the theory of snow: see below, Chapter 14 (p. 143).
18 See below, Chapter 13 (pp. 133, 136).
19 Theiler in his *Poseidonios* (1982) F336–343 includes the whole of *De mundo* Chapter 4 (the meteorological chapter), with other texts. For a detailed criticism of the view that *De mundo* 4 is derived from Posidonius, with particular reference to Theiler, see Strohm (1987).
20 *De mundo* 395a32–35. On Posidonius' view see below, Chapter 16 (p. 160).
21 Theiler (1982) F336a emends νιφετοῦ, "snowstorm", to ὑετοῦ, "rain", which is close to Seneca's report of Posidonius; but one can hardly use an emended text as evidence.
22 *De mundo* 394b2, with Furley's translation, in Forster and Furley (1955).
23 Diogenes Laertius VII.154. On this see below, Chapter 12 (p. 119).
24 On the meaning of these terms see below, Chapter 12 (p. 119).
25 *De mundo* 392a5ff. On Posidonius' view, see below, Chapter 5 (p. 37).

26 *De mundo* 397b8 ἄφθαρτον δι' αἰῶνος, "eternally indestructible" in Furley's translation (Forster and Furley [1955]). For Posidonius' view see Diogenes Laertius VII.142.
27 *De mundo* 400a7.
28 *De mundo* 397b22–24.
29 Betegh and Gregoric (2014) 577.
30 *De mundo* 396a25–27.
31 On Aristotle and tides see below, Chapter 13 (pp. 131, 137).
32 *Mete.* 368b23–25. See below, Chapter 12 (p. 118–9).
33 *De mundo* 393b14.
34 For a recent review see Thom (2014) 3–8.
35 See below, Chapter 13 (p. 129).
36 So Pliny *Nat.* VI.81. Burri (2014) 90 takes this view, without citing Pliny; Strabo XV.1.14–15, which he cites, does not say when Taprobane became known to the Greeks.
37 E.g., Strohm (1987) 80–81.
38 See below, Chapter 12 (p. 118–9).
39 Thom (2014) 7.
40 Roos/Wirth (1968) 186–95.
41 So Capelle (1905).
42 von Wilamowitz-Möllendorff (1906).
43 So (e.g.) Roos/Wirth (1968) xxvii–xxviii; Stadter (1980) 30 and 202 n.19.
44 Theiler (1982) prints the meteorological fragments of Arrian in his *Poseidonios* as F336b, 337b, 338b, 340b.
45 See below, Chapter 9 (p. 81).
46 See below, Chapter 6 (p. 48).
47 Cleomedes II.7.126 (II.7 lines 11–14 Todd) = Posidonius T57 EK, with Kidd's (1999) translation.
48 Kidd's (1999) translation.
49 Cleomedes I.11.65 (I.8 lines 158–62 Todd) = Posidonius F19 EK.
50 Cleomedes I.6.31–3 (I.4 lines 90–131 Todd) = Posidonius F210 EK. On the relation of Cleomedes to Posidonius see Kidd (1988) 138.
51 See Diogenes Laertius VII.39, 87, 91, 134, 142–43 (Posidonius named in three lists), and 148.
52 See Diogenes VII.92, on the number of kinds of virtue; VII.103, Posidonius dissented from the view that wealth and health are not good things; VII.128, Panaetius and Posidonius did not accept the view that virtue is sufficient in itself for well-being (αὐτάρκη ... πρὸς εὐδαιμονίαν). I do not count very minor differences, such as the account of the rainbow at VII.152, where Posidonius' view seems more like an elaboration of an earlier Stoic view than a real difference.
53 See below, Chapter 4 (pp. 21 and 31 n.4), and Chapter 15 (pp. 153–5) for evidence suggesting Posidonius knew the Greek original.
54 For instance, Xenophanes frs. 30 and 32, where the verse form shows them to be verbatim.
55 *Mete.* 342b25ff, 345a11ff, 365a14ff, 369b12ff.
56 Diogenes Laertius VII.153 (*SVF* I.117) and Seneca *NQ* VII.19.1 (*SVF* I.122).
57 *SVF* II.703 = Aëtius III.3.13 in Diels (1879) 369–70.
58 *SVF* II.693 and 701 = Arius Didymus fr. phys. 26 and 35 in Diels (1879) 461–62, 468; for the reasons for attributing these and other fragments to Arius see Diels (1879) 69–88, also Mansfeld and Runia (1997) 238–65.
59 Aëtius may be as late as the early 2nd century A.D. (Diels [1879] 101, Mansfeld and Runia [1997] 319), Arius Didymus as early as the late 1st century B.C. But this early dating depends on Diels's identification of Arius Didymus with the Arius who was

court philosopher to the Emperor Augustus (Diels [1879] 80ff), and this identification has been shown to be doubtful (see discussion in Mansfeld and Runia [1997] 240–2).

60 Published with English translation and commentary in Daiber (1992).
61 Bakker (2016) 70.
62 Daiber (1992) 283ff.
63 See Bakker (2016) 67–72; also 145–6, where Bakker draws attention to other features of the Syriac Meteorology which are inconsistent with other evidence for Theophrastus.
64 Bakker (2016) 67 suggests that this use of analogy is characteristic also of Epicurus' meteorology; but familiar analogies are mentioned relatively rarely in Epicurus' *Letter to Pythocles*.
65 Much of this paragraph is derived from Bakker (2016) 147–52. (He includes some evidence for which I do not have space.)
66 Daiber (1992) 268 (section 13(21) of his English translation).
67 Theophrastus *De ventis* 22; Alexander *In Mete.* 93.35ff Hayduck; Olympiodorus *In Mete.* 97.5ff Stüve.
68 *Mete.* 342a24ff, on the horizontal motion of shooting stars, and, tentatively, *GA* 782b18ff, on the curliness of hair.
69 See section 6(42)-(47) of the Syriac Meteorology.
70 Sections 6(42)-(59) and 13(21) in the Syriac Meteorology. Contrast Aristotle *Mete.* 341b6–10.
71 Van Raalte (2003).

4   The history of Greek meteorology before Posidonius

Pre-philosophical origins

The beginnings of Greek meteorology – the first hints of the attempt to explain meteorological events without attributing them to the action of a divinity – are older than the beginning of Greek philosophy. Hesiod says in *The works and days*, after urging the wearing of warm clothes in winter:

Ψυχρὴ γάρ τ᾽ ἠὼς πέλεται Βορέαο πεσόντος
ἠώιος δ᾽ ἐπὶ γαῖαν ἀπ᾽ οὐρανοῦ ἀστερόεντος
ἀὴρ πυροφόρος τέταται μακάρων ἐπὶ ἔργοις·
ὅστε ἀρυσσάμενος ποταμῶν ἄπο αἰεναόντων,
ὑψοῦ ὑπὲρ γαίης ἀρθεὶς ἀνέμοιο θυέλλῃ
ἄλλοτε μέν θ᾽ ὕει ποτὶ ἕσπερον, ἄλλοτ᾽ ἄησι
πυκνὰ Θρηικίου Βορέου νέφεα κλονέοντος.

For the dawn is chill when Boreas has once made his onslaught, and at dawn a fruitful mist is spread over the earth from starry heaven upon the fields of blessed men; which, drawn from the ever flowing rivers and raised high above the earth by wind-storm, sometimes rains towards evening and sometimes blows [sc. as wind], when Thracian Boreas huddles the thick clouds.

(Hesiod, *Op.* 547–53, with trans., modified, of
H.G. Evelyn-White [1914, Harvard University Press, Loeb ed.] p. 43)

In this sentence the only possible grammatical function of ὅστε, "which", meaning the mist, is to be the subject of the verbs ὕει, "rains", and ἄησι, "blows"; apparently it is the mist which is raining or blowing.

At *Theogony* 687ff Hesiod describes the effects of the thunderbolts hurled by Zeus in the gods' battle with the Titans, including

Σὺν δ᾽ ἄνεμοι ἔνοσίν τε κονίην τ᾽ ἐσφαράγιζον
βροντήν τε στεροπήν τε καὶ αἰθαλόεντα κεραυνόν

DOI: 10.4324/9780429399930-4

Also winds noisily stirred up earthquake and dust-storm, and thunder and lightning and blazing thunderbolt.[1]

Here a natural force, wind, causes other meteorological phenomena. As we shall see, this idea, and that quoted from *Works and days*, was to be of great importance in the philosophers' meteorological theories.[2]

Philosophers before Aristotle

Greek philosophy is traditionally regarded as beginning with Thales, around 600 B.C. or a little later. Very little is known about his "philosophical" views, but it seems quite likely that they included ideas about meteorology. It seems generally accepted, on Aristotle's authority, that Thales in some sense regarded water as the origin of all things, and believed that the earth floats on water as a ship does[3]; and if, as many but by no means all scholars believe, the Latin *Liber Aristotelis de inundatione Nili*[4] is a translation of a genuine work of Aristotle's,[5] then we also have Aristotle's authority for attributing to Thales the view that the Nile floods in summer because northerly Etesian winds hold the river back in its course.[6] That Thales also believed earthquakes to be caused by the earth being rocked by the water on which it floats is plausible but depends on Seneca's sole authority.[7]

With Anaximander (Thales' pupil), and Anaximander's pupil, Anaximenes, we are on rather surer ground. We know they both wrote books which were known to later writers: a quotation survives from Anaximander's book,[8] and Diogenes Laertius comments on Anaximenes' literary style.[9] Unlike Thales, Anaximander and Anaximenes were not men whose names were familiar in Greece in the centuries that followed their deaths: no extant author before Aristotle mentions either of them.[10] There would have been no reason to father on them theories which were not their own; so we may accept that, as later authors state, they put forward explanations of meteorological phenomena, though our records of those explanations may have been distorted in the course of transmission. For Anaximander we have explanations of wind, rain, thunder and lightning, earthquakes and the origin of the sea, recorded by Alexander of Aphrodisias (attributing to Anaximander and another a theory described, with no author's name, in Aristotle's *Meteorologica*),[11] by Hippolytus,[12] by Aëtius,[13] by Seneca[14] and by Ammianus Marcellinus.[15] Explanations are attributed to Anaximenes of wind, cloud, rain, hail, snow, thunder and lightning, rainbows, and earthquakes, recorded by Aristotle,[16] Hippolytus,[17] Aëtius,[18] Seneca,[19] Galen,[20] Simplicius[21] and the Scholia to Aratus.[22]

Meteorology, therefore, was a major interest to Anaximander and Anaximenes, and after them it was a subject which later pre-Socratics – those, at least, who followed the Ionian rather than the Western tradition[23] – could not ignore, if they sought to describe the workings of the natural world. Xenophanes wrote about wind, cloud, rain and rainbows, as we know from verbatim fragments,[24] and also, according to Aëtius, about comets, shooting stars, lightning and St Elmo's fire[25]; though we may suspect that what he said was very brief. Among later pre-Socratics, the one most interested in meteorology seems to have been Anaxagoras: our sources attribute to him theories of thunder and lightning, along with *keraunos* ("thunderbolt", i.e., lightning-strike), *typhōn* and *prēstēr* (kinds of tornado or similar phenomena[26]); of cloud, rain, hail and snow; of wind; of rainbows and mock-suns; of shooting stars, meteorites, comets and the Milky Way; of earthquakes; of the Nile floods; and of the origin of the sea and its salt.[27] At least three others seem to have had a serious interest in meteorology. To Diogenes of Apollonia ideas are attributed about the seasons, rain, wind, thunder and lightning, meteorites, the origin of the sea and its salt, and the Nile floods.[28] Democritus, we are told, had theories about thunder, lightning, and related phenomena; wind; comets; the Milky Way; earthquakes; the Nile floods; and the sea; and he seems also to have been interested in weather prediction, drawing up a *"parapēgma"*, a calendar with weather predictions for particular days, and discussing weather-signs.[29] Metrodorus of Chios is reported to have held theories about thunder, lightning and related phenomena; clouds; wind; rainbows; shooting stars; the Milky Way; the sea and its salt; and earthquakes.[30]

Other pre-Socratics, without including meteorology among their main interests, yet touched on it occasionally. Heraclitus is said to have had a theory of two exhalations,[31] arguably important for its influence on later thinkers,[32] and to have proposed explanations of thunder and lightning,[33] rain, wind and similar phenomena[34]; though one may suspect these to be re-interpretations of cryptic expressions by Heraclitus, which to him may have had some other meaning: surviving fragments use meteorological terms, *prēstēr* and *keraunos*, where he evidently means his archetypal principle, *pyr* (fire), or some form of it.[35] Other important pre-Socratics reported to have proposed explanations of thunderstorms are Empedocles[36] and Leucippus[37]; minor figures to whom thunderstorm theories are attributed are Cleidemus[38] and Anaxagoras' pupil Archelaus.[39] Other thinkers with more astronomical interests proposed theories of comets and the Milky Way: such were "those called Pythagoreans",[40] and Hippocrates of Chios and his pupil Aeschylus.[41] Oinopides seems to have held a theory about the Milky Way (mythological, like one of the "Pythagorean" ones),[42] and also a theory of the summer floods of the Nile.[43] A very obscure thinker, Thrasyalces, is known only for his explanation of the latter phenomenon, and for the idea that there are only two principal winds, Boreas and Notus.[44]

The details of all these theories I shall discuss, as needed, in later chapters, but two general points are worth making here. First, the earliest philosophers were attempting to explain the nature of the world we live in and the causes of phenomena occurring in it in terms which avoided the involvement of direct divine action, and so needed to deal with meteorological events because gods were thought to cause them: Zeus caused thunder and lightning, hurled thunderbolts and also sent hail, rain and snow[45]; Poseidon was "earthshaker" – his shaking of the earth is graphically described at *Iliad* 20.57–63; the winds are divinities,[46] or are controlled by divinities,[47] even minor ones.[48]

A surviving line from Xenophanes[49] illustrates the philosophers' denial of such beliefs:

ἥν τ' Ἶριν καλέουσι, νέφος καὶ τοῦτο πέφυκε

"She whom men call Iris [i.e., the rainbow], this too is by nature a cloud"

Iris was a goddess to Homer,[50] but only a cloud to Xenophanes. We do not know (lacking verbatim texts) if other early philosophers were so explicit, but their attitude was surely similar.

To provide a convincing counter to the mythological view, the philosophers had to seek natural, non-divine, materials and forces which could produce meteorological phenomena: in Aristotelian terms, they had to seek the material and efficient causes of them. No idea of purpose, of an Aristotelian final cause, was needed, and this remained the normal pre-Socratic attitude. I know only two exceptions: Aristotle mentions a Pythagorean belief that the purpose of thunder is to frighten the spirits of the dead in Tartarus[51]; less weirdly, Diogenes of Apollonia speaks of the regularity of rain, wind and fine weather among the proofs that νόησις, intelligence, directs the world.[52] Apart from this, so far as we know, pre-Socratic explanations of meteorological phenomena dealt solely with material and efficient causes.

My second general point is that, although some pre-Socratics wrote about many meteorological phenomena, few if any of them can have written about them at length, since they wrote on other subjects also, and were mostly not voluminous writers. Apart from Democritus, for whom a long list of works is preserved by Diogenes Laertius,[53] none of them is known to have written more than a very few works, and it is doubtful if any of them wrote a separate work on meteorology. One who may have done so is Diogenes of Apollonia, who is said by Simplicius[54] to have written a "Meteorology" (Μετεωρολογία); but this may have been just one section of a more general work.[55] Aristotle did not know it, or thought little of it, for he never names Diogenes in his *Meteorologica*. To Democritus is attributed a work entitled Αἰτίαι ἀέριοι, "Aerial causes"[56]; but this work was possibly spurious,[57] and possibly not about meteorology.[58] Few, if any, pre-Socratic philosophers can have written much in defence of their meteorological theories, or to explain why rival theories must be rejected.

The fifth century Sophists had no interest in meteorology (apart from Antiphon, to whom an account of hail is attributed[59]), nor had Socrates, at least in his maturity. Plato's lack of interest is manifest: even in the *Timaeus*, where he gives an account of the physical world, there seem to be only three places where he offers even partial explanations of meteorological phenomena: at 49C, writing about the changes of the four elements into each other, he mentions "cloud and mist" (νέφος καὶ ὀμίχλην) as a stage in the change from air to water; at 59E he explains hail and ice as formed by the complete freezing of water, and snow and hoar-frost as formed by its being half-frozen (ἡμιπαγές); at 80C he mentions the falling of thunderbolts (τὰ τῶν κεραυνῶν πτώματα) in the course of an explanation of how movements occur although there is no void.

Aristotle, other Peripatetics and Epicurus

Aristotle deals with the natural world at much greater length than Plato, and could not ignore meteorology. I doubt it was central to his interests. Unlike his astronomy, it was not crucial to his understanding of the nature of the cosmos and of the divine[60]; and another aspect of the natural world, biology, was one that he wrote on at far greater length than on meteorology.[61] Nevertheless, *Mete.* I–III was, almost certainly, the most detailed study of meteorology that any Greek thinker had yet written, and one which was in one respect strikingly original, in that Aristotle bases his account of almost all the phenomena on a new theory of two exhalations, one dry and one wet.[62] In another respect Aristotle follows the tradition of almost all the pre-Socratics: there is no mention of final causes. I suspect that he wrote *Mete.* I–III more because meteorology was a traditional part of a thorough study of the natural world than because he felt deeply interested in the subject; and so he follows the tradition of describing only material and efficient causes, and does not mention final causes at all. He very likely thought that the final causes of meteorological events are non-existent,[63] or else obscure, but it is surprising that he does not discuss the matter, as he does in rather similar circumstances in *De generatione et corruptione* and *Mete.* IV.[64] In *Physics* II, Chapter 8, he speaks of rain in winter (as opposed to rain in summer) being normal, and therefore natural; and so, on Aristotle's principles, it should have a final cause – causing the crops to grow is mentioned.[65] There is nothing like this in *Mete.* I–III.

Aristotle's successor, Theophrastus, also wrote at some length about meteorology. The extant *De ventis* discusses and endeavours to explain the properties of particular winds and kinds of wind: the book begins with a reference to an earlier work which dealt with the nature and causes of winds in general, and *De ventis* 5 indicates that rain was dealt with in another work. Diogenes Laertius' list of Theophrastus' works includes two which were evidently about meteorology in general: Μεταρσιολεσχία, in one book, and Μεταρσιολογικά, in two books.[66] The Syriac Meteorology claims to be by Theophrastus, and deals with thunder, lightning, thunderbolts, clouds, rain, snow, hail, dew, hoar-frost, wind (including *prēstēr*), lunar haloes and earthquakes, but there are serious grounds for doubting whether this work accurately represents Theophrastus.[67] There is independent confirmation of the breadth of his study of meteorology: Proclus Diadochus mentions that he wrote on the causes of thunder, wind, thunderbolts, lightning, *prēstēr*, rain, snow and hail,[68] and Seneca speaks of his theory of earthquakes (as identical with Aristotle's)[69]; but for most phenomena there is no reliable information about what view Theophrastus held.

Both *De ventis* and the Syriac Meteorology are concerned exclusively, or almost exclusively, with the material and efficient causes of meteorological phenomena. *De ventis* says nothing about final causes; the Syriac meteorology has only the two sentences "Neither the thunderbolt nor anything that has been mentioned has its origin in God. For it is not correct (to say) that God should be the cause of disorder in the world".[70] This effectively denies purpose to meteorological phenomena (and ignores the purpose which was seen in seasonal rain) – but it is doubtful whether this statement in the Syriac Meteorology is really Theophrastus' view.[71] In his *Metaphysics* Theophrastus expresses doubt about final causes in general, but seems to allow that the seasons of the year may have a purpose, and says that enquiry is needed[72]; a later passage includes influxes and refluxes of the sea, and droughts and ὑγρότητες (excessive humidity? flooding?), among phenomena which appear purposeless.[73] In effect he left intact the tradition of confining the study of meteorology to material and efficient causes.

Theophrastus was not the only pupil or successor of Aristotle to put forward meteorological ideas. We have records of his pupil Callisthenes' ideas about earthquakes[74] and the Nile floods.[75] More important was Strato, Theophrastus' successor as head of the Peripatetic school, who was known as φυσικός, "student of nature".[76] Strabo attributes to him an elaborate theory about the Black Sea, Mediterranean and Atlantic: about their past and possible future changes, and flows of water into and between them.[77] We also have reports of Strato's theories of earthquakes,[78] of comets, and of thunder, lightning and related phenomena.[79] But his interest in the subject seems to have been limited: no work definitely on meteorology is attributed to him.[80]

Another thinker seriously concerned with meteorology was Epicurus (a contemporary of Strato). No obviously meteorological title is attributed to him, but we may be sure that he dealt with it at some length in his 37 books *On nature*.[81] His teaching on astronomy and meteorology is summarised in his *Letter to Pythocles*.[82] In addition to the sun, moon, stars and planets, the letter deals with cloud, rain, thunder, lightning, thunderbolt, *prēstēr*, earthquake, wind, hail, snow, dew, hoar-frost, ice, rainbow, lunar halo, comets, shooting stars and weather-signs; that is, nearly all the phenomena with which I am concerned. Sometimes his atomic theory suggests theories to him which were unavailable to non-atomists, as when he says that clouds may be formed παρὰ περιπλοκὰς ἀλληλούχων ἀτόμων, "by entanglements of atoms that hold on to each other",[83] presumably by being hook-shaped; but generally his meteorological explanations are similar in character to those of writers of other philosophical schools. I discuss his approach to meteorology in more detail, and compare it with that of the Stoics and Posidonius, in Chapter 19.

Epicurus died in 271/70 B.C., Strato in 270/69 or 269/68.[84] After that, for over 150 years no-one wrote at length about the causes of meteorological phenomena (or, if anyone did, the record of it is lost). The philosophers of that period had little or no interest in the subject. Among the Peripatetics, Strato's successors generally did not share his interests, though there are minor Peripatetic works which deal with meteorological causes and which may date from this period: Book XXVI, on winds, of the Pseudo-Aristotelian *Problemata*, and the meteorological section of the *De mundo* (though I have argued above, and it is the commonly held view, that the latter work is probably later than Posidonius).[85] Among the Epicureans no-one other than Epicurus is recorded as having written about meteorology.[86] (Lucretius is an exception, but his meteorology was almost certainly a little later than that of Posidonius.[87]) The philosophers of the Academy and of the lesser schools of philosophy (Megarics, Cyrenaics, Cynics, the earliest Sceptics) had little or nothing to say on the subject. Among philosophers, there remain the Stoics: they had little interest in meteorology, but did not ignore the subject completely, and, as Posidonius was a member of their school, it is important to see what they are recorded as having said about it.

The Stoics before Posidonius

Lists of the works of six early Stoics (Zeno, Aristo, Herillus, Dionysius, Cleanthes and Sphaerus) given in Diogenes Laertius Book VII include no titles on meteorology; however, Diogenes Laertius reports Zeno's view of lightning, citing his Περὶ τοῦ ὅλου, *On the whole* (i.e., "on the universe"), and Seneca reports Zeno's theory of comets. Diogenes' list of Chrysippus' works breaks off before reaching works on physics, but meteorology cannot have been a major interest of his, since opinions on meteorological phenomena are attributed to him by name only in three passages of Stobaeus, on the seasons of the year, on mist, rain and related phenomena, and on lightning and related phenomena.[88]

The Stoics of course did not share Epicurus' conviction that no god is concerned in the workings of our world. Cleanthes in his *Hymn to Zeus* writes in traditional terms about Zeus as controller of the thunderbolt:

Τοῖον ἔχεις ὑποεργὸν ἀνικήτοις ὑπὸ χερσὶν
ἀμφήκη πυρόεντα ἀειζώοντα κεραυνόν

So serviceable a thunderbolt you have in your unconquerable hands, forked, fiery, ever-living.

This is allegorical, however, for the writer goes on:

τοῦ γὰρ ὑπὸ πληγῆς φύσεως πάντ' ἔργα <τελεῖται>
ᾧ σὺ κατευθύνεις κοινὸν λόγον, ὃς διὰ πάντων
φοιτᾷ

for by its blows all the works of nature are accomplished, and by it you direct the common plan, which goes through all things.[89]

The thunderbolt is here a means by which Zeus directs the world. There is other evidence, which I quote below, that Cleanthes did speak of actual meteorological events, such as thunderbolts, as caused by the gods; but the Stoics would not have found it difficult to combine the idea of god as directing literal thunder, lightning and other such phenomena with a naturalistic explanation of them. In their view there is "part of god which pervades all things" (τὸ μέρος αὐτοῦ [sc. θεοῦ] τὸ διῆκον διὰ πάντων)[90]; if, then (for example), thunder and lightning are due to wind bursting a cloud – a naturalistic explanation reportedly given by the Stoics,[91] and similar to views held by other ancient thinkers[92] – there is, to a Stoic, something of the divine pervading the wind and the cloud, which could, from one point of view, be regarded as causing the thunder and lightning.

Cleanthes' *Hymn to Zeus* was apparently not the only place where the early Stoics linked meteorology with allegory. The scholiast on Hesiod, *Theogony* 139,[93] indicates that Zeno interpreted the *cyclopes* of that line, Brontes, Steropes and Arges, as thunder, lightning and thunderbolt, and says that they are said to be sons of *ouranos*, the sky, because these happenings occur περὶ τὸν οὐρανόν, "about the sky".

There is evidence that the early Stoics were reluctant to discuss the causes of natural phenomena. Strabo, in a passage already referred to, implies that, as a Stoic, he need not concern himself with Posidonius' un-Stoic discussions of natural phenomena (ὅσα φυσικώτερα): πολὺ γάρ ἐστι τὸ αἰτιολογικὸν παρὰ αὐτῷ καὶ τὸ Ἀριστοτελίζον, ὅπερ ἐκκλίνουσιν οἱ ἡμέτεροι διὰ τὴν ἐπίκρυψιν τῶν αἰτιῶν, "for there is much enquiry into causes in him, that is, 'Aristotelising', a thing which our school [the Stoics] sheers off from because of the concealment of causes".[94] To some extent our records of early Stoic meteorology bear this out. There is an account in Stobaeus of Chrysippus' account of the seasons[95]:

Χρυσίππου … Ἔαρ δὲ ἔτους ὥραν κεκραμένην ἐκ χειμῶνος ἀπολήγοντος καὶ θέρους ἀρχομένου· ἢ τὴν μετὰ χειμῶνα ὥραν πρὸ θέρους. Θέρος δὲ ὥραν ἔτους τὴν μάλισθ' ὑφ' ἡλίου διακεκαυμένην. Μετόπωρον δὲ ὥραν ἔτους τὴν μετὰ θέρος μέν, πρὸ χειμῶνος δὲ κεκραμένην. Χειμῶνα δὲ ὥραν ἔτους τὴν μάλιστα κατεψυγμένην ἢ τὸν περὶ γῆν ἀέρα κατεψυγμένον.

Chrysippus … Spring is a season of the year mixed from winter ceasing and summer beginning; or, the season after winter before summer. Summer is a season of the year most burnt by the sun. Autumn is a mixed season of the year, after summer but before winter. Winter is the season of the year that is most chilled, or the air chilled around the earth.

The sun here causes the summer's heat, but no cause is mentioned for the other three seasons. Contrast the account of the seasons in Diogenes Laertius' summary of Stoic meteorology[96]: a summary which cites Posidonius three times (and Zeno only once), making it likely that the account of the seasons, for which no individual is named, is that of Posidonius:

Χειμῶνα μὲν εἶναί φασι τὸν ὑπὲρ γῆς ἀέρα κατεψυγμένον διὰ τὴν τοῦ ἡλίου πρόσω ἄφοδον, ἔαρ δὲ τὴν εὐκρασίαν τοῦ ἀέρος κατὰ τὴν πρὸς ἡμᾶς πορείαν, θέρος δὲ τὸν ὑπὲρ γῆς ἀέρα καταθαλπόμενον τῇ τοῦ ἡλίου πρὸς ἄρκτον πορείᾳ, μετόπωρον δὲ τῇ παλινδρομίᾳ τοῦ ἡλίου ἀφ’ ἡμῶν γίνεσθαι.

They say that winter is the air above the earth having been chilled because of the sun's departure to a distance [sc. from us], spring is the pleasant temperature of the air due to [sc. the sun's] movement towards us, summer is the air above the earth heated by the sun's movement to the north, and autumn occurs by the sun's going back again away from us.[97]

A cause is here given for every season.[98]

The early Stoics, however, could not entirely avoid mention of causes when they wrote of meteorological phenomena. Another passage of Stobaeus shows, I think, an attempt to omit causation:

Χρύσιππος ἔφησε τὴν ὁμίχλην νέφος διακεχυμένον, ἢ ἀέρα πάχος ἔχοντα· δρόσον δὲ ἐξ ὁμίχλης καταφερόμενον ὑγρόν· ὑετὸν δὲ φορὰν ὕδατος ἐκ νεφῶν· ὄμβρον δὲ λάβρου ὕδατος καὶ πολλοῦ ἐκ νεφῶν φοράν· χάλαζαν δὲ ὑετοῦ πεπηγότος διάθρυψιν· χιόνα δὲ νέφος πεπηγὸς ἢ νέφους πῆξιν· τὸ δ’ ἐπὶ τῆς γῆς πεπηγὸς ὕδωρ κρύσταλλον· πάχνην δὲ δρόσον πεπηγυῖαν.

Chrysippus said that mist is thinned cloud or air that has become thick. Dew is moisture carried down from mist. Rain is movement of water from cloud. A rainstorm is water from clouds moving violently and in great quantity. Hail is the breaking up of frozen rain. Snow is frozen cloud or freezing of a cloud. Water frozen on the earth is ice. Hoar-frost is frozen dew.[99]

One might say that no causes are mentioned here: there is no suggestion, such as we find in other ancient authors, that cold or compression is causing these phenomena.[100] What is said of rain, rainstorm, ice and hoar-frost is purely descriptive of what is obvious to everyday observation. But in what is said of mist, dew, hail and snow there is, if not a cause, at least a process which we cannot observe. The text implies, as one would expect, that hail and snow are produced by different processes (what these processes are I discuss later[101]), and to postulate the process of production is to postulate something of the cause.

Aëtius III.3.12[102] describes Chrysippus' theory of thunder, lightning and related phenomena:

Χρύσιππος ἀστραπὴν ἔξαψιν νεφῶν ἐκτριβομένων ἢ ῥηγνυμένων ὑπὸ πνεύματος, βροντὴν δ’ εἶναι τὸν τούτων ψόφον ... ὅταν δ’ ἡ τοῦ πνεύματος φορὰ σφοδροτέρα γένηται καὶ πυρώδης, κεραυνὸν ἀποτελεῖσθαι, ὅταν δ’ ἄθρουν ἐκπέσῃ τὸ πνεῦμα καὶ ἧττον πεπυρωμένον πρηστῆρα γίγνεσθαι, ὅταν δ’ ἔτι ἧττον ᾖ πεπυρωμένον τὸ πνεῦμα, τυφῶνα.

Chrysippus (says that) lightning is the ignition of clouds that are rubbed or broken by wind, and thunder is the sound of this. ... But when the motion of the wind is more violent and fiery a thunderbolt is produced; when the wind falls [sc. from the cloud] in a mass and burning less fiercely a *prēstēr* occurs; when the wind is burning still less fiercely, a *typhōn*.

Diogenes Laertius reports Zeno as giving almost exactly the same explanation of lightning (and of thunder and thunderbolt, if Diogenes is still reporting him).[103] When writing of these phenomena, Zeno and Chrysippus, it seems, could not avoid adopting a theory of their cause, and the same evidently applied to Zeno when writing about comets.[104] The early Stoics apparently tried to avoid speculative explanations of the causes of meteorological phenomena, but, when giving their accounts of the natural world, they did not entirely succeed in doing so; and when they did they gave, as all previous philosophers who dealt with the subject had done, naturalistic explanations.

The early Stoics, then, did speak about the material and efficient causes of at least some meteorological phenomena; had they anything to say about final causes, about how such phenomena fitted into their concept of a world controlled by divine providence? Among their predecessors, Diogenes of Apollonia regarded the seasonal changes from winter rain to spring and summer heat, which enable plants to grow, as proof that intelligence governs the world, the same idea occurs in Plato,[105] and Aristotle and Theophrastus mention seasonal weather changes as possible examples of a final cause of meteorological events. There is evidence that the first Stoics spoke of these changes as the work of a beneficent providence. In *De natura deorum* Book I Cicero makes Velleius, as defender of Epicureanism, complain that Zeno has represented as gods entities which are clearly not gods in the ordinary sense; one instance is that Zeno attributed *vis divina*, divine power, "annis mensibus annorumque mutationibus", "to years, months, and changes of the years", presumably meaning seasons.[106] Eusebius (perhaps following Arius Didymus) says that Cleanthes regarded the sun as the ruling principle (ἡγεμονικόν) of the cosmos, in part, διὰ τὸ ... πλεῖστα συμβάλλεσθαι πρὸς τὴν τῶν ὅλων διοίκησιν, ἡμέραν καὶ ἐνιαυτὸν ποιοῦντα καὶ τὰς ἄλλας ὥρας, "because it contributes most to the running of the world, causing day and year and the other divisions of time"[107]: ὥρας here cannot mean "seasons" exclusively, but it surely includes them. Cicero, in his account of Stoic theology, reports Cleanthes as describing four causes of human belief in the gods. One of them we take "ex magnitudine commodorum quae percipiuntur caeli temperatione fecunditate terrarum aliarumque commoditatum complurium copia", "from the magnitude of the benefits which we derive from our temperate climate, from the earth's fertility, and from a vast abundance of other blessings".[108] Here "caeli temperatione", "temperate climate", must be a reference to the climate known to the Greeks and its seasonal changes. The argument from seasonal changes as a proof of divine providence recurs in Cicero's account of Stoic theology at *De natura deorum* II.19, but then it is combined with mention of tides, so is very likely derived from Posidonius.[109]

As another cause of human belief in the gods, Cleanthes mentioned "quae terreret animos fulminibus tempestatibus nimbis nivibus grandinibus ...", "that which terrifies our minds by thunderbolts, storms, cloud-bursts, snow, hail ..." (and other phenomena, including earthquakes and comets).[110] This may well be true as an historical statement; given its context, we must also assume that Cleanthes, as a believer in the gods, thought it a valid argument in proof of their existence, suggesting (though this is not in Cicero's text) that he thought the purpose of these frightening phenomena was to inspire awe of the gods in human minds, or to warn evildoers of coming divine punishment. Whatever Cleanthes' thought was, it cannot have convinced his successors in the Stoic school, since we do not find this argument attributed to them; perhaps it seemed to them, as Sandbach says, "inconvenient for believers in Providence".[111]

In the surviving accounts of early Stoic thought, these seem to be the only hints of a divine purpose in meteorological events, and they are from accounts of Stoic theology; texts which are actually about early Stoic meteorology, if they mention causes, confine themselves to material and efficient ones.

If the early Stoics had little interest in meteorology, their successors down to the time of Posidonius seem to have had even less.[112] No work on meteorology is attributed to any of them. We have only a report that Antipater interpreted a myth about Apollo slaying a dragon as referring to the sun dispersing dark, pestiferous exhalations[113]; two brief reports of views attributed to Boethus, on comets[114] and on weather-signs[115]; and two fragments of Panaetius, on comets and on the habitability of equatorial regions.[116] Posidonius, when he undertook a major study of meteorology, was, for a Stoic, definitely an innovator.

The contribution of non-philosophers

So far I have dealt only with men counted as philosophers who concerned themselves with meteorology, and I have dealt only with men who discussed the causes of meteorological phenomena. It was, almost always, philosophers who offered general explanations of phenomena such as thunder and lightning, or wind, or rain; I discuss in a later chapter passages in ancient authors (possibly influenced by Posidonius) in which the causes of meteorological phenomena are spoken of as specially, or to a great extent, the concern of philosophers.[117] But writers in other fields did also concern themselves with aspects of meteorology. Eudoxus of Cnidus lived circa 390 to circa 340 B.C. and is best known as mathematician and astronomer; he drew up, like Democritus, a calendar which included the weather to be expected on particular days of the year,[118] and offered a theory, said to have been learned from Egyptian priests, about the summer floods of the Nile.[119] The latter subject was also a concern of historians: we have a lengthy discussion of it in Herodotus,[120] and several reports of the theory proposed by Ephorus.[121] Euthymenes of Massilia, a sailor who had voyaged along the west coast of Africa, put forward a theory of the floods on the basis of what he had seen there[122]; Nicagoras, thought to have been a "paradoxographer", also had ideas on the subject[123] – my list is not exhaustive. All these writers were, or are believed to have been, earlier than or older contemporaries of Aristotle, and it is doubtful if any of them influenced Posidonius' meteorology. In the generations which followed, besides the few men I have mentioned who discussed meteorological causes, there were those who studied other aspects of the subject. We have the poetry of Aratus about weather-signs, and the remains of a number of *parapēgmata*, the calendars, like those of Democritus and Eudoxus, which included the weather to be expected on particular days.[124] There were also men who certainly, or probably, influenced Posidonius' meteorology although they themselves were not concerned, or only marginally concerned, with the subject, and who are regarded not as philosophers, but rather as geographers, or astronomers, or historians. Their contribution will be one of the subjects of the next chapter.

Notes

1 Hesiod, *Th.* 706–7.
2 West (1971) 97 draws attention to the importance of the ideas found in these two passages for later meteorological theorising.
3 On Thales and his ideas about water see Kirk, Raven and Schofield (1983) 76–99 (on water pp. 88ff); Guthrie (1962) 45–71 (on water pp. 54ff).
4 In Gigon (1987), no. 695, cf. nos. 688, 690.
5 Burstein (1976) 136 n.3 lists scholars who have accepted this work as Aristotle's (as Burstein does himself). Sharples (1998) 197–8, especially note 571, lists both scholars who have accepted and those who have rejected Aristotle's authorship, including some who maintain that the author is Theophrastus (i.e., an author scarcely later than Aristotle). More recently, Jakobi and Luppe (2000) treat the work as genuinely by Aristotle; Fowler (2000) argues against Aristotelian authorship; Beullens (2014) 322–4 is tentatively in favour of it.
6 Herodotus II.20 (printed in DK 11A16) describes the theory but names no author. Later authors who attribute the theory to Thales include Diogenes Laertius I.37 (DK 11A1); Aëtius IV.1.1 (DK 11A16); Diodorus I.38.2; Seneca *NQ* IVa.2.22.
7 Seneca *NQ* III.14.1 (DK 11A15) and VI.6.1.
8 In Simplicius, *In Ph.* 24.13ff (DK 12A9). See Kirk, Raven and Schofield (1983) 102, 105–7.
9 Diogenes Laertius II.3 (DK 13A1).
10 Guthrie (1962) 72.
11 Alexander, *In Mete.* 67.3ff Hayduck, commenting on Aristotle, *Mete.* 353b6 ff (DK 12A27): on wind and the sea.
12 Hippolytus, *Haer.* I.6.7 (DK 12A11): on wind, rain and lightning.
13 Aëtius III.3.1, III.7.1 and III.16.1 (DK 12A23, 24, 27): on thunder and lightning, wind and the sea.
14 Seneca, *NQ* II.18 (DK 12A23): on thunder and lightning.
15 Ammianus XVII.7.12 (DK12A28): on earthquakes.
16 Aristotle, *Mete.* 365b6 ff (DK 13A21).
17 Hippolytus, *Haer.* I.7.3 and 7–8 (DK 13A7): on wind, cloud, rain, hail, snow, lightning, rainbows and earthquakes.
18 Aëtius III.3.2, III.4.1, III.5.10. III.15.3 (DK 13A17, 18 and citation at 13A21): on thunder and lightning, clouds, rain, hail, snow, rainbows and earthquakes.
19 Seneca *NQ* II.17, on thunder and lightning, and VI.10 (cited at DK 13A21) on earthquakes.

20 Galen in Kühn (1821–33) XVI.395.16–396.1(DK 13A19): on wind.
21 Simplicius, *In Ph.* 24.26 (DK 13A5): on wind and cloud.
22 Scholia to Aratus p. 515.27 M. (DK 13A18): on rainbows.
23 On this distinction see, for instance, Kirk, Raven and Schofield (1983) 213.
24 DK 21B30 and 32.
25 Aëtius II.18.1, III.2.11, III.3.6 (DK 21A39, 44, 45)
26 On these phenomena see Graham, Herzog and Williams (2021).
27 See fragments DK 59B16 and B19 and testimonia DK 59A1 (section 9), A42 (sections 10–12), A12, A80–6a, and A89–91.
28 Fragment DK 64B3 and testimonia DK 64A12 and A15–18.
29 DK 68A91–3a and A97–9a, and for weather-prediction DK 68B14 (the "*parapēgma*") and B147 (weather-signs).
30 DK 70A13–19 and A21.
31 Diogenes Laertius IX.9 (DK 22A1).
32 See below, Chapter 10 (p. 90).
33 Aëtius III.3.9 (DK 22A14).
34 Diogenes Laertius IX.10 (DK 22A1).
35 See DK 22B31 and 64.
36 Aristotle *Mete.* 369b12ff and Aëtius III.3.7 (DK 31A63).
37 Aëtius III.3.10 (DK 67A25).
38 Aristotle *Mete.* 370a10ff (DK 62,1).
39 Aëtius III.3.5 (DK 60A16).
40 Aristotle *Mete.* 342b29ff (DK 42,5) and 345a13–18; Aëtius III.1.2 (DK 58B37b and c).
41 Aristotle *Mete.* 342b35ff (DK 42,5), for comets, and *Mete.* 345b9ff with Alexander *In Mete.* p. 38.28ff Hayduck (DK 42,6), for the Milky Way.
42 Compare Achilles *Isag.* p. 55.18–21 Maass with Aristotle *Mete.* 345a13–16 (both in DK 41,10).
43 Diodorus I.41.1 (DK 41,11).
44 Strabo XVII.1.5 and I.2.21 (DK 35, 1 and 2)
45 Thunderstorms and thunderbolts: e.g. *Iliad* 13.796; 21.198–9; *Odyssey* 12.415–6. Rain, etc.: e.g., *Iliad* 10.5–8.
46 Most notably in Homer, at *Iliad* 23.194ff.
47 E.g., *Odyssey* 5.292ff (Poseidon); 3.289 (Zeus); 5.383ff (Athene).
48 *Odyssey* 5.268 (Calypso); 10.20ff (Aeolus); 12.149f (Circe).
49 DK 21B32, line 1.
50 E.g., *Iliad* 23.198ff.
51 *APo.* II.11, 94b33.
52 See fragment DK 64B3.
53 Diogenes Laertius IX.45–9 (DK 68A33).
54 Simplicius *In Ph.* 151.20 (DK64A4).
55 See Kirk, Raven, and Schofield (1983) 435–6.
56 Diogenes Laertius IX.47 (DK 68A33).
57 It is not included in Thrasylus' "tetralogies" of Democritus' works, so may be inauthentic: see Kirk, Raven, and Schofield (1983) 405–6.
58 Another author, Ion of Chios, is said by the *Suda* to have written περὶ μετεώρων (DK 36A3), but if this is meant to be a book-title it is surely wrong: other authors who give details of Ion's works mention nothing like this (see DK 36).
59 DK 87B29 (from Galen *In Epid.* III.32).
60 So most clearly at *Met.* Λ, chapter 7,1072a19ff: the unmoved mover, identified with god, causes the revolution of the heavens, and so on it depend ὁ οὐρανὸς καὶ ἡ φύσις, "the heavens and nature", or, as Tredennick (1933–5) renders it, "the sensible universe and the world of nature" (1072b14).

61 *Mete*. Books I–III fill 40 Bekker pages, the biological works more than 300.
62 First described in detail at *Mete*. 341b6ff. See further Chapter 10 below (p. 91–2).
63 As Wilson (2013) 93ff argues.
64 *GC* II.9 335b6f, II.10 336b25ff; *Mete*. IV.12 389b28–390a17.
65 On rain see *Physics* II.8, 198b18–20 and 198b36–199a5.
66 Diogenes Laertius V.43 and 44.
67 See above, Chapter 3 (p. 16–17).
68 Proclus Diadochus, *In Platonis Timaeum commentaria* 176E (p. 121 lines 3–5 Diehl). (I owe this reference to Daiber [1992] 284).
69 Seneca *NQ* VI.13.1.
70 Syriac Meteorology 14.(14)–(16) (Daiber (1992) 270).
71 See Van Raalte (2003).
72 *Metaphysics* 4, 7a19–b5.
73 *Metaphysics* 9, 10a28ff.
74 Seneca *NQ* VI.23.2–4.
75 See below, Chapter 15 (p. 153–4).
76 Polybius XII.25c3, Cicero *De finibus* V.5.13, Diogenes Laertius V.58, etc.
77 See Strabo I.3.4–5 (= Strato fr. 91 Wehrli).
78 Seneca *NQ* VI.13.2 (=Strato fr. 89 Wehrli).
79 Aëtius III.2.4 and III.3.15 (= Strato frs. 86 and 87 Wehrli). Wehrli (1969a) also prints, as Strato fr. 88, a passage of Hero's *Pneumatics* about changes between the four elements which includes explanations of dew and of wind. But there is no direct evidence that this is from Strato.
80 See list of his works in Diogenes Laertius V.58–60. (The work Περὶ τοῦ πνεύματος in that list may have been about wind but was perhaps more likely about breath. Wehrli [1969a] has no information about this work except its title.)
81 See list in Diogenes Laertius X.27–28.
82 Preserved in Diogenes Laertius X.83–116. It has been doubted whether this letter is genuinely by Epicurus, but there is no reason to doubt that it contains Epicurean doctrine. See Bailey (1926) 275.
83 *Letter to Pythocles* 99.
84 Dorandi (1999) 36, 43, 50–51.
85 On the date of the *De mundo* see above, Chapter 3 (p. 13–14).
86 There are no titles of works likely to be about meteorology in the lists of works by Metrodorus of Lampsacus and Hermarchus in Diogenes Laertius X.24–5, nor are any titles on this subject known from the Herculaneum papyri (see Gigante [1979] 45ff).
87 George Boys-Stones suggested to me two likely Epicurean writers on meteorology, probably contemporary with Lucretius: Egnatius (on whom see Uden [2021]) and Catius, who wrote on the Epicurean theory of vision (Cicero, *Fam.* 15.16). If they wrote on meteorology from an Epicurean viewpoint, we have no evidence for what they said.
88 On these reports of Zeno and Chrysippus see above, Chapter 3 (p. 16).
89 *SVF* I.537, lines 5–9. There is doubt about the text of line 7, but the point is clear.
90 Diogenes Laertius VII.147.
91 Diogenes Laertius VII.153–4. (See below, p. 28–9.)
92 E.g. Epicurus, *Letter to Pythocles* 100–3.
93 *SVF* I.118.
94 Strabo II.3.8 (Posidonius T85 EK) with the translation of Kidd (1999) 58–9. (I have already referred to this passage above, Chapter 3 [p. 11]).
95 Stobaeus, *Ecl.* I p. 106.24 W. (= Arius Didymus fr. phys. 26 Diels, part of *SVF* II.693).
96 Diogenes Laertius VII.151–4.

97 Diogenes Laertius VII.151–2 (part of *SVF* II.693).
98 Posidonius' explanation – if it was his – was not new: Aristotle *Mete.* 361a12–14 gives a similar explanation of summer and winter.
99 Stobaeus, *Ecl.* I p. 245.23 W. = *SVF* II.701 (Arius Didymus fr. phys. 35 Diels).
100 Cold: e.g., Aristotle *Mete.* 346b24–32, 347b18, 348b2–22; compression: e.g., Epicurus, *Letter to Pythocles* 99.
101 See Chapter 14, pp. 143–6.
102 Aëtius III.3.12 in Mansfeld and Runia (2020) = *SVF* II.703 (Stobaeus, *Ecl.* I, p. 233,9 W).
103 Diogenes Laertius VII.153–4 (*SVF* I.117).
104 Seneca *NQ* VII.19.1 (*SVF* I.122). See below, Chapter 9 p. 78.
105 *Philebus* 30C.
106 *De natura deorum* I.36 (*SVF* I.165).
107 Eusebius, *Praeparatio evangelica* XV.15.7 (= Arius Didymus fr. 29 Diels; *SVF* I.499).
108 Cicero, *De natura deorum* II.14, with the translation of Rackham (1933).
109 See below, Chapter 18 p. 181–2.
110 Cicero, *De natura deorum* II.14.
111 Sandbach (1975) 70.
112 I base this paragraph on an examination of *SVF* III p. 209ff ("Chrysippi discipuli et successores") and, for Panaetius, van Straaten (1962).
113 *SVF* III p. 250 (Antipater text 46 = Macrobius I.17.57.).
114 Aëtius III.2.7 (*SVF* III p. 267). See below, Chapter 9 n.30 (p. 86).
115 Cicero, *Div.* I.13 and II.47 (*SVF* III p. 265). See below, Chapter 17 (p. 175–6).
116 Fragments 75 and 135 Van Straaten.
117 Galen, *Institutio logica*, XIII, and Diogenes Laertius VII.132–3, discussed below, Chapter 20 (p. 192–3).
118 Eudoxus fragments 146–267 in Lasserre (1966).
119 Eudoxus fragments 287 and 288 in Lasserre (1966).
120 Herodotus II.19–26.
121 E.g., Diodorus I.39.7; Seneca as summarised by Johannes Lydus *De mensibus* IV (from the lost ending of *NQ* IVa; printed with Seneca's text in Oltramare [1961]); Aëtius IV.1.6.
122 Seneca *NQ* IVa.2.22; cf. Aëtius IV.1.2. On Euthymenes see Jacoby (1907).
123 Aristotle(?), *De inundatione Nili*; *Scholia in Apollonium Rhodium vetera* IV.269. On Nicagoras see Nicagoras (3) in *Brills' New Pauly* (2002–10) 9, 705 (article by F. Lasserre).
124 See Taub (2003) 20ff.

5 Earlier authors on meteorology used by Posidonius

Posidonius rejected the view of the earlier thinkers of his school, that meteorology was not worth serious attention, and made it the object of a major study. Why he did this it will be best to discuss after examining the recorded details of his ideas on the subject, but it is worthwhile to consider now which earlier writers are likely to have been important to him as he developed his own work on meteorology.

He was a Stoic, and accepted the principles of Stoic physical theory, so would, we may assume, have looked first at what had been said on meteorology by his Stoic predecessors; but, as we have seen, that did not amount to much. There was a wide choice among thinkers of other schools who had written on meteorology, and he might have taken ideas about particular phenomena from almost any of them. If an ancient explanation of a meteorological phenomenon was to make sense to those who read it, it had to be, at least to some extent, expressed in terms of other phenomena familiar to ancient readers – of the evaporation and condensation of water in domestic contexts, of friction, of the behaviour of bodies when heated or cooled, and so on – and these features of a meteorological theory could have been adopted by Posidonius even if the theory had been devised by a thinker, such as an atomist, whose general physical system was very different from Posidonius' own.

If, however, he sought a model on which to base a meteorological system, his choice was much more limited. The physical systems of most of the pre-Socratics were very different from that of the Stoics; the one pre-Socratic whom they took seriously and used as a source for ideas of their own, Heraclitus, seems to have had, as we have seen, little to say about meteorology.[2] Also, although several pre-Socratics had proposed explanations of a variety of meteorological phenomena, it is doubtful if any of them had written on the subject in detail.[3] Epicurus had written in some detail, but his atomism would have prevented Posidonius from basing a system on his ideas, and he would also have found unacceptable Epicurus' insistence on multiple explanations for each phenomenon. Posidonius may occasionally have accepted multiple explanations for a meteorological phenomenon (thunder and lightning is a possible example[4]), but this was unusual for him; there is something unsatisfactory in a view that several different causes all produce the same result, unless one can

DOI: 10.4324/9780429399930-5

show that each different cause produces a characteristic variation in the result, which Epicurus, to judge from the *Letter to Pythocles*, seems generally not to have done. (Epicurus' denial of Providence would have been less important in the study of meteorology, since, as we have seen, ancient writers, even if believers in Providence or final causes, generally ignored them when writing on this subject – Posidonius followed this tradition.)

Two other writers had written extensively on meteorology: Aristotle and Theophrastus. It is unlikely that Posidonius would have based his meteorology on that of Theophrastus: if Theophrastus ever discussed the relation of his meteorological ideas to the first principles of physics, it is likely that he did as he does when discussing first principles in his *Metaphysics* and the beginning of *De igne*, and raised doubts and asked questions without suggesting clear answers. That would have made him a difficult authority for a thinker to follow who hoped, as the evidence suggests Posidonius did, to devise a systematic meteorology. Posidonius could certainly have used Theophrastus as a source of answers to particular meteorological problems; however, such little evidence as there is does not suggest that he used Theophrastus' extant *De ventis* when discussing details of wind phenomena.[5]

I have already argued that the Syriac Meteorology cannot be regarded as a reliable source for Theophrastus, although it purports to be by him.[6] Its ideas are generally ones which existed in antiquity; but, assuming that a Greek original (whoever wrote it) existed in Posidonius' day, Posidonius could not have based his meteorology on it, with its multiple explanations of most phenomena and lack of reference to first principles. He might have taken from it answers to particular meteorological questions.

The case is different with Aristotle – a more distinguished philosopher than his pupil Theophrastus, and one whose importance, after a period of neglect, was coming to be generally recognised by philosophers of the later 2nd and of the 1st centuries B.C., and in particular by Posidonius' teacher Panaetius and by Posidonius himself.[7] Aristotle's *Meteorologica* did not have the disadvantages which, I have suggested, the meteorological works of Epicurus and Theophrastus had, or are likely to have had. It gives a detailed account of meteorology which starts from the first principles of Aristotle's element theory and which generally avoids multiple explanations. It was therefore a much more convenient work for Posidonius to consult; and note that Strabo, in a sentence already quoted,[8] specifically associates Posidonius' "enquiry into causes" (αἰτιολογικόν) with "Aristotelising" (Ἀριστοτελίζον). There is direct evidence that Posidonius knew the *Meteorologica*. In two places Strabo tells us that Posidonius spoke of Aristotle as holding views which Aristotle puts forward in *Meteorologica*. According to Strabo I.2.21 (F137a EK), a passage about wind names, Posidonius correctly reported the sense of Aristotle's account, at *Mete.* 363a25–b26, of the directions from which a number of winds blow. At II.2.2 (F49.10–43 EK) Strabo says that Posidonius criticised Aristotle's theory of the temperate zones: the sense (though with updated vocabulary) of the theory which Aristotle expounds at *Mete.* 362a32ff is accurately reported.[9] I discuss these passages in detail below.[10] Further evidence of Posidonius' knowledge of *Mete.* comes from theories held by Posidonius which *Mete.* has evidently influenced, most obviously his theory of thunder and lightning.[11]

Posidonius' use of Aristotle's *Mete.* must have been facilitated by the fact that, in several relevant respects, their world views were similar. To both of them, the elements which make up the earth and the region around it are earth, water, air and fire,[12] and these elements are, at least indirectly, transformable into each other[13]; the cosmos is not infinite, is spherical, and is unique[14]; the earth is also spherical, and is motionless.[15] To Aristotle,[16] to the Stoics,[17] and presumably to Posidonius, the earth is surrounded by concentric regions of water, air and fire.

A difficulty for Posidonius' use of Aristotle was the major difference in their concept of the heavens. For Aristotle, the material of the stars, planets, sun and moon, and of the heavens in which they exist, is not fire, air, water or earth, but an indestructible and changeless, eternally revolving, fifth element, which is not itself hot, but generates, by friction, heat in the sublunary elements below.[18] For the Stoics there is no such fifth element; for them, ἀνωτάτω μὲν οὖν εἶναι τὸ πῦρ, ὃ δὴ αἰθέρα καλεῖσθαι, ἐν ᾧ πρώτην τὴν τῶν ἀπλανῶν σφαῖραν γεννᾶσθαι, εἶτα τὴν τῶν πλανωμένων, "highest is fire, which is also called *aithēr*, in which the sphere of the fixed stars is first formed, then that of the planets" (and after that air, water and earth).[19] To Posidonius the sun is εἰλικρινὲς πῦρ, "pure fire",[20] and an ἄστρον (a star in a wide sense, including sun and moon) is λαμπρὸν καὶ πυρῶδες, "bright and fiery".[21] (The moon, however, is ἀερομιγῆ, "mixed with air",[22] or, in another source, μικτὴν ἐκ πυρὸς καὶ ἀέρος, "mixed from fire and air"[23] – presumably it is at the boundary of the regions of air and fire.[24]) This difference, as we shall see, affected the way in which Posidonius used some of Aristotle's theories.

As I have said, little was written that was specifically about meteorology between the death of Epicurus on 271/270 B.C. and the birth of Posidonius, probably in the 130s; but there were men who made important contributions on matters peripheral to the subject, which Posidonius had to take account of. Writers on geography had studied the climatic zones of the earth, advancing knowledge of the size of the zones and of their climates. The polymath Eratosthenes and the historian Polybius (both cited by Strabo together with Posidonius as writers on geography[25]) will be cited in the next section of this study for their work on these subjects. Dicaearchus, a pupil of Aristotle, had made the first attempts to measure the height of Greek mountains; as we shall see, these measurements were important to Posidonius' view of the atmosphere.[26] Seleucus – otherwise known for his denial that the earth is at rest[27] – was, in his study of the tides, an important predecessor of Posidonius, who, we know, cited him on this subject.[28] Another astronomer, Hipparchus, dealt with non-meteorological subjects which also interested Posidonius, such as the distance of the sun and the moon from the earth[29] and the shape of the οἰκουμένη (the inhabited world),[30] and also expressed an opinion about the tides.[31] The work of all these men on climatic zones, on mountain heights and on tides must have been known to Posidonius (even if, as is conceivable, only indirectly).

Posidonius could have found, and presumably did find, meteorological information in sources of other kinds also. One likely source is narrative history. Strabo speaks of the extension of geographical knowledge due to those who conquered previously unknown lands – Alexander the Great, the Romans, Mithridates Eupator, the Parthians[32]; Posidonius, historian as well as philosopher, must have read many historical narratives, including accounts of these conquests and of exploratory journeys which these and other rulers ordered, and will surely have derived meteorological information from them. Some of his accounts of particular earthquakes may well be derived from such sources[33]; his knowledge of the Nile floods is another possible example.[34]

Posidonius of course did not depend solely on written sources. He also acquired information from his own observations and experience, and from what people told him by word of mouth. His knowledge of the tides is one example which combines his own observation with the statements of others[35]; what he said of easterly "etesian" winds in the western Mediterranean is another[36]; and it seems highly probable that his knowledge of the "stony plain" in what is now southern France is partly derived either from personal autopsy or from local informants.[37] The possibility must be borne in mind in other instances also, that the authority for something said by Posidonius is either his own personal observation or the verbal statements of witnesses.

As we shall see, the facts which Posidonius learned from geographers and others who were not primarily writing about meteorology, or which he observed for himself of heard of from witnesses, are the basis of his most interesting and important contributions to the subject.

Notes

1 Cleanthes wrote four books of "Interpretations of Heraclitus" and a work on Heraclitus was among the works of Cleanthes' pupil Sphaerus (see Diogenes Laertius VII.174, 178 = *SVF* I.481 and 620). For comments on Heraclitus' importance to the Stoics see Guthrie (1962) 404; Kirk, Raven and Schofield (1983) 185.
2 See above, Chapter 4 (p. 22).
3 See above, Chapter 4 (p. 23).
4 See below, Chapter 8 (p. 70).
5 See below, Chapter 11 (p. 107).
6 See above, Chapter 3 p. (p. 16–17).
7 See above, Chapter 2 (p. 7).
8 Strabo II.3.8, quoted Chapter 4 (p. 27).
9 Sandbach (1985) 60 cites F49.10–43 EK, but not F137a EK, as evidence that Posidonius knew the *Mete.* Sandbach also cites, as evidence suggesting Posidonius' knowledge of *Mete.*, Alexander's statement that Posidonius accepted Aristotle's explanation of haloes (see below, Chapter 16, p. 164–5), and two passages from the scholia on Aratus and one from Strabo which cite Posidonius immediately before or after a citation of Aristotle: it is likely, Sandbach argues, that the writers took their knowledge of Aristotle from Posidonius. (These three passages are, from the

scholia, F131a EK, on comets, and F121 EK, on mock-suns; and, from Strabo, F229 EK, on the "Stony Plain". I discuss these passages below, see Chapter 9 [p. 79], Chapter 16 [p. 166–7] and Chapter 12 [p. 123].)

10 See below, Chapter 11 (p. 102), and Chapter 7 (p. 55–7).

11 See below, Chapter 8 (pp. 67, 69–71). On Posidonius' knowledge of Aristotle's *Mete.* see also Kidd (1988) 85; Pajón Leyra (2013) 726: in addition to passages referred to in the last paragraph and note 9, she cites Strabo III.1.5 (Posidonius F119 EK), on sunsets, which she compares with *Mete.* 342b5ff, and Strabo III.3.3 (F220 EK), on tides, on which see below, Chapter 13 (p. 131). In my opinion, these two passages do not add much to the argument, since what Strabo says does not closely resemble the related passages in *Mete.* (and Strabo III.1.5 does not mention Aristotle); but the evidence for Posidonius' use of *Mete.* is sufficient without them.

12 For Aristotle see (e.g.) *De caelo* IV.4, 311a15ff, *Mete.* I.2, 339a11ff.; for the Stoics, Diogenes Laertius VII.136. For Posidonius, Simplicius (*In cael.*, IV 3 = F93a EK) summarises various aspects of Aristotle's theory of the four elements and adds καὶ Ποσειδώνιος ὁ Στωϊκὸς παρὰ τούτων λαβὼν πανταχοῦ χρῆται, "and Posidonius the Stoic taking it from them [Aristotle and Theophrastus] uses it everywhere". It is not clear how much of Aristotle's theory Simplicius is attributing to Posidonius, but he surely implies that Posidonius accepted the four elements.

13 For Aristotle see *GC* II.4, 331a7–b2. For the Stoics, Diogenes Laertius VII.141 says τὰ δὲ μέρη τοῦ κόσμου φθαρτά· εἰς ἄλληλα γὰρ μεταβάλλει· φθαρτὸς ἄρα ὁ κόσμος, "the parts of the cosmos are perishable; for they change into each other; so the cosmos is perishable". Posidonius is mentioned in VII.142 as one of several authors who spoke about the coming-to-be and perishing of the cosmos, so he presumably accepted this view. (See further below, Chapter 10 [p. 92–?].)

14 Aristotle: cosmos not infinite, *De caelo* I.5, 273a5ff; spherical, *De caelo* II.4, 286b10ff; unique, *De caelo* I.8, 276a18ff. Posidonius: all three qualities, Diogenes Laertius VII.140 (F8 EK); spherical, Strabo II.2.1 (F49.7 EK); unique, Diogenes Laertius VII.138 (F14 EK).

15 Aristotle: earth spherical, *De caelo* II.14, 297a8ff; motionless, *De caelo* II.8, 289b6. Posidonius: earth spherical, Strabo II.2.1 (F49.7 EK) and Simplicius, *In Ph.* II.2 (F18.18–20 EK). The same passage (F18.39–45) implies that, for Posidonius, the earth is at rest (see below, Chapter 21 [pp. 202, 204]); Diogenes Laertius VII.145 says that was the Stoic view.

16 *De caelo* IV.4, 311a15ff.

17 Diogenes Laertius VII.137; also, for earth, water and air, VII 155.

18 On the properties of the fifth element see *De caelo* I 2–3, 268b11–270b32. On its being the material of the heavenly bodies, and its heating the sublunary regions by friction (which indicates that it is not itself hot), *De caelo* II 7, 289a11–35, also *Mete.* I.3, 340b4ff.

19 Diogenes Laertius VII.137.

20 Diogenes Laertius VII.144 (F17 EK).

21 Arius Didymus, *Epitome*, fr. 32 (F127 EK).

22 Diogenes Laertius VII.145 (F10 EK).

23 Aëtius II.25.5 (F122 EK).

24 Cf. Stobaeus, *Eclogae* I, p. 184.8 W. = *SVF* II 527: according to Chrysippus, the sphere of the moon is πλησιάζουσαν τῷ ἀέρι. Διὸ καὶ ἀερωδεστέραν φαίνεσθαι, "coming near to the air. For which reason it [the moon] appears more airy".

25 Strabo I.1.1; see also the end of I.2.1.

26 See below, Chapter 6 (p. 46–9).

27 Aëtius III.17.9.

28 See below, Chapter 13 (p. 131–2).

29 See below, Chapter 6 (p. 43–4).

30 Agathemerus, *Geographiae informatio* I.2 (= Posidonius F200a EK).

31 Strabo I.1.9 (= Posidonius F214 EK). It is unclear what Hipparchus' tidal theory was.
32 Strabo I.2.1.
33 See below, Chapter 12 (p. 119–20).
34 See below, Chapter 15 (especially p. 154).
35 See below, Chapter 13 (p. 130–3).
36 See below, Chapter 11 (p. 108).
37 See below, Chapter 12 (p. 123).

6 The region in which meteorological phenomena occur

Meteorological events occur in the space above the surface of the earth – this is generally true even with the wide definition of "meteorology" which I am using, though rivers, seas and earthquake are exceptions. What view did the Greeks, and Posidonius in particular, have of that space?

The region of air and the space above it

At latest by the second half of the 5th century B.C. the Greeks were agreed that the space immediately above the earth's surface is filled with a substance, normally invisible, which they usually called ἀήρ, "air". Philosophers take it for granted, as Empedocles does in his account of breathing and the *clepsydra*,[1] and Democritus in his account of vision.[2] The same is true of non-philosophical authors: we find typically, for example, in Herodotus αἰθρίου ... ἐόντος τοῦ ἠέρος, "the air being clear"[3]; in Sophocles ὦ φάος ἁγνὸν καὶ γῆς ἰσόμοιρ' ἀήρ, "O holy light and air that shares earth equally with it"[4]; in Euripides δι' ἀέρος ... ποτανοὶ ... οἰωνοί, "birds flying through air".[5]

It was, then, established long before Posidonius' time that ἀήρ, "air", fills the space immediately above the earth. It was also widely held, by the Stoics among others, that the heavenly bodies and the region in which they are located are different in nature from the earth and the air immediately around it, and that the heavenly bodies are divine. The divinity of the heavenly bodies was familiar from Greek myth,[6] and was thought to be confirmed by observation and calculation. The construction of *parapēgmata*, which began in the 5th century B.C., and tried to relate weather phenomena to astronomical events such as the risings and settings of stars,[7] must have driven home an obvious truth, that astronomical events, unlike the weather, occur with complete regularity, and appear to be doing so eternally. The theories of Eudoxus and Callippus seemed to show that the apparently irregular movements of sun, moon and planets might be produced by a combination of regular circular motions.[8] Eternal, regular movements were thought to be evidence of divinity.[9]

These views were adopted by Plato. In the *Timaeus* he says that the creator τοῦ ... θείου τὴν πλείστην ἰδέαν ἐκ πυρὸς ἀπειργάζετο, "made the form of the divine mostly from fire", and placed it περὶ πάντα κύκλῳ τὸν οὐρανόν,

DOI: 10.4324/9780429399930-6

"in a circle about the whole sky".[10] From this cause there came to be ὅσ' ἀπλανῆ τῶν ἄστρων ζῷα θεῖα ὄντα καὶ ἀΐδια καὶ κατὰ ταὐτὰ ἐν ταὐτῷ στρεφόμενα ἀεὶ μένει, "the fixed stars, being animals, divine and eternal, and they always continue turning in the same way in the same position [sc. in the sky]".[11] In the *Laws*, if not with complete clarity in the *Timaeus*, Plato affirms that the planets, too, have constant, regular circular motions.[12] In the *Timaeus* it is not just the heavenly bodies themselves that are fiery. Plato indicates that each of the four elements has its own region: τὰ πλήθη τῶν γενῶν τόπον ἐναντίον ἄλλα ἄλλοις κατέχειν, "the main masses of the Kinds occupy regions opposite to one another".[13] He implies that there is a region of fire high above the earth, with a region of air between the earth and the fire.[14]

Aristotle expressed more clearly this idea of the natural places of the four elements, placing water above earth, air above water, and fire above air.[15] He also accepted that the heavenly bodies are divine,[16] but held that they are formed from an element different from these four,[17] which he liked to call "the first body" or "first element".[18] This element is "eternal, has neither increase nor diminution, but is ... unalterable and affected by nothing"[19]; its motion is circular,[20] it "cannot have weight or lightness" and "it is not possible for it to be moved towards the centre or away from the centre [sc. of the cosmos]".[21] It is not itself hot, nor are the stars, sun and moon within it hot, but by friction it heats the air or elemental fire which is immediately below it, and so causes the movements and changes that occur in the sublunary region.[22]

What is important to Aristotle's concept of μετεωρολογία – and consequently to the concept of meteorology which was taken up by at least some Stoics, and which I am using in this book[23] – is the almost complete separation of this "first element" from the sublunary world of earth, water, air and fire in which we live. If the heavens are changeless, with an eternal circular motion, then no body or phenomenon can be in the heavens unless it is eternal, and has a circular motion (or a motion composed of a combination of circular movements, as Aristotle believed that the sun, moon and planets have[24]); bodies and phenomena which lack these properties must be sublunary. Therefore, not only phenomena such as shooting stars and the aurora borealis, but even comets, are sublunary.[25] (This does not apply to the Milky Way, and it is not clear to me why Aristotle included that among sublunary phenomena.[26])

We have seen already that the Stoics, presumably including Posidonius, agreed with Aristotle that concentric regions of water, air and fire surround the earth. Like Aristotle, they regarded the region of the heavenly bodies as divine; but they rejected Aristotle's "first element",[27] holding that the highest parts of the cosmos are fire, which they regarded as divine and which (unlike Aristotle) they also called *aithēr*.[28] That Posidonius shared this Stoic view is evident from Arius Didymus, who says: ἄστρον δὲ εἶναί φησιν ὁ Ποσειδώνιος σῶμα θεῖον ἐξ αἰθέρος συνεστηκός, λαμπρὸν καὶ πυρῶδες, "Posidonius says that a star is a divine body, bright and fiery, formed from *aithēr*".[29]

The Stoics accordingly did not share Aristotle's view of the nearly total separation of the heavens from the sublunary world. Their sun is itself hot, so there is no need of the hypothesis of frictional heating to explain how it heats the earth, and its fire and that of the other heavenly bodies is fed by exhalations from the earth.[30] This means that matter is constantly passing from the region of air to that of fire, and consequently there was no difficulty in a theory such as that which Posidonius seems to have held about comets, that a comet is caused by an unusual mass of air passing into the region of fire.[31]

Aristotle holds that at least some of the heavenly bodies are much larger than the earth and are a vast distance from it, as is evident, he says, if we consider "what is now sufficiently demonstrated by mathematics".[32] There seems to be no evidence to show what these mathematical calculations were or what sizes or distances of heavenly bodies had been computed from them. The earliest surviving calculation is that of Aristarchus (3rd century B.C.), who computed that (to use modern notation) the sun's diameter is between 6.33 and 7.17 times that of the earth.[33] As Neugebauer showed, it follows from Aristarchus' calculation (though Aristarchus himself does not draw the conclusion) that the distance of the sun from the earth is 400 times the radius of the earth.[34]

Other ancient thinkers put forward calculations of their own, among them Posidonius. Pliny *Nat.* II 85[35] states:

Posidonius minus [or "non minus"] XL stadiorum a terra altitudinem esse, in quam nubila ac venti nubesque perveniant, inde purum liquidumque et inperturbatae lucis aera, sed a turbido ad lunam viciens C milia stadiorum, inde ad solem quinquiens miliens, et [or "eo"] spatio fieri ut tam inmensa eius magnitudo non exurat terras.

According to Posidonius, the distance the atmospheric region, winds and clouds reach above the earth is less than [or "not less than"] 40 stades, and from there the air is clear, translucent and calm; but from the turbulent region to the moon, the distance is 2,000,000 stades, and from there to the sun, 500,000,000 stades; and that interval of distance ensures that the sun, despite its huge size, does not burn up the earth.[36]
(Trans. I.G. Kidd [1999, Cambridge University Press], p. 175–6.)

For the moment I disregard the "turbulent region". I also disregard the question whether Pliny is right to indicate that 500,000,000 stades is the distance of the moon from the sun rather than of the earth from the sun: this seems insignificant when dealing, as we clearly are here, with extremely round numbers.

Between the times of Aristarchus and Posidonius, further attempts had been made, notably by Archimedes in *The Sand-Reckoner* and Hipparchus, to calculate such quantities as the size and the distance of the sun and the moon (but Archimedes' aim was to calculate, not the most accurate values, but the largest possible values, in order to illustrate the notation he had devised for very large numbers).[37] Posidonius was interested in such calculations: we know he attempted to calculate the circumference of the earth and the diameter of the sun.[38] So it is not surprising that he gives figures for the distances of sun and moon from the earth: Kidd shows that Pliny's figures are plausible round numbers for Posidonius to have given. Cleomedes reports Posidonius' attempt to calculate the diameter of the sun, in which he assumed that the sun's orbit as it revolves around the earth is 10,000 times the circumference of the earth, possibly more, possibly less[39]: Kidd shows that on this assumption, if the earth's circumference is 240,000 stades,[40] then the sun's orbit is 2,400,000,000 stades; if, for a rough calculation, we assume $\pi = 3$, then the radius of its orbit is 400,000,000 stades (which, in a geocentric system, is the distance of the sun from the earth); and it may be more.[41] It seems plausible that Posidonius should have said 500,000,000 stades, as a very round number. For the moon's distance we know of no calculation by Posidonius, but Hipparchus calculated it as rather more than 2,000,000 stades.[42]

Posidonius' astronomical calculations were based partly on very approximate empirical data and partly on guesswork, since he had no evidence for his estimate of the sun's orbit; Cleomedes only says, presumably quoting him, that it is plausible (πιθανόν) that the sun's orbit is not less than 10,000 times the earth's circumference, σημείου γε λόγον τῆς γῆς πρὸς αὐτὸν ἐχούσης, "at least if the earth is a mathematical point in relation to it [sc. the sun's orbit]"[43] – a common assumption among ancient astronomers,[44] but one that suggests no actual figure for the distance from earth to sun. Nor can the fact that the sun does not burn the earth have given Posidonius a basis for calculating the distance. The only known ancient calculation in which the figure 10,000 appears in this context is one by Archimedes in *The Sand-Reckoner*, which concludes that the diameter of the sun's orbit is *fewer* than 10,000 times the earth's diameter.[45] This was presumably the source of Posidonius' figure of 10,000; but he ignores the fact that the calculated result was *fewer than* 10,000.[46] Nevertheless, so far as the distance of the sun is concerned, Posidonius' hypothesis is much nearer to the truth than were the distances calculated, not only by Aristarchus, but also by Hipparchus, who apparently calculated that the sun's distance from the earth is 490 times the earth's radius, and later by Ptolemy, who made the sun's average distance 1,210 times the radius of the earth.[47] If, as Posidonius thought, the sun's orbit is 10,000 times the earth's circumference, then the radius of the sun's orbit, approximately its distance from the earth, is 10,000 times the earth's radius. By modern calculation the sun is about 150,000,000 kilometres from the earth, more than 23,000 times the earth's radius of 6,371 kilometres.[48] But Posidonius' relative success was the result of chance – unless one likes to call it intuition.[49]

These calculations have consequences for meteorology, by making incredible theories in which happenings in the *aēr*, close to the earth, affect the heavenly bodies, for instance, a theory attributed to Anaxagoras, that the sun's "turning" to the south, the summer solstice, occurs ἀνταπώσει τοῦ πρὸς ταῖς ἄρκτοις ἀέρος, "by counter-pressure of the *aēr* in the north",[50] and a suggestion by Herodotus, who speaks of the sun ἀπελαυνόμενος ἐκ μέσου τοῦ οὐρανοῦ ὑπὸ τοῦ χειμῶνος καὶ τοῦ βορέω, "driven from the middle of the sky by winter storm and north wind".[51] Some in later antiquity, for instance Epicurus, refused to accept the astronomical calculations; hence Epicurus says not only that the size of the sun "relative to us is as great as it appears",[52] but also that solstices may be due "to a counterthrust of the air".[53]

The height up to which weather phenomena occur

Pliny, as quoted above, gives it as Posidonius' view that, above the "turbulent region" and reaching up to the moon, there is a region of clear, calm air. This is consistent with other reports about Posidonius' view of the moon. Diogenes Laertius VII.144–5 reports him as saying that the sun is pure fire, but the moon is ἀερομιγῆ ... καὶ πρόσγειον, "mixed with air and near the earth"[54]; Aëtius II.25.5 says he thought that the moon is μικτὴν ἐκ πυρὸς καὶ ἀέρος, "mixed from fire and air".[55] It is evidently at the boundary of the region of air and the region of elemental fire which is above it,[56] but far above the "turbulent region" in which weather phenomena occur. This reflects a widely held Greek view.

From Aristotle's time onwards it was agreed, by those who accepted the astronomers' calculations, that the sun, moon and stars are enormous distances away. But clouds and the phenomena associated with them (rain, thunderstorms and so on) appear to ordinary observation to be quite close to us, and there were no calculations to prove them otherwise. Up to what height above the earth do such phenomena occur? Pliny, quoted above, reports Posidonius' answer: "Posidonius minus [or "non minus"] XL stadiorum a terra altitudinem esse, in quam nubila ac venti nubesque perveniant", "according to Posidonius, the distance the atmospheric region, winds and clouds reach above the earth is less than [or "not less than"] 40 stades".

Posidonius probably said that clouds and the like occur up to a height of less than 40 stades,[57] but presumably thought that they might occur nearly up to that height, or he would not have mentioned the figure 40.[58] Where did he get the figure of 40 stades, or nearly 40 stades, for the height of clouds and wind? The answer, I believe, goes back to, and beyond, Aristotle.

Aristotle discusses the sublunary region at *Mete.* 340a24ff. He begins: Εἰ δὴ γίγνεται ὕδωρ ἐξ ἀέρος καὶ ἀὴρ ἐξ ὕδατος, διὰ τίνα ποτ' αἰτίαν οὐ συνίσταται νέφη κατὰ τὸν ἄνω τόπον; "If water comes to be from air and air from water, why are clouds not formed in the upper region?"[59] He presumably thought it a matter of everyday observation that clouds do not occur at a very great height above the earth. His explanation is that the region down as far as the moon is filled with his "first element", and that this heats by its motion the upper part of the sublunary region, which contains not air but rather οἶον πῦρ, "a sort of fire" – fire as an element.[60] A further reason is that the fire and air is carried round by the motion of the heavens and that this motion prevents condensation: ῥεῖν γὰρ ἀναγκαῖον ἅπαντα τὸν κύκλῳ ἀέρα, ὅσος μὴ ἐντὸς τῆς περιφερείας λαμβάνεται τῆς ἀπαρτιζούσης ὥστε τὴν γῆν σφαιροειδῆ εἶναι πᾶσαν, "for the whole encircling mass of air must necessarily be in motion, except that part of it which is contained within the circumference that makes the earth a perfect sphere".[61] By this he evidently means that the only air not to be carried round by the rotation of the heavens is that below the height of the highest mountains.

No reason is given why this should be so; and within the passage there is a parenthesis: φαίνεται γὰρ καὶ νῦν ἡ τῶν ἀνέμων γένεσις ἐν τοῖς λιμνάζουσι τόποις τῆς γῆς, καὶ οὐχ ὑπερβάλλειν τὰ πνεύματα τῶν ὑψηλῶν ὀρῶν, "thus in fact we find that winds rise in low marshy districts of the earth, and do not blow above the highest mountains".[62] No cause is suggested, and nothing in Aristotle's system requires him to say that that there are no winds above the highest mountains. The reason for his saying it is surely that it was a widely held Greek belief, based ultimately on a mythological conception of the home of the gods, such as we find in the description of Mount Olympus in the *Odyssey*:

ὅθι φασὶ θεῶν ἕδος ἀσφαλὲς αἰεὶ
ἔμμεναι· οὔτ' ἀνέμοισι τινάσσεται, οὔτε ποτ' ὄμβρῳ
δεύεται οὔτε χιὼν ἐπιπίλναται, ἀλλὰ μάλ' αἴθρη
πέπταται ἀνέφελος, λευκὴ δ' ἐπιδέδρομεν αἴγλη.

where, they say, is the home of the gods which ever stands fast. It is neither shaken by winds nor wetted by rain, nor does snow come near it, but clear, cloudless sky is spread over it and bright radiance covers it.[63]

Aristotle had many successors who accepted this belief. Capelle (1916), on which work this paragraph and the next two are largely based, quotes as upholding it the Pseudo-Aristotelian *Problemata*,[64] Plutarch,[65] Geminus,[66] Arrian,[67] Pomponius Mela,[68] the commentators on Aristotle's *Mete*. (Alexander,[69] Philoponus[70] and Olympiodorus[71]), Gregory of Nyssa,[72] Isidore of Seville[73] and others. Most of them[74] support this doctrine by a tale that the remains of a sacrifice performed on one of the highest peaks are found there undisturbed – in most versions of the story, a year later; there is a similar tale in Solinus.[75] This story is told about four different mountains (Cyllene, Oeta, Athos and Olympus), which surely indicates that it was a widespread popular story.

This, then, was a generally held view; but how high were the highest mountains thought to be? From the Hellenistic period and later we hear of attempts to measure, or at least to estimate, the height of the highest mountains. The earliest were by Dicaearchus, a pupil of Aristotle. We know little about how the measurements were made, but there are references to instruments being used in making them, particularly the *dioptra*.[76]

We have in ancient sources a number of figures – measured, estimated or guessed – for the heights of particular mountains. I disregard two, recorded by Pliny, in Roman miles, which seem obviously fantastic: 50 miles (74,000 metres) for the height of the Alps,[77] and 6 miles (8,880 metres) for Mount Haemus (the Balkan mountain range, in modern Bulgaria).[78] Other, more reasonable, heights are given in stades (sometimes with a number of feet added or subtracted).

A stade is not necessarily a precise measurement. Scholars have long been aware of the different lengths of a stade that can be derived from different ancient sources.[79] I mention just some examples from the evidence that scholars have cited. Bauslaugh (1979) examined distances in stades given by Thucydides and compared them with modern measurements of the same distances: on this basis, the shortest stade in Thucydides was, he found, 130 or 140 metres, the longest was 260 metres, or possibly 290. Later ancient writers sometimes indicate how many stades they reckon as equal to a Roman mile, or give a distance in stades between two places which can be compared with the same distance in miles as given in another ancient source. Strabo VII.7.4 speaks of "reckoning, as most people do, eight stades to the mile" (λογιζομένῳ … ὡς μὲν οἱ πολλοί, τὸ μίλιον ὀκταστάδιον), but adds that Polybius reckoned eight and one-third stades. He repeats these figures at VII fragment 56(57); but at V.3.12 he says it is 160 stades from Rome to Aricia, a distance given as 16 miles by the *Antonine itinerary*,[80] so Strabo's figure here implies 10 stades to the mile. After Strabo, Pliny reckons 8 stades to the mile,[81] but later writers, such as Dio Cassius, tend to reckon 7.5 stades to the mile.[82] Occasionally, we find 7 stades to the mile.[83] A Roman mile has been calculated to be almost exactly 1,480 metres.[84] If this is right, then at 10 stades to the mile, one stade equals 148 metres; at 7.5 stades to the mile, 1 stade equals 197 metres; at 7 stades to the mile, 1 stade equals 211 metres.

Eight ancient texts give nine heights in stades for five mountains, all of them in Greece. (For simplicity, in what follows the "equivalent" in metres of a height in stades is the range between a stade of 150 metres and one of 200 metres. Actual heights, unless stated otherwise, are from *Times comprehensive atlas* (2014).)

1 Strabo VIII.6.21: Acrocorinth: 3.5 stades (equivalent to 525–700 metres. Actual height 575 metres.[85])
2 Geminus, *Isagoge* XVII.5: Atabyrius (Rhodes): less than 10 stades (probably), measured by Dicaearchus.[86] (10 stades is equivalent to 1,500–2,000 metres. Actual height 1,215 metres.)
3 Geminus, *Isagoge* XVII.5: Cyllene: less than 15 stades, measured by Dicaearchus. (15 stades is equivalent to 2,250–3,000 metres. Actual height 2376 metres.)
4 Strabo VIII.8.1: Cyllene: two heights in one text: some say 15 stades, some say 20 stades. (20 stades is equivalent to 3,000–4,000 metres.)
5 Apollodorus: Cyllene: 9 Olympic stades less 80 feet[87] (equivalent to 1,325–1,775 metres).
6 Plutarch, *Aem.*15: Olympus: 10 stades plus 96 feet, measured by Xenagoras (equivalent to 1,530–2,030 metres. Actual height 2,911 metres).
7 Martianus Capella II. 149: Olympus: 10 stades (equivalent to 1,500–2,000 metres).
8 Pliny, *Nat.* II.162: Pelion: 10 stades, measured by Dicaearchus (equivalent to 1,500–2,000 metres. Actual height 1,624 metres).

Round numbers of 10, 15 or 20 stades cannot have been intended as very precise figures. Only in passage 5 is it stated what sort of stade is meant. Dicaearchus' three heights, the height of Acrocorinth, and the height of 15 stades for Cyllene, though imprecise, are accurate: the actual height is within the range of possible interpretations of the height in stades. The mountains concerned, apart from Cyllene, are close to the sea, so the natural course would have been to measure the height as that above sea-level. The highest and the lowest heights for Cyllene and the two heights for Olympus are not accurate, even in the sense described. Neither is close to the sea, so the height above sea-level would be hard to calculate; for the smaller heights we can, I think, be confident that sea-level was not the level above which the height was measured.

From evidence like this, ancient thinkers had to derive their views about the height of the highest mountains, and consequently, as we have seen, their views of the height up to which weather phenomena (winds, clouds and so on) occur.

Eratosthenes, in the 3rd century B.C., concluded that the height of the highest mountains is 10 stades. I suggest that he based this on Dicaearchus' one recorded height not qualified by "less than", that is, 10 stades for Pelion. His view seems to have been widely held throughout antiquity. Geminus (1st century B.C. or A.D.) probably said that weather phenomena do not reach a height of 10 stades.[88] Plutarch, around 100 A.D., said that according to οἱ γεωμετρικοὶ (presumably here "the earth-measurers") the height of no mountain exceeds 10 stades.[89] Theon of Smyrna in the 2nd century A.D.,[90] the mathematician Theon of Alexandria in the 4th century,[91] and Simplicius in the 6th century[92] all quote Eratosthenes' view with approval.

Other ancient authors suggested greater heights. Cleomedes (2nd century A.D.) said that the height-limit for mountains is 15 stades,[93] Arrian (in the same century) said that the height-limit for weather phenomena is 20 stades,[94] Philoponus (6th century A.D.) said that the limit for both mountains and weather phenomena is 12 stades.[95] But Posidonius, according to Pliny, held that clouds and wind occur up to a height of less than (or, possibly, not less than) 40 stades. Other figures in this passage are plausible, as we have seen, so there seems no good reason to doubt this one.[96] So, either Posidonius rejected the common view, that clouds and wind do not occur above the highest mountain peaks, or he accepted it, but thought that the height of the highest peaks had been underestimated. The first alternative is possible: the theories discussed above were not the only ones: Pliny[97] mentions a view that clouds extend to a height of 900 stades. But the second is likelier. As we shall see, Posidonius is fairly conservative in his meteorological theories, and tends to follow Aristotle where Aristotle's theories do not conflict with Stoic physics; but he will correct an earlier theory where more extensive geographical knowledge enabled him to do so.[98] He had travelled in western Europe, so was probably more aware than most earlier Greek writers of the height of the Alps: the same is true of Polybius, who, we know, commented on the great height of the Alps compared with Greek mountains.[99] Posidonius himself may not have travelled through the Alps,[100] but he would surely have been told of their exceptional height. The highest recorded ancient estimate for the height of a Greek mountain is 20 stades for Cyllene. It would have been an assumption as reasonable as any that the highest of the Alps might be as much as but not more than twice the height of the highest Greek mountain. Hence, I suggest, Posidonius assumed up to 40 stades as the height of the highest mountains, and accepted the view that the height of the highest mountains is a limit beyond which clouds and wind do not occur.

Forty stades would be between 6,000 and 8,000 metres. This would be a considerable over-estimate for the Alps: Mont Blanc, the highest of them, is only 4,810 metres. However, Posidonius probably said "less than 40"; and also, though Posidonius could not have known it, the highest mountain of all, Mount Everest, is 8,848 metres.[101]

To sum up this section: there had been development in Greek ideas between the times of Aristotle and Posidonius, as is shown by the fact that Posidonius, unlike Aristotle, gave a figure for the height up to which clouds and winds occur. Probably, that figure is based on an estimate of the height of the highest mountains, and Posidonius was not the first man to try to measure or estimate that height: Eratosthenes, at least, had done so before him, and it is highly likely that he too accepted the common Greek view that clouds and wind do not occur above the highest mountains. Posidonius' contribution seems to have been the realisation that Greek mountains are far from being the highest in the world. As regards weather, Posidonius had by chance reached a result closer to the truth than those of other Greek thinkers. Modern meteorologists have found that most weather phenomena occur in the troposphere, the lowest region of the atmosphere, the height of which varies, with latitude and season, between 6 and 18 kilometres.[102] Posidonius' 40 stades, between 6 and 8 kilometres, is within this range; the figures given by other ancient thinkers are well below it.[103]

No-one in antiquity is recorded as accepting Posidonius' 40 stades (Pliny, after quoting Posidonius' figures, says "Inconperta haec et inextricabilia", "These things are not reliably known and cannot be disentangled"), and Eratosthenes' 10 stades seems to have remained the orthodox, or at least a widely held, view; but we should give credit to Posidonius for reinforcing, if not originating, the correct view – accepted by Arrian, Cleomedes and Philoponus – that 10 stades is too low a figure.

Notes

1 DK 31B100.
2 Theophrastus, *De sensu* 50, 54, 74, 80, 81 (DK 68A135).
3 Herodotus II.25.
4 Sophocles, *Electra* 86–7.
5 Euripides, *Helen* 1478–80.
6 Examples: *Iliad* 3.277, the Sun is called on to witness Agamemnon's oath; *Odyssey* 12.374ff, the Sun complains to Zeus when Odysseus' men slaughter his cattle. In Hesiod's *Theogony*, line 19, Sun and Moon are included in a list of divinities; lines 371–4, they are children of Theia and Hyperion; lines 381–2, Erigeneia gives birth to the stars.

7 On *parapēgmata* see Taub (2003) 20ff (for the date of their origin, page 26), and earlier authors there cited.
8 See, for example, Heath (1913) 193–216.
9 Cf. Aristotle, *De caelo* 270b6–17: πάντες τὸν ἀνωτάτω τῷ θείῳ τόπον ἀποδιδόασι ... δῆλον ὅτι ὡς τῷ ἀθανάτῳ τὸ ἀθάνατον συνηρτημένον, "all assign the highest place to the divine ... supposing, obviously, that immortal is closely linked with immortal" (270b6–9, tr. Guthrie [1939]), i.e. the immortal, unchanging heavens must be linked to the immortal gods. Sense-perception, Aristotle goes on, confirms this: in the whole of human memory there has never been any change in the outermost heaven (τὸν ἔσχατον οὐρανόν) or any part of it.
10 *Timaeus* 40A.
11 *Timaeus* 40B.
12 *Laws* VII, 822A.
13 *Timaeus* 63D, with translation of Bury (1929).
14 This is clear from Plato's speaking of bodies both of earth and of fire drawn εἰς ἀνόμοιον ἀέρα, "into air, which is unlike it" (*Timaeus* 63B–C – water is not mentioned in this passage).
15 *De caelo* 311a15ff, *Mete.* 339a11ff.
16 *De caelo* 270b5ff (also 269a320, *Met. XII* 1074a39ff).
17 *De caelo* 268b27ff, *Mete.* 339b16ff.
18 E.g., *De caelo* 270b2 τὸ πρῶτον τῶν σωμάτων, "the first of bodies", *Mete.* 339b17 τοῦ πρώτου στοιχείου, "the first element".
19 *De caelo* 270b1–2, ἀΐδιον καὶ οὔτ᾽ αὔξησιν ἔχον οὔτε φθίσιν, ἀλλ᾽ ... ἀναλλοίωτον καὶ ἀπαθές.
20 *De caelo* 269a2ff, 269b30.
21 *De caelo* 269b31–3, ἀδύνατον ἔχειν βάρος ἢ κουφότητα· οὔτε ... ἐνδέχεται αὐτῷ κινηθῆναι ἐπὶ τὸ μέσον ἢ ἀπὸ τοῦ μέσου.
22 *De caelo* 289a11–35; *Mete.* 340b6–341a37.
23 See above, Chapter 1 (p. 000).
24 See, in *De caelo*, 292b32–293a8.
25 *Mete.* I, chapters 4–7 (341b1–345a10).
26 *Mete.* I, chapter 8 (345a11–346b15).
27 For the agreement of the Stoics generally, and of Posidonius, with Aristotle about earth, water, air and fire as elements, but rejection of the "first element", see above, Chapter 5 (p. 37). On the divinity of *aithēr* see the next note.
28 See, for example, Diogenes Laertius VII.137 ἀνωτάτω ... εἶναι τὸ πῦρ, ὃ δὴ αἰθέρα καλεῖσθαι, ἐν ᾧ πρώτην τὴν τῶν ἀπλανῶν σφαῖραν γεννᾶσθαι, εἶτα τὴν τῶν πλανωμένων, "highest is fire, which indeed they called *aithēr*, in which first the sphere of the fixed stars is formed, then that of the planets"; Cicero, *Academica priora* II.126 (*SVF* I.154) "Zenoni et reliquis fere Stoicis aether videtur summus deus", "to Zeno and almost all the other Stoics *aether* seems the highest god". On Aristotle's usage see Ross (1936) 578. (I have checked what he says by a search of *Thesaurus linguae Graecae*). At *De caelo* 270b6–22 and *Mete.* 339b20–21 Aristotle cites what he says was the traditional meaning of *aithēr* in support of his theory of a "first element", but he never calls this element *aithēr* when describing his own theory.
29 Arius Didymus, *Epitome* fr. 32 (Posidonius F127 EK).
30 See below, Chapter 10 (p. 93).
31 On Posidonius' comet theory see below, Chapter 9 (p. 79–80).
32 *Mete.* 339b33 (τὰ νῦν δεικνύμενα διὰ τῶν μαθημάτων). This means, he says (339b37–340a3) that εἰ ... τά τε διαστήματα πλήρη πυρὸς καὶ τὰ σώματα συνέστηκεν ἐκ πυρός, πάλαι φροῦδον ἂν ἦν ἕκαστον τῶν ἄλλων στοιχείων, "if both the intervals [sc. between earth and heavenly bodies] were full of fire and the bodies were composed of fire, each of the other elements would long ago have vanished [sc. burnt up

by the vast quantity of fire]". He has already said of the earth, at 339b8–9, ὦπται διὰ τῶν ἀστρολογικῶν θεωρημάτων ... ὅτι πολὺ καὶ τῶν ἄστρων ἐνίων ἐλάττων ἐστίν, "astronomical researches have now made it clear that the earth is far smaller even than some of the stars" (tr. Lee [1952]). Cf. *De caelo* 298a19–20, τὸν ὄγκον ... τῆς γῆς ... μὴ μέγαν πρὸς τὸ τῶν ἄλλων ἄστρων μέγεθος, "the bulk of the earth is not great in relation to the size of the other heavenly bodies". Such statements imply they are at a great distance, since they appear small when seen from the earth. Aristotle quotes a calculated figure at *De caelo* 298a16: the earth's circumference calculated as 400,000 stades.

33 Aristarchus, *On the sizes and distances of the sun and moon*, proposition 15; see Heath (1913) 403–9; Neugebauer (1975) 634–43.

34 Neugebauer (1975) 637.

35 Posidonius F120 EK.

36 For comments see Kidd (1988) 465–6.

37 See Kidd (1988) 444–54 on Posidonius F115 EK; Heath (1913) 341–3 and Neugebauer (1975) 325–9 on Hipparchus; Heath (1913) 347–8 and Neugebauer (1975) 643–51 on Archimedes.

38 On the sun see next note; on the earth, Cleomedes I.10.50–2 (I.7 lines 1–50 Todd) = Posidonius F202 EK.

39 Cleomedes II.1.79–80 (II.1 lines 269–86 Todd) = Posidonius F115 EK.

40 The most likely figure for Posidonius' calculation of the earth's circumference: see below, Chapter 7 (p. 56).

41 See Kidd (1988) 466 (on F120) and 444–7 (on F115).

42 Pappus states that Hipparchus calculated the mean distance of the moon as 67⅓ times the radius of the earth (Swerdlow (1969) 289); if the radius is about 40,000 stades, then the distance of the moon is about 2,700,000 stades. (Besides Swerdlow's work, see Kidd [1988] 450.)

43 Posidonius F115 lines 17–19 EK. There is some doubt about the text, but this must be the meaning. (So given by Kidd [1988] 445. Kidd (1999) 172 does not translate the words σημείου – ἐχούσης.)

44 See (e.g.) Heath (1913) 308–10.

45 See Heath (1913) 348; Neugebauer (1975) 646.

46 Heath (1913) 348 and Kidd (1988) 447 point out the relation of Archimedes' calculation to Posidonius.

47 For the results calculated by Hipparchus and Ptolemy, see Neugebauer (1975) 325–6 and 110.

48 For these modern figures, see Wikipedia articles Earth and Earth radius (read 4 July 2021).

49 I also discuss Posidonius' calculations, from a different viewpoint, in Chapter 21 (p. 202).

50 Aëtius II.23.2 (DK 59A72). Compare Hippolytus, *Haer.* I.8.9 (DK 59A42).

51 Herodotus II.26. He is describing a counterfactual, but it only makes sense if the force which moves the sun is the same force which, in his view, actually moves it. Compare II.24, the sun ἀπελαυνόμενος ... ὑπὸ τῶν χειμώνων, "driven away ... by winter storms".

52 Κατὰ τὸ πρὸς ἡμᾶς τηλικοῦτόν ἐστιν ἡλίκον φαίνεται (*Letter to Pythocles*, 91, tr. Mensch and Miller (2018) 524).

53 Κατὰ ἀέρος ἀντέξωσιν (*Letter to Pythocles*, 93, tr. Mensch and Miller [2018] 524).

54 Posidonius F17 and F10 EK.

55 Posidonius F122 EK.

56 See above, Chapter 5 (p. 37).

57 The best codices omit "non" before "minus XL": Kidd (1988) 465.

58 On the figure forty ("XL"), which has been doubted, see below.

59 *Mete.* 340a24–5.
60 *Mete.* 340b5–32.
61 *Mete.* 340b32–341a5. My quotation is 340b34–36, with the translation of Lee (1952); for the interpretation see Lee's note.
62 *Mete.* 340b36–341a1, with the translation of Lee (1952).
63 *Odyssey* 6.42–45.
64 *Problemata* XXVI.36 (944b12–16).
65 *De primo frigido* 951B.
66 *Isagoge* XVII.2ff.
67 *Fragmenta de rebus physicis* 4, in Roos/Wirth (1968).
68 II.2.31.
69 *In Mete.* I.3 (p. 16.12ff Hayduck).
70 *In Mete.* I.3 (p. 26.32ff Hayduck).
71 *In Mete.* I.3 (p. 22.26ff Stüve).
72 *Hexahemeron* p. 96C3ff.
73 *De natura rerum* 30.5.
74 All those named here, apart from Aristotle's *Mete.*, Plutarch, Gregory and Isidore.
75 8.4 (p. 62.4ff M.)
76 Plutarch, *Aem.* 15; Philoponus, *in Mete.* p. 27.9–12 Hayduck; Theon Smyrnaeus p. 124.19 Hiller (= Dicaearchus fr. 107 Wehrli); Hero *Dioptr.* 13, p. 234.3–18 Schöne.
77 Pliny, *Nat.* II.162. (However, Theiler [1982] vol. II p. 162 would emend this to 5 Roman miles, i.e., 40 stades.)
78 Pliny, *Nat.* IV.41. The highest peak in the Balkan range is 2,376 metres (*Times comprehensive atlas* [2014]).
79 For detailed discussions see Hultsch (1882), especially pp. 48–73; Lehmann-Haupt (1929), columns 1931–63; recently, Arnaud (2005) 84–7.
80 *Antonine itinerary* 106.5–107.2.
81 Pliny, *Nat.* II.85 (a stade = 125 Roman "passus", so 8 stades = 1,000 "passus" = 1 Roman mile); XII.53 (40 stades = 5 miles.)
82 See Dio Cassius 38.18.7 and 51.19.6: both passages give distances in stades obviously obtained by multiplying by 7.5 a distance in Roman miles. Other passages in which 7.5 stades are reckoned to equal one mile are in the *Geometrica* which claims to be by Hero (Heiberg [1912] 194); in two anonymous geographical texts printed by Müller (1855–61) vol. 1, 423 and 426; and in Julian of Ascalon (of the 5th century?), who says that "according to the custom now prevailing" (κατὰ τὸ νῦν κρατοῦν ἔθος) a mile = 7.5 stades (see Geiger [1992], with text on p. 43).
83 Hesychius, article μίλιον ("Mile"), has μέτρον ὁδοῦ, σταδίων ζ, οἱ δὲ ζ ἥμισυ ("a measure of a journey, 7 stades, but some say 7 and a half").
84 See Hultsch (1882) 91–8 and 700–701, concluding that the Roman mile is 1478.5 metres. *Brill's new Pauly* (2002–10). 8, 881–4, article "Milestones" (by M. Rathmann), says 1,481 metres, *Oxford classical dictionary* (2012), article "Measures", says 1,480 metres.
85 Capelle (1916) 31 n.4.
86 There is some doubt about the text: see the note of Aujac (1975).
87 See Jacoby (1929) p. 1078: FGrH 244 Apollodoros von Athen frg. 130.
88 Geminus, *Isagoge* XVII.2.
89 Plutarch *Aem.*15.
90 Theon Smyrnaeus p. 124–5 Hiller. In this passage δέκα σταδίων ('ten stades') for the height of the highest mountain is a conjecture, but proved correct not only by parallels in other authors but also by Theon's deduction from Eratosthenes' figure of 252,000 stades for the earth's circumference, that the earth's diameter is 80,182 stades and the highest mountain therefore one eight-thousandth of that.
91 Theon of Alexandria's commentary on the *Almagest*, p. 394.17–395.2 and 398.1–5 in ed. of A. Rome (1936).

92 Simplicius, *In Cael.* p. 550.1–4 Heiberg. Capelle (1916) 16 says that the evidence for ten stades as the view of Eratosthenes is incompatible with Geminus' report of Dicaearchus' measurement of Cyllene (see above) and that the figure in Geminus' text must be corrupt. But if Dicaearchus said "less than 15 stades" for Cyllene, then Eratosthenes could correctly say that the highest definite measured height was the 10 stades for Pelion.

93 Cleomedes I.7 lines 123–4 Todd.

94 Arrian, *Fragmenta de rebus physicis* 4 in Roos/Wirth (1968).

95 Philoponus, *In Mete.* 26.32ff and 27.9ff Hayduck.

96 It is accepted by modern editors of Pliny: by Mayhoff (1906), by Rackham (1938), by Beaujeu (1950), and by König and Winkler (1974), and it is accepted by EK (their F120) and by Vimercati (2004), his A78. However, Capelle (1916) 29 suggests that it is an error, and Theiler (1982) I p. 221 (his F297) and vol. II p. 177 (cf. also p. 161) emends "XL" in the MSS. of Pliny to "XV", regarding as from Posidonius the figure of 15 stades as the height of the highest mountains given by Cleomedes I.7 lines 123–4 Todd (Posidonius F288 in Theiler [1982]). But this passage of Cleomedes does not name Posidonius, and I have argued above (Chapter 3 [p. 14]) that Cleomedes does not simply copy Posidonius. We cannot assume that Cleomedes is following Posidonius where he does not name him.

97 *Nat.* II 85, following his report of Posidonius.

98 See, for example, Chapter 7 (p. 55–7) below, on climatic zones.

99 Polybius XXXIV.10.15 = Strabo IV.6.12, quoted by Capelle (1916) 23 nn. 1 and 7.

100 On the likely extent of Posidonius' travels in western Europe see Kidd (1988) 16–20.

101 Mont Blanc and Mount Everest: see *Times comprehensive atlas of the world* (2014), plates 72 and 32.

102 Wikipedia, 'Troposphere' (read 21 April 2022).

103 The relation of Posidonius' view to the modern concept of the troposphere is pointed out by Virmercati (2004) 537–8.

7 Climatic zones

This chapter concerns the ancients' knowledge and beliefs about how climate and weather differ in different regions of the earth, and how the sun controls this – which tended to mean how the different climatic regions are related to the equator, tropics and poles of an earth which, as ancient thinkers had realised, is spherical.

Strabo (II.2.2[1]) reports Posidonius' own view about the origin of Greek theories on this subject:

> φησὶ δὴ ὁ Ποσειδώνιος τῆς εἰς πέντε ζώνας διαιρέσεως ἀρχηγὸν γενέσθαι Παρμενίδην ἀλλ' ἐκεῖνον μὲν σχεδόν τι διπλασίαν ἀποφαίνειν τὸ πλάτος τὴν διακεκαυμένην, ὑπερπίπτουσαν ἑκατέρων τῶν τροπικῶν εἰς τὸ ἐκτὸς καὶ πρὸς ταῖς εὐκράτοις·[2]

> Posidonius says that it was Parmenides who was the founder of the division into five zones [i.e., arctic and antarctic, northern and southern temperate, and tropical or torrid], but that he represented the torrid [literally "burnt through"] zone as virtually double in width, falling beyond the two tropics outwards, and overlapping the temperate zones.
> (Trans. I.G. Kidd [1999, Cambridge University Press], p.109)

That Parmenides had something to say about tropical zones and the inhabited parts of the earth is also reported by Aëtius[3] and by Achilles,[4] and Diogenes Laertius, in different passages, attributes to him the belief that the earth is "spherical" (σφαιροειδῆ) and "round" (στρογγύλην),[5] which is possibly relevant because a ζώνη, literally "belt", should go *round* something – if the word is Parmenides' own, and not a later interpreter's (see below). There is controversy about Parmenides' view, and I cannot go into the details here; Kidd concludes that Parmenides may have had a theory of zones "in a very elementary sense" and that "Posidonius' evidence should not be rejected out of hand".[6] There is also evidence suggesting that some "Pythagoreans" of the 5th or 4th century B.C. had a theory of climatic zones.[7]

DOI: 10.4324/9780429399930-7

Whatever the truth about these theories, there certainly was, in the 5th century B.C., a Greek belief that parts of the earth south of Greece are drier and hotter than Greece, that parts of the earth north of Greece are colder, and that parts both of the hotter and of the colder region are uninhabitable. Herodotus tells that, compared with Greece, Egypt is rainless and dry,[8] that the country south of Egypt is hotter than Egypt,[9] and that a region of Africa is "terribly waterless and desolate of all" (ἄνυδρος δεινῶς καὶ ἔρημος πάντων), apparently meaning "desolate of all living things".[10] In and beyond the country of the Scythians, Herodotus tells of a region frozen for eight months of the year,[11] and of a region full of snow and uninhabited.[12]

The first Greek thinker who, we can say for certain, related these ideas about climate to the tropics and poles of the earth was Aristotle, who discusses the subject at *Mete.* 362a32–b30. This is part of his account of the winds, and his immediate aim is to show that the south wind, as known to the Greeks, blows from the "summer tropic", the Tropic of Cancer, not from the south pole. At *Mete.* 362a32–5 he begins a sentence: δύο γὰρ ὄντων τμημάτων τῆς δυνατῆς οἰκεῖσθαι χώρας, τῆς μὲν πρὸς τὸν ἄνω πόλον, καθ' ἡμᾶς, τῆς δὲ πρὸς τὸν ἕτερον καὶ πρὸς μεσημβρίαν, "there being two sectors of habitable land, the one towards the upper [i.e., north] pole, where we are, the other towards the other [sc. pole] and the south ...". Each of these sectors is bounded on one side by "the tropic" (τὸν τροπικόν), in our terms the Tropic of Cancer or of Capricorn, and on the other by "the ever visible" (τὸν διὰ παντὸς φανερόν)[13] – presumably the terrestrial latitude corresponding to the celestial latitude above which the stars around the pole never set. Places ὑπὸ τὴν ἄρκτον, "under the Bear", as a polar constellation, are uninhabitable through cold.[14] Aristotle also says: νῦν δ' ἀοίκητοι πρότερον γίγνονται οἱ τόποι πρὶν ἢ ὑπολείπειν ἢ μεταβάλλειν τὴν σκιὰν πρὸς μεσημβρίαν, "as it is, places become uninhabitable before the shadow either ceases or turns to the south"[15] (at the summer solstice the sun is directly over the Tropic of Cancer, so at noon objects on the tropic cast no shadow, and objects south of the tropic cast a shadow southwards). This suggests that some places north of the tropic are uninhabitable, but presumably Aristotle thought that this applied only to places a negligible distance north of the tropic. The excessive heat (καῦμα, ἀλέαν) of the tropics, as well as the cold of the arctic, is mentioned at 362b17 and b27, but Aristotle does not call the tropical zone διακεκαυμένη, as Strabo and Aëtius do, nor does he in this passage use the noun ζώνη: presumably this use of these two words is post-Aristotelian.

Posidonius criticised Aristotle's theory, as well as that of Parmenides. Strabo, following the sentence I quoted about Parmenides, continues (in indirect speech, so still reporting what Posidonius said):

Ἀριστοτέλη δὲ αὐτὴν καλεῖν τὴν μεταξὺ τῶν τροπικῶν, <τὰς δὲ μεταξὺ τῶν τροπικῶν> καὶ τῶν ἀρκτικῶν εὐκράτους. ἀμφοτέροις δ' ἐπιτιμᾷ δικαίως. διακεκαυμένην γὰρ λέγεσθαι τὴν ἀοίκητον διὰ καῦμα· τῆς δὲ μεταξὺ τῶν τροπικῶν πλέον ἢ τὸ ἥμισυ τοῦ πλάτους <οὐκ> οἰκήσιμόν ἐστιν ἐκ τῶν ὑπὲρ Αἰγύπτου στοχαζομένοις Αἰθιόπων[16]

Aristotle called it [i.e., called "the torrid zone"] the zone between the tropics, and the zones between the tropics and the arctic circles he called temperate zones. He [Posidonius] criticises both, and rightly so. For the torrid zone is defined as the zone that is uninhabitable because of the heat; and of the zone between the tropics, more than half of the breadth is uninhabitable [or, omitting οὐκ, "is habitable"], to make a conjecture from the Ethiopians beyond Egypt.

(Trans. of I.G. Kidd[1999, Cambridge University Press], p.109, modified)

Kidd[17] suggests that the mention of "conjecture" is an example of Strabo echoing "guarded statements" by Posidonius about evidence he thought uncertain; but this seems to me doubtful, as the calculations which follow appear to be those of Strabo, not Posidonius.

Strabo then shows that Aristotle is wrong about the extent of the uninhabitable torrid zone.[18] He says that the distance from Syene, on the summer tropic (the Tropic of Cancer), to the "Cinnamon-producing parallel", is measured as 8,000 stades, and that that parallel is the beginning of the torrid zone; and from that he calculates the size of the torrid zone in relation to the distance between the tropics. First, he uses Eratosthenes' estimate of the earth's circumference, 252,000 stades. From the equator to the tropic is four-sixtieths of the earth's circumference[19]; four-sixtieths of 252,000 is 16,800; therefore, from the "cinnamon-producing parallel" to the equator is 16,800 minus 8,000, that is, 8,800 stades, and the ratio of the distance between the two tropics to the width of the torrid zone is 16,800:8,800, or 21:11. (This is the calculation that Strabo has made, though he does not spell it out in his text.)

Strabo next says that if we use the shortest measurement of the earth's circumference, such as that of Posidonius, that is (according to our MSS. of Strabo), about 180,000 stades, περὶ ἥμισύ που ἀποφαίνει τὴν διακεκαυμένην τῆς μεταξὺ τῶν τροπικῶν, ἢ μικρῷ τοῦ ἡμίσους μείζονα, "it renders the torrid zone as somewhere about half the zone between the tropics, or a little more than half".[20] This conclusion does not accord with a circumference of 180,000 stades: four-sixtieths of 180,000 is 12,000 stades; subtract the measured 8,000 stades from Syene to the "cinnamon-producing parallel", and only 4,000 stades are left for the distance from that parallel to the equator. On this hypothesis, the torrid zone is only one-third of the distance between the tropics. If, however, we take the figure given by Cleomedes for Posidonius' estimate of the earth's circumference, 240,000 stades,[21] then from the equator to the tropic is four-sixtieths of 240,000, that is, 16,000 stades, the distance from the equator to the "cinnamon-producing parallel" is 16,000 minus 8,000, that is, 8,000 stades, and the torrid zone is, as Strabo says, one half of the distance between the tropics. We may suspect, therefore, that our MSS. of Strabo misreport Posidonius' estimate, and that it was really 240,000 stades.[22]

When Strabo says, as quoted above, "he criticises both, and rightly so",[23] it is clear that "rightly so" (δικαίως) is Strabo's comment, and that what precedes is a report of Posidonius. In what comes after δικαίως the infinitive λέγεσθαι suggests that the words διακεκαυμένην – καῦμα are indirect speech, reporting Posidonius, and one might suppose that in what follows those words, although it is not in indirect speech, at least the view that there is a small uninhabitable torrid zone, is Posidonius' view[24]; but it cannot be, because Posidonius' view of the zones (described by Strabo a few lines later) is different. Posidonius did agree with Strabo in not regarding the whole region between the tropics as uninhabitable, and no doubt criticised Aristotle for saying that it was, but he did not agree with Strabo about the torrid zone.

Strabo goes on:

τοῖς τε ἀρκτικοῖς, οὔτε παρὰ πᾶσιν οὖσιν, οὔτε τοῖς αὐτοῖς πανταχοῦ, τίς ἂν διορίζοι τὰς εὐκράτους, αἵπερ εἰσὶν ἀμετάπτωτοι; τὸ μὲν οὖν μὴ παρὰ πᾶσιν εἶναι τοὺς ἀρκτικούς, οὐδὲν ἂν εἴη πρὸς τὸν ἔλεγχον... τὸ δὲ μὴ πανταχοῦ τὸν αὐτὸν τρόπον, ἀλλὰ μεταπίπτειν, καλῶς εἴληπται. [25]

How could anyone determine the limits of the temperate zones, which are fixed and non-variable, by arctic circles, which are not available to all observers and are not the same everywhere? Well, his point about the arctic circles not being available to all observers would be irrelevant to his criticism. ... But his other point about the arctic circles not being similar everywhere, but changing, is well taken.
(Trans. I.G. Kidd [1999, Cambridge University Press], p. 110–11)

In the first sentence of this quotation, Strabo mentions two objections to the use of the "arctic circles" to define the boundary between temperate and arctic (or antarctic) zones; he then denies the validity of one of them, so they cannot be his own objections, but are presumably those of Posidonius. Aristotle[26] defined the boundary towards the poles of the temperate zones as τὸν διὰ παντὸς φανερόν, "the ever visible" (sc. circle), evidently meaning (as explained above) the terrestrial latitude corresponding to the celestial latitude above which the stars nearest the pole never set. Posidonius' apparent objections were, first, that in some places there is no "ever visible circle" (since there is none at the equator, nor, for observers in the northern hemisphere is there such a circle in the southern hemisphere, or vice versa); second, that this circle, where it exists, varies with the latitude of the observer, which cannot apply to an uninhabitable zone of the earth.[27]

Thus Posidonius accurately reports statements made by Aristotle in *Mete.* 362a32–b30, and pertinently criticises them.

The zone between the tropics is the only zone for which there is evidence for Posidonius' views of the climate.[28] Following my last quotation, Strabo says: Αὐτὸς δὲ διαιρῶν εἰς τὰς ζώνας, πέντε μέν φησιν εἶναι χρησίμους πρὸς τὰ οὐράνια, "In his own division into zones, he [obviously Posidonius] says that five are useful in relation to celestial phenomena"[29]: two, one from each pole, to the arctic or antarctic circle; two, one from each of those circles to the northern or southern tropic; and one between the tropics. Strabo then goes on, clearly still describing Posidonius' theory:

Πρὸς δὲ τὰ ἀνθρώπεια ταύτας τε καὶ δύο ἄλλας στενὰς τὰς ὑπὸ τοῖς τροπικοῖς καθ᾽ ἃς ἥμισύ πως μηνὸς κατὰ κορυφήν ἐστιν ὁ ἥλιος, δίχα διαιρουμένας ὑπὸ τῶν τροπικῶν. Ἔχειν γάρ τι ἴδιον τὰς ζώνας ταύτας, αὐχμηράς τε ἰδίως καὶ ἀμμώδεις ὑπαρχούσας καὶ ἀφόρους πλὴν σιλφίου καὶ πυρωδῶν τινων καρπῶν συγκεκαυμένων. Ὄρη γὰρ μὴ εἶναι πλησίον, ὥστε τὰ νέφη προσπίπτοντα ὄμβρους ποιεῖν, μηδὲ δὴ ποταμοῖς διαρρεῖσθαι ... Ὅτι δὲ ταῦτ᾽ ἴδια τῶν ζωνῶν τούτων δηλοῦν φησι τὸ τοὺς νοτιωτέρους αὐτῶν ἔχειν τὸ περιέχον εὐκρατότερον καὶ τὴν γῆν καρπιμωτέραν καὶ νοτιωτέραν καὶ εὐυδροτέραν.

These zones are also related to human geography along with two other zones, narrow strips which lie under the tropics, where they have the sun directly overhead for about half a month [i.e., half a month each year], since they are cut in two by each tropic. He said these two zones have peculiarities of their own; they are peculiarly parched and sandy, and produce nothing but silphium and some fiery burnt-up fruits; for no mountains are near for clouds to hit and produce rain, nor is there any irrigation from rivers. ... [Details are given of the men and animals of these zones.] And further evidence, he says, that all this is peculiar to these zones is that the people to the south of them have a more temperate climate, and a country that is more fertile and better watered.

(Strabo II.2.3, trans. I.G. Kidd [1999, Cambridge
University Press], p. 111–12, slightly modified)

This scheme leaves no room for an uninhabitable region at the equator: that would surely be another zone, and would bring the total to nine.[30] That Posidonius denied that there is an uninhabitable zone at the equator is confirmed by Cleomedes[31]: ἀπὸ τούτων ὁ Ποσειδώνιος τὸ ἐνδόσιμον λαβὼν καὶ πᾶν τὸ ὑπὸ τὸν ἰσημερὸν κλίμα εὔκρατον εἶναι ὑπέλαβε, "Posidonius, taking his key note from these facts [that Syene and Ethiopia, on or near the summer tropic, are inhabited], assumed that the whole latitude at the equator also was temperate."[32] That the earth is uninhabitable south of the "cinnamon-producing parallel" is Strabo's view (see Strabo I.4.2, II.1.15 and II.5.7, as well as II.2.2), but not the view of Posidonius.

That the earth at the equator is temperate in climate and habitable was not an original view of Posidonius: Eratosthenes (though with him there is some doubt)[33] and Polybius[34] are both reported to have held it, as is Posidonius' own teacher, Panaetius[35] (although other Stoics thought the equatorial zone uninhabitable[36]). In the generations following Aristotle, Greek knowledge of the upper Nile must have increased greatly following Alexander the Great's conquest of Egypt, and it must have been widely known that lands south of the summer tropic were inhabited, though some, like Strabo, thought that there were uninhabitable lands still further south. Others, like Posidonius, presumably felt that they had enough information (which surely might have been obtained by travellers on the upper Nile or voyagers on the Indian Ocean) to abandon entirely the hypothesis of an uninhabitable zone at the equator. As Geminus says, ἐπὶ πολλοὺς τόπους τῆς διακεκαυμένης ζώνης ἐληλύθασί τινες, καὶ τὰ πλεῖστα οἰκήσιμα εὕρηται, "people have gone to many places of the torrid zone, and most have been found habitable".[37]

That the climate at the northern and the southern tropic is drier and less temperate than the climate at the equator is a view in which Posidonius, according to Geminus,[38] was anticipated by Polybius. The original reason why Polybius and Posidonius held this view was presumably that they thought they had information enough, from travellers in Africa and Arabia, to be sure that the land there at the latitude of the northern tropic was mostly desert, and that land further south was less dry and more fertile: Geminus says here, of Polybius: ἱστορίας φέρει τῶν κατωπτευκότων τὰς οἰκήσεις καὶ ἐπιμαρτυρούντων τοῖς φαινομένοις, "he brings forward accounts of those who have seen the inhabited lands and bear witness to what can be seen". He adds that Polybius also reasoned from the sun's movements; but the accounts of first-hand witnesses surely came first.

Several causes were proposed to explain the temperate climate of the equatorial region. Panaetius[39] is reported as saying that the air is cooled by strong etesian (northerly) winds there, and by an exhalation of cold brought by a breeze from the ocean. According to Strabo,[40] Polybius held that the land at the equator ὑψηλοτάτη ἐστί· διόπερ καὶ κατομβρεῖται, "is very high; because of this it is also rained upon" (because the etesian winds blow clouds from the north against the high ground); to which, says Strabo, Posidonius objected that οὐδὲν... εἶναι κατὰ τὴν σφαιρικὴν ἐπιφάνειαν ὕψος διὰ τὴν ὁμαλότητα, οὐδὲ δὴ ὀρεινὴν εἶναι τὴν ὑπὸ τῷ ἰσημερινῷ, ἀλλὰ μᾶλλον πεδιάδα ἰσόπεδόν πως τῇ ἐπιφανείᾳ τῆς θαλάττης, "there is no high point in a spherical surface, because of its evenness [i.e., all spheres have a uniform surface]; nor, in fact, is the land under the equator mountainous, but rather flat, on a level more or less with the surface of the sea". The objection about the uniformity of the surface of a sphere is a strange one, unless Polybius meant something different from ordinary mountains; but Posidonius apparently also asserted, as an observed fact, that land at the equator is not mountainous: for this he very likely had information (he said that he had enquired about the voyages in the Indian Ocean of Eudoxus of Cyzicus,[41] and he could have made other enquiries) which justified him in saying that *some* land near the equator is not much above sea level.[42]

Posidonius' explanation of the "temperate" climate of the equatorial region is given by Cleomedes[43]:

ἔπου, γὰρ, φησίν, ἐπὶ πλέον τοῦ ἡλίου περὶ τοὺς τροπικοὺς διατρίβοντος, οὐκ ἔστιν ἀοίκητα τὰ ὑπ' αὐτοῖς, οὐδὲ τὰ ἐπὶ τούτων ἐνδοτέρω, πῶς οὐκ ἂν πολὺ πλέον τὰ ὑπὸ τῷ ἰσημερινῷ εὔκρατα εἴη, ταχέως τῷ κύκλῳ τούτῳ καὶ προσιόντος τοῦ ἡλίου καὶ πάλιν ἴσῳ τάχει ἀφισταμένου αὐτοῦ καὶ μὴ ἐγχρονίζοντος περὶ τὸ κλίμα, καὶ μὴν διὰ παντός, φησίν, ἴσης τῆς νυκτὸς τῇ ἡμέρᾳ οὔσης ἐνταῦθα καὶ διὰ τοῦτο σύμμετρον ἐχούσης πρὸς ἀνάψυξιν τὸ διάστημα; καὶ τοῦ ἀέρος τούτου ἐν τῷ μεσαιτάτῳ καὶ βαθυτάτῳ τῆς σκιᾶς ὄντος, καὶ ὄμβροι γενήσονται καὶ πνεύματα δυνάμενα ἀναψύχειν τὸν ἀέρα

for, he [Posidonius] says, when places at the tropics [sc. of Cancer and Capricorn] are not uninhabitable, although the sun spends more time at them, nor are places further within [i.e. within the zone between the tropics] uninhabitable, how could places at the equator fail to be much more temperate, when the sun swiftly approaches this circle [i.e. the equator] and again with equal speed departs from it, and does not stay long about this latitude – and when, besides, the night is always equal to the day there, and because of this has the interval that is exactly suitable for cooling? Also, this air [sc. at the equator] being in the most central and deepest part of the shadow, there will be both rain and winds able to cool the air.

Three causes are here suggested to explain why the climate at the equator should be temperate: first, the sun spends more time at or near the two tropics than it does at or near the equator (which in a way is true: it is a familiar fact that day length changes more slowly at midsummer and midwinter, when the sun is over one or other tropic, than it does at the equinoxes, when it is over the equator), and so, Posidonius claims, the sun heats the tropics more than the equator; second, day and night at the equator are always equally long, so that (Posidonius suggests) the air there at night has the ideal length of time in which to cool; third, air at the equator is in the deepest shadow – he presumably means that, at night, air at the equator cools because there it is more fully screened from the sun than it is anywhere else, that is, the full diameter of the earth is between it and the sun, which is above the equator on the opposite side of the earth. In support of this, Posidonius cited the rains that are reported to fall in Ethiopia, that is, not far from the equator.[44]

Not all of this was original to Posidonius. Geminus[45] attributes to Polybius the idea that the sun spends longer over the two tropics than over the equator and so heats the tropics more. Strabo, too, mentions this theory: just after his report that Polybius believed in a temperate equatorial region, and that the cause was high ground at the equator, he says[46]:

συνηγορεῖ δὲ τούτοις καὶ τὰ τοιαῦτα, ὧν μέμνηται καὶ Ποσειδώνιος, τὸ ἐκεῖ τὰς μεταστάσεις ὀξυτέρας εἶναι τὰς εἰς τὰ πλάγια, ὡς δ' αὔτως καὶ τὰς ἀπ' ἀνατολῆς ἐπὶ δύσιν τοῦ ἡλίου· ὀξύτεραι γὰρ αἱ κατὰ μεγίστου κύκλου τῶν ὁμοταχῶν κινήσεων.

This is supported by the following sorts of argument, mentioned too by Posidonius: the fact that there the oblique changes of course of the sun are more rapid, and in the same way also its movement from rising to setting; for in motions completed in the same time those over the greatest circle of circumference are more rapid.

(Trans. I.G. Kidd [1999, Cambridge University Press], p. 114)

The rapid "oblique changes" seem to be the apparent movements of the sun across the equator at the equinoxes, also mentioned by Cleomedes and Geminus. The argument from the movement of the sun from sunrise to sunset is one they do not mention: I take the idea to be that every daily revolution of the sun around the earth takes the same amount of time, but the equator is longer than a tropic, so that the sun, when over the equator, passes over each individual stade of it in a shorter time than it takes to pass over a stade of a tropic when it is over that, so that each stade at the equator is heated less. Strabo seems to imply that these arguments were not original to Posidonius.

One other idea is attributed to Posidonius concerning the climate of different areas of the world. Strabo,[47] after criticising a statement by Posidonius about the small number and size of rivers in north-west Africa, reports him as saying τὰ μὲν ἀνατολικὰ ὑγρὰ εἶναι, τὸν γὰρ ἥλιον ἀνίσχοντα ταχὺ παραλλάττειν, τὰ δ' ἑσπέρια ξηρά, ἐκεῖ γὰρ καταστρέφειν, (in Kidd's[48] translation) "eastern areas are wet, for the sun in rising passes by quickly, while the west is dry, because there the sun retires". This appears to mean that, over the *oikoumenē* as a whole, the inhabited world as known to the ancients, the east is wetter than the west[49]; but the words quoted provide no plausible explanation of how this happens, since (as Strabo comments[50]) the speed of the sun's revolution is clearly constant. Kidd finds the words in question "incomprehensible".[51] However, if Posidonius was talking about the whole *oikoumenē*, he had a precedent, since Aristotle reaches the opposite conclusion by a similar argument: at *Mete.* 364a24ff he says that westerly winds are colder and easterly winds hotter, ὅτι πλείω χρόνον ὑπὸ τὸν ἥλιόν ἐστι τὰ ἀπ' ἀνατολῆς· τὰ δ' ἀπὸ δυσμῆς ἀπολείπει τε θᾶττον καὶ πλησιάζει τῷ τόπῳ ὀψιαίτερον, "because the winds from the sunrise are beneath the sun for a longer time; but it leaves more quickly those from the sunset and reaches that region later"- on which E.W. Webster commented "a poor argument even for a flat-earth man; and for Aristotle with his round earth lamentable".[52]

We can, however, excuse Posidonius if we accept the interpretation of Shcheglov (2006), who argues that the words I quoted from Strabo refer not to the *oikoumenē* as a whole, but just to Africa, and to the theory, discussed above, that the sun in its annual movements heats the two tropics more than the equator, so that north-west Africa, near the northern tropic, is drier than east Africa at the equator.

There is other evidence that Posidonius thought at least one part of the east, that is, India, to be wetter than Africa. Strabo says elsewhere[53] that, according to Posidonius, τοὺς Ἰνδοὺς τῶν Αἰθιόπων διαφέρειν τῶν ἐν τῇ Λιβύῃ· εὐερνεστέρους γὰρ εἶναι καὶ ἧττον ἔψεσθαι τῇ ξηρασίᾳ τοῦ περιέχοντος, "Indians differ from the African Ethiopians. ... Indians are more developed physically, less burnt by the dryness of the atmosphere". That India has the wetter climate is true,[54] and Pliny seems to provide[55] a plausible explanation, namely, that in Posidonius' view, "eius venti adflatu iuvari Indiam salubremque fieri", "India is helped and made healthy by the blowing of that wind", the wind being Favonius, the west wind, mentioned just before - which looks likes a reference to the rain-bearing south-west monsoon.[56]

On climatic zones generally, as with the discussion of the height to which clouds and rain occur, we can see that Greek knowledge and ideas had developed between the time of Aristotle and that of Posidonius. Astronomers and geographers had calculated, with approximate correctness, the circumference of the earth and the degrees of latitude which separate the tropics from the equator. From the reports of travellers, serious students had gained a better knowledge of the geography and climate of North Africa, the Middle East and India, and had seen that Aristotle was wrong to regard the equatorial zone as uninhabitable. With their wider geographical knowledge, they realised that people do live on the equator, and believed (which is true at least of North Africa) that land at the tropic is drier and more burnt up than that further south, nearer the equator – a realisation of which there is already a hint in Herodotus, with his tale of the Nasamones, who travelled through the African desert and found beyond it a marshy country inhabited by small men.[57] These advances in knowledge were, it is clear, set forth in the works of Posidonius, even if his original contribution was not great: Eratosthenes (probably), Polybius and Panaetius had held similar views before him. At the most, Posidonius had added some facts, as that there is (at least some) flat land at sea-level at the equator, and some additional – ingenious, but quite speculative – explanations.

Posidonius' correct view that land at the equator is habitable was never universally held in antiquity. Geminus agreed with Posidonius about this,[58] as did Ptolemy in his *Geography*[59]; but Achilles seems to have regarded the matter as doubtful, saying of the torrid zone ταύτην δὲ οἳ μὲν ἀοίκητον, οἱ παλαιοί, τινὲς δὲ οἰκεῖσθαι λέγουσιν, "some, the ancients, say it is uninhabitable, but some say it is inhabited".[60] Strabo, as already mentioned, thought the equator uninhabitable; Cleomedes reports Posidonius' view, but also criticises his arguments and concludes that he was wrong[61]; the Anonymus who reports Panaetius' view has asserted immediately before that the torrid zone is uninhabitable, but then adds that "some people" (τινές), including Panaetius, think otherwise[62]; to Virgil and Pliny the torrid zone is uninhabitable.[63]

Notes

1 Posidonius F49.10–14 EK. For commentary on Strabo II.2.2–2.3 see Kidd (1988) 222–31; Theiler (1982) II, 22–5.

2 I print the text as given by EK, and Vimercati (2004) 126, omitting the words which they put in square brackets, τῆς μεταξὺ τῶν τροπικῶν after τὴν διακεκαυμένην. (Theiler [1982] I, 28, in his F13, prints a slightly different text, retaining these words; but the difference does not affect the present discussion.)

3 Aëtius III.11.4. This and the relevant words from Strabo II.2.2 are printed as DK 28A44a.

4 Achilles, *Isagoga in Aratum* 31 (Maass [1898] 67.27–33; Posidonius F209 EK).
5 Diogenes Laertius IX.21 (DK 28A1) and VIII.48 (DK 28A44).
6 Kidd (1988) 224–5.
7 Aëtius III.14.1 (see Diels [1879] 378) attributes a theory of zones to Pythagoras himself. This is incredible, and Diels [1879] 181 may well be right to regard the story as a Hellenistic or later invention; but Guthrie [1962] 294 regards it as evidence that the theory was "known as a Pythagorean tenet").
8 Herodotus II.19 and 26.
9 Herodotus II.22.
10 Herodotus II.32.
11 Herodotus IV.28.
12 Herodotus IV.31.
13 *Mete.* 362b2–3.
14 *Mete.* 362b9.
15 *Mete.* 362b7–8.
16 Posidonius F49.14–20 EK. I continue to print the text as given by EK, and by Jones (1917–32).
17 Kidd (1988) 76–7.
18 For this paragraph and the next see Posidonius F49.20–36 EK.
19 See Strabo II.5.7.
20 Posidonius F49. 32–36. I quote lines 34–6, with the trans. of Kidd (1999) 110.
21 Cleomedes I.7 lines 44–5 Todd (Posidonius F202.46–7 EK).
22 See Taisbak (1974) 261–2; Kidd (1988) 226–7.
23 Posidonius F49.17 EK.
24 As Kidd (1988) 223 assumes. (Elsewhere, e.g., Kidd [1988] 750, he accepts a report that, for Posidonius. land at the equator is habitable. Theiler [1982] II, 23 accepts that, in Posidonius' view, land at the equator is habitable, as does Vimercati [2004] 585–6).
25 The final lines of Strabo II.2.2 (Posidonius F49.37–43 EK).
26 *Mete.* 362b3.
27 See Jones (1917–32) 364–5, note; Kidd (1988) 228–9.
28 Besides Strabo, Posidonius' theory of zones is also mentioned by Achilles, *Isagoga in Aratum* 31 (Maass [1898] 67.27–33; Posidonius F209 EK), who says: οἱ μὲν γὰρ ἓξ αὐτὰς εἶπον ὡς Πολύβιος καὶ Ποσειδώνιος τὴν διακεκαυμένην εἰς δύο διαιροῦντες, "some said that there are six [sc. zones), like Polybius and Posidonius, dividing the torrid one into two". Strabo II.3.1(Posidonius F49.62–4 EK) confirms this for Polybius, but Strabo's detailed account of Posidonius' different theory is surely far more probable than Achilles' brief statement. See Kidd (1988) 748.
29 Strabo II.2.3 = Posidonius F49.44–5 EK. Trans. from Kidd (1999) 111, slightly modified.
30 Since there would then be two temperate equatorial zones, one each side of the equator, plus an uninhabitable zone at the equator.
31 Cleomedes I.6.31 (I.4 lines 94–5 Todd) = Posidonius F210.5–7 EK. Symeon Seth, *De utilitate corporum caelestium*, 44 (Posidonius F211.1–3 EK.) also says that, according to Posidonius, the climate is temperate at the equator; but, as an 11th century author (Kidd [1988] 753), his testimony may not be independent of our other sources.
32 Tr. Kidd (1999) 275.
33 Strabo II.3.2 (Posidonius F49.118–20 EK) says he thought the earth temperate at the equator; but surviving verses of Eratosthenes (see Maass [1898] 63–4) imply that it is not. See Kidd (1988) 236 for discussion, with citation of earlier views.
34 See Strabo II.3.2 (Posidonius F49.118–20 EK); Geminus, *Isagoge*, XVI.32–3.
35 Anonymus I, *Isagoge*, 6, in Maass (1898) 97 = Panaetius fr. 135 Straaten.

36 Diogenes Laertius VII.156.
37 Geminus, *Isagoge*, XVI.31.
38 Geminus, *Isagoge*, XVI.32–33.
39 Panaetius fr. 135 Straaten (see note 35 above).
40 Strabo II.3.2–3 (Posidonius F49.121–3 and 130–4 EK); with the translation of Kidd (1999) 114–15.
41 Strabo II.3.4–5 (Posidonius F49.150–228 EK, especially 173–80: Eudoxus was blown off course, evidently to East Africa). See Kidd (1988) 254–6 for a discussion of likely ancient knowledge of the Indian Ocean in Posidonius' time.
42 Strabo II.3.3 (Posidonius F49.136–43 EK) says Posidonius was inconsistent, and said elsewhere that he suspected (ὑπονοεῖν) that there were mountains at the equator, and also that there was ocean there. Here we may suspect that Strabo, keen to criticise Posidonius, is being tendentious. His second point is not a valid objection to Posidonius' claim about land at sea level, unless Posidonius said that *all* the equator is covered by ocean, which he clearly did not (see Kidd [1988] 239, 458–61). As to mountains, Kidd (1988) 238 points out that there is no other evidence for Strabo's statement.
43 Cleomedes I.6.31–2 (I.4 lines 98–107 Todd) = Posidonius F210.10–20 EK. For discussion of this passage see Kidd (1988) 750–1.
44 Cleomedes I.6.32 (I.4 lines 107–9 Todd) = Posidonius F210.20–23 EK. For comments see Kidd (1988) 749–52; Vimercati (2004) 586–7.
45 Geminus, *Isagoge*, XVI.34–6.
46 Strabo II.3.2 (Posidonius F49.125–9 EK). For discussion see Kidd (1988) 237.
47 Strabo XVII.3.10 (Posidonius F223.9–11 EK).
48 Kidd (1988) 802 and (1999) 297.
49 So interpreted by, for instance, Thomson (1948) 214.
50 Strabo XVII.3.10 (Posidonius F223.21–3 EK).
51 Kidd (1988) 802. Theiler (1982) II, 70 suggests that, for Posidonius, the east is wet but the west dry because the sun, when it passes over eastern lands, has just passed over the ocean and drawn moisture from it, but when it passes over the western οἰκουμένη it has been passing over dry land: this is a theory Posidonius might well have held, but is surely not a possible interpretation of Strabo's words. Vimercati (2004) 598 suggests that Strabo omits part of Posidonius' theory.
52 Webster (1931), in a note on this passage; quoted by Lee (1952) 193.
53 Strabo II.3.7 (Posidonius F49.329–331 EK), with the translation of Kidd (1999) 123.
54 See online World Bank data for average annual rainfall (checked August 2021).
55 Pliny, *Nat.* VI.57–8 = Posidonius F212 EK.
56 See below, Chapter 11 (p. 108–9).
57 Herodotus II.32.
58 Geminus, *Isagoge*, XVI.25 and 31.
59 In book IV.7.11, 12, 24 and 26 and in IV.8 he mentions places which he says have latitudes on or south of the equator, and he must have thought these places accessible on or across the equator, or he could not have had knowledge of the latitudes. At IV.7.31 he mentions a district which extends μεχρὶ τῆς Κολόης λίμνης, μεθ᾽ ἣν οἱ Μαστῖται μεχρὶ τῶν τοῦ Νείλου λιμνῶν, "up to Lake Coloë, after which are the Mastitae up to the lakes of the Nile". He has previously placed Lake Coloë on the equator and the lakes of the Nile 6° and 7° south of it (IV.7.23–4): clearly the Mastitae find the equator habitable. (I use the edition of Ptolemy by Stückelberger and Grasshoff [2006]).
60 Maass (1898) 63, lines 4–5.
61 Cleomedes I.6.32–3 (I.4 lines 109–31 Todd) = Posidonius F210.24–48 EK.
62 Anonymus, *Isagoge*, in Maass (1898) 96–7.
63 Virgil, *Georgics*, I.233–8; Pliny, *Nat.* II.172.

8   Thunder and lightning

The last two chapters have dealt with subjects for which increased geographical knowledge, and techniques of measurement and calculation which were available to the ancients, could and did give them a better understanding of the world. Modern critics may castigate the avoidable inaccuracy of many of their measurements, but they did make a real advance towards what we now know to be the truth. For thunder and lightning this does not apply. The ancients had no understanding of electricity, and so had no way to investigate the true nature of these phenomena, though many thinkers had theories about them. (I shall concentrate on thunder and lightning, and largely ignore phenomena which the ancients connected with these, such as thunderbolts – i.e., lightning strikes – and tornadoes, since we lack evidence for Posidonius' views of these.)

The problem to be solved was to find a way by which a loud noise and a flash of light might be produced from a cloud. To explain the noise was not too difficult. We all know that wind can cause a noise – a very loud noise when a strong wind overturns a heavy object. Also, assuming the lightning flash to be fiery, we all know that fire is noisy when it burns a fuel or boils a liquid fiercely, or when a mass of liquid puts a fierce fire out. An explanation of thunder might require the supposition that a thundercloud has properties not possessed by the cloud-like bodies we see at close quarters, such as the smoke and "steam" produced by domestic fires and boiling water; for instance, the supposition, best known from Aristophanes' parody,[1] that a thundercloud is like a bladder, with wind trapped within it. Granted such assumptions, a reasonably plausible explanation of thunder was not too hard to find. To explain the lightning-flash was another matter.

Theories before Posidonius

I have already quoted two lines from Hesiod's *Theogony*, which speak of winds (ἄνεμοι) causing thunder, lightning and thunderbolt.[2] The earliest philosophical views, from the 6th century B.C., are known only from reports in much later authors, but were apparently similar to this. Aëtius[3] says that, according to Anaximander,

DOI: 10.4324/9780429399930-8

ἐκ τοῦ πνεύματος ταυτὶ πάντα συμβαίνειν· ὅταν γὰρ περιληφθὲν νέφει παχεῖ βιασάμενον ἐκπέσῃ τῇ λεπτομερείᾳ καὶ κουφότητι, τόθ᾽ ἡ μὲν ῥῆξις τὸν ψόφον, ἡ δὲ διαστολὴ παρὰ τὴν μελανίαν τοῦ νέφους τὸν διαυγασμὸν ἀποτελεῖ.

all these phenomena happen from wind; for when it [i.e., wind] has been surrounded by thick cloud, and forcibly escapes by its fineness and lightness, then the breaking [sc. of the cloud] produces the noise, and the spreading out [of the wind?] beside the blackness of the cloud produces the flash of light.

If this is right, Anaximander initiated the analogy between thunder and the bursting of a bladder. Aëtius[4] adds that Anaximenes (Anaximander's pupil) held the same theory,[5] adding a comparison of lightning to phosphorescence in the sea – an idea adopted by few later thinkers.[6]

These theories do not explain the fieriness of lightning, which is evident from its brightness, and the fires sometimes caused by the obviously related *keraunos* (lightning-strike). In the 5th century B.C., Empedocles and Anaxagoras (thinkers who believed that matter does not really change, so that what is fire now must always have been fire) explained thunder and lightning as caused by pre-existing fire enclosed in a cloud. Aristotle says:

Καίτοι τινὲς λέγουσιν ὡς ἐν τοῖς νέφεσιν ἐγγίγνεται πῦρ· τοῦτο δ᾽ Ἐμπεδο-κλῆς μέν φησιν εἶναι τὸ ἐμπεριλαμβανόμενον τῶν τοῦ ἡλίου ἀκτίνων, Ἀναξαγόρας δὲ τοῦ ἄνωθεν αἰθέρος, ὃ δὴ ἐκεῖνος καλεῖ πῦρ κατενεχθὲν ἄνωθεν κάτω. τὴν μὲν οὖν διάλαμψιν ἀστραπὴν εἶναι τὴν τούτου τοῦ πυρός, τὸν δὲ ψόφον ἐναποσβεννυμένου καὶ τὴν σίξιν βροντήν.

And yet some say that fire is present in the clouds. This Empedocles says is some of the sun's rays trapped [sc. in the clouds], Anaxagoras says it is some of the upper *aithēr* (which he calls fire) which has been carried down from above. Lightning they then say is the flash of this fire, thunder the noise and hissing of it as it is quenched.[7]

That thunder is the sound of the lightning-fire being quenched by the water of the thundercloud (analogous to the sound of red-hot stones being put into cold water[8]) seems a reasonable view, and is one which Aristotle himself adopts, at least as a hypothesis, at *Analytica posteriora* II.8 (93a1–94a19).[9] In *Mete.* 370a6–10 he criticises it, at least as presented by Empedocles and Anaxagoras.

Democritus used his atomic theory to explain lightning. According to him (says Aëtius) lightning is σύγκρουσιν νεφῶν, ὑφ᾽ ἧς τὰ γεννητικὰ τοῦ πυρὸς διὰ τῶν πολυκένων ἀραιωμάτων ταῖς παρατρίψεσιν εἰς τὸ αὐτὸ συναλιζόμενα διηθεῖται, "a collision of clouds, by which those [sc. "atoms", presumably] which generate fire are filtered by friction through interstices which contain much void, being brought together into the same place."[10]

Presumably, friction between the colliding clouds causes the fire-atoms (spherical, and so mobile[11]) which are in them to collect together, and this by itself produces what we perceive as fire; and presumably a similar process occurs when men light fires by friction between two pieces of wood. Democritus' lightning theory is another which attributes to thunderclouds properties – in this case, solidity and abrasiveness – which the cloudlike bodies we have contact with manifestly do not have.

After Democritus, the next important theory of thunder and lightning is Aristotle's, in *Mete.* II.9 (369a13ff). The natural movement of the dry exhalation, he implies, is to rise, and he goes on (369a25ff):

ὅση δ' ἐμπεριλαμβάνεται τῆς ξηρᾶς ἀναθυμιάσεως ἐν τῇ μεταβολῇ ψυχομένου τοῦ ἀέρος, αὕτη συνιόντων τῶν νεφῶν ἐκκρίνεται, βίᾳ δὲ φερομένη καὶ προσπίπτουσα τοῖς περιεχομένοις νέφεσι ποιεῖ πληγήν, ἧς ὁ ψόφος καλεῖται βροντή...(b5ff) τὸ δὲ πνεῦμα τὸ ἐκθλιβόμενον τὰ πολλὰ μὲν ἐπυροῦται λεπτῇ καὶ ἀσθενεῖ πυρώσει, καὶ τοῦτ' ἐστιν ἣν καλοῦμεν ἀστραπήν...(371a18ff) ἐὰν δ' ἐν αὐτῷ τῷ νέφει πολὺ καὶ λεπτὸν ἐκθλιφθῇ πνεῦμα, τοῦτο γίγνεται κεραυνός.

but as much of the dry exhalation as is trapped during the change as the air cools, is ejected as the clouds condense, and, being forcibly borne along and striking against the clouds which surround it, it produces an impact, the noise of which we call thunder ... (b5ff) and the wind which is squeezed out usually burns with a fine and weak fire, and this is what we call lightning ... (371a18ff) but if in the cloud itself much fine wind is squeezed out, this becomes a *keraunos* (lightning-strike).

To Aristotle, dry exhalation is the ἀρχὴ καὶ φύσις, "source and nature" of wind,[12] so he is not inconsistent in speaking, in the extracts I quote, first of dry exhalation and then of wind. Aristotle diverges from earlier versions of the theory of wind trapped in a cloud, by speaking of wind being squeezed out by a cloud, rather than escaping by its own force: his theory of natural motion requires him to explain why hot, dry exhalation descends as lightning, instead of rising. He compares this squeezing out with fruit-stones jumping out when squeezed between the fingers: they often rise, though naturally heavy, just as thunderbolts and the like descend, though naturally light.[13] He also says that thunder is produced by a process similar to that which produces a crackle when wood burns.[14]

This is reasonable, if we allow that thunderclouds have properties that we normally observe only in solid bodies. He does not explain how the fire is ignited, but we can perhaps explain this from what he has said earlier in his *Meteorologica*. The dry exhalation, Aristotle has said, is dry and hot,[15] that is, it has the qualities which in *De generatione et corruptione* are those of the element fire[16]; in the *Meteorologica* Aristotle explains that hot, dry matter is οἷον ὑπέκκαυμα ... ὥστε μικρᾶς κινήσεως τυχὸν ἐκκαίεσθαι πολλάκις, "like fuel ... so that if it happens to be moved slightly it often catches alight".[17] Presumably, when there is lightning, the dry exhalation is moved enough for it to catch alight. In this way Aristotle combines a plausible origin for the fire of lightning with the earliest view, not abandoned by some of the later pre-Socratics,[18] that thunder and lightning are due to wind.

There is some evidence for the thunder and lightning theories of Aristotle's successor, Theophrastus, even disregarding the questionable testimony of the Syriac Meteorology. In Chapter 1 of *De igne*, he says that the ignition of fire is usually "with force" (μετὰ βίας), and to other instances of this he adds ἐκ δ' αὐτοῦ τοῦ ἀέρος ἐν τοῖς νέφεσι συστροφῇ καὶ θλίψει, "and from the air itself in clouds by concentration and squeezing"; he names *keraunoi* as an example, presumably lightning would be another. Other evidence suggests that he and his successor Strato both wrote of these phenomena as caused by the interaction of the hot (*thermon*) and the cold (*psychron, psychrotēs*) without specifying the type of matter involved (such as fire, wind, air).[19] I do not discuss these ideas here, as there is no sign that they influenced Posidonius.

One other relevant text which certainly antedates Posidonius is Epicurus' *Letter to Pythocles*, 100–2. This describes various possible causes of thunder and lightning, among which much is familiar: for instance, "breakings of clouds" (ῥήξεις νεφῶν) in Epicurus recalls the "burst bladder" theory, and "rubbing and collision of clouds" (παράτριψιν καὶ σύγκρουσιν νεφῶν) recalls the theory of Democritus. Other ideas are not recorded from earlier authors (though this may be due to defects in our evidence). One such idea which seems to have been used by Posidonius is that wind may rotate in a cloud: thunder may occur κατὰ πνεύματος ἐν τοῖς κοιλώμασι τῶν νεφῶν ἀνείλησιν, καθάπερ ἐν τοῖς ἡμετέροις ἀγγείοις, "by the rolling of wind in the hollows of the clouds, as in our jars",[20] presumably referring to a sound produced in a jar by blowing into it. Ἀνείλησιν might just mean "confinement", but the confined wind must be moving, so the ειλ root surely has its sense of "revolving" here.

More important to Posidonius than Epicurus' theories would have been what was said by earlier members of his own school, the Stoics. According to Diogenes Laertius, Zeno said that lightning is ἔξαψιν νεφῶν παρατριβομένων ἢ ῥηγνυμένων ὑπὸ πνεύματος ... βροντὴν δὲ τὸν τούτων ψόφον ἐκ παρατρίψεως ἢ ῥήξεως, "a catching fire of clouds which are rubbed against or broken by wind ... and thunder is the sound of them from rubbing or breaking".[21] Stobaeus describes Chrysippus' lightning theory in exactly the same words, except that he has ἐκτριβομένων, I suppose "rubbed hard", for παρατριβομένων, and says of thunder simply βροντὴν δ' εἶναι τὸν τούτων ψόφον, "and thunder is the sound of them".[22] Aëtius says: οἱ Στωϊκοὶ βροντὴν μὲν συγκρουσμὸν νεφῶν, ἀστραπὴν δὲ ἔξαψιν ἐκ παρατρίψεως, "the Stoics [sc. say that] thunder is a collision of clouds, and lightning is the setting on fire of them from rubbing [against each other?]".[23] It looks as though the early Stoics laid some stress on friction as a cause, presumably because friction is a cause of fire known to everyone; but it seems not to have been essential, since breaking of a cloud can also cause lightning.

The theory of Posidonius

Against this background, what was the theory of Posidonius? His view is reported by Seneca *NQ* II.54.1–3[24]:

> Nunc ad opinionem Posidonii revertar. E terra terrenisque omnibus pars umida efflatur, pars sicca et fumida; haec fulminibus alimentum est, illa imbribus. Quidquid in aera sicci fumosique pervenit, id includi se nubibus non fert, sed rumpit claudentia; inde est sonus quem nos tonitrum vocamus. In ipso quoque aere quidquid extenuatur simul siccatur et calfit. Hoc quoque si inclusum est aeque fugam quaerit, et cum sono evadit, ac modo universum eruptionem facit, eoque vehementius intonat, modo per partes et minutatim. Ergo tonitrua hic spiritus exprimit dum aut rumpit nubes aut pervolat; volutatio autem spiritus in nube conclusi valentissimum est adterendi genus.[25]

> Let me now return to the opinion of Posidonius. From the earth and everything terrestrial something moist is exhaled, and something dry and smoky. The latter is nourishment for thunderbolts, the former for rain. Whatever dry and smoky matter comes into the air does not tolerate being shut up within clouds but breaks the materials that are enclosing it; hence the sound which we call thunder. Also, whatever is rarefied in the air itself is at the same time dried and heated; this too if it has been shut up [sc. in a cloud] equally seeks escape and comes out with noise, and sometimes bursts out all as one body and so causes thunder more violently, but sometimes escapes bit by bit and gradually. Therefore this wind forces out thunder while it either breaks clouds or flies through them. The rotatory movement of wind shut up in a cloud is the strongest kind of rubbing.[26]

This theory is unlike that of Zeno and Chrysippus, in that friction plays a subordinate part, but it has some confirmation in Cicero's description of a "Stoic" theory; it would be natural for Cicero to describe the theory of his former teacher, Posidonius:

> Placet …Stoicis eos anhelitus terrae, qui frigidi sint, cum fluere coeperint, ventos esse; cum autem se in nubem induerint eiusque tenuissimam quamque partem coeperint dividere atque dirrumpere idque crebrius facere et vehementius, tum et fulgores et tonitrua existere.

> The Stoics believe that those exhalations of the earth which are cold, when they begin to flow, are winds; when they have plunged into a cloud and begun to divide and break up all its finest parts, and to do that more constantly and strongly, then both lightning and thunder occur.[27]

Here, as in Seneca's account of Posidonius' theory, thunder occurs when exhalation within a cloud breaks it. We cannot rely on Cicero's details. He is arguing that a naturalistic explanation of thunderstorms, such as the Stoics accepted, is incompatible with their view that they are signs of future events; he has no need for accuracy in detail, but what he says should be recognisable as a Stoic theory – most likely that of Posidonius.

Seneca, writing a long discussion of thunder and lightning, ought to be more attentive to details. At first sight he seems to be describing here at least two causes of thunder: the dry, smoky exhalation from the earth and the rarefaction, drying and heating of air. (This may well mean the rarefaction of part of the cloud, to judge by Seneca's theory at *NQ* II.57, which is evidently based on this one.[28]) The action of wind might be a third cause, but probably is not; the mention of "this wind" ("hic spiritus") implies that the air, or the exhalation, *becomes* wind as it bursts out of the cloud; and the rotatory rubbing of the wind against the cloud seems likely to be part of the process of breaking out, as it seems to be in the explanation of thunder in the *De mundo*.[29] Also, Posidonius' idea may have been that the dry exhalation not only bursts out of the cloud but also rarefies and heats part of the cloud, or air within it; in this case the passage is describing different stages in a single process for the production of thunder. Lightning is not mentioned, but the rubbing that is mentioned surely caused lightning as well as thunder in Posidonius' theory, as it does in earlier theories, quoted above. This is confirmed by *NQ* II.55.2, where Seneca seems to be answering possible objections to Posidonius' theory, and says that, granting that fire is ignited in a cloud, "spiritu nascitur et attritu", "it originates from wind and rubbing". It is also confirmed by Seneca's related theory in *NQ* II.57, where friction explicitly causes the fire of lightning and thunderbolts.[30]

Seneca begins *NQ* II.54 by saying that he is *returning* to the opinion of Posidonius; but he has not previously named Posidonius in *NQ* II, except for some incidental details at II.26.4. Presumably, he is referring back to some passage where he gave Posidonius' view without naming him. Of the several suggestions that have been made about what this passage is,[31] the likeliest, in my view, is that of Kidd,[32] that Seneca is referring to the whole of II.21–30, and that that passage is in fact largely based on Posidonius, although Seneca begins it by saying that he is now giving his own exposition of the problem.[33] As Kidd argues,[34] the structure of *NQ* II suggests this: in II.21–30 Seneca discusses the causes of thunder and related phenomena, with special attention to thunder in II.27–30; II.31–53 is a long digression, mainly about the significance of thunderbolts as portents; at II.54 he returns to the discussion of causes, and one would naturally expect him to be picking it up from the point where he left it at II.30; and, which confirms that he is doing this, at the end of II.30 he defends his view of thunder by citing the theory of dry and wet exhalations: II.30.3, "diximus enim utriusque naturae corpora efflare terras et sicci aliquid et umidi in toto aere vagari", "for we have said[35] that the earth exhales bodies of both natures and that some dry matter and some wet circulates in the air as a whole". (The point, in the context of *NQ* II.30, is that this explains why thunder occurs when a volcanic eruption or a sandstorm has sent clouds of dry matter into the air – thunder requires clouds, but clouds are not just water-vapour.)

Probably, then, *NQ* II.54 refers back to II.21–30; but this is not certain, and in any case II.21–30 cannot all be derived from Posidonius: Seneca, judging from how he begins II.21, has surely introduced ideas of his own, and he twice cites Posidonius' pupil, Asclepiodotus.[36] Analysis of Posidonius' theory of thunder and lightning must be based on *NQ* II.54.

Most of what is attributed to Posidonius in II.54 is familiar from earlier writers. That thunder is due to material becoming trapped in a cloud and bursting out of it is a view attributed to Anaximander, Anaximenes, Empedocles, Anaxagoras and Leucippus,[37] and, though this is second-hand evidence, Aristophanes' parody guarantees that this concept is at least as old as the 5th century B.C.; that wind is a cause of such phenomena is an idea found in Hesiod, in several pre-Socratic writers, and in Aristotle. The theory of two exhalations, one dry and smoky and one wet, of which the former causes thunder and related phenomena, is clearly derived from Aristotle; we know of no other thinker before Posidonius who held this theory and applied it to thunder and lightning.[38] The idea of wind rotating within a cloud is found in Epicurus (see p. 68 above) and may have occurred in other authors earlier than Posidonius: there are parallels in the Syriac Meteorology,[39] and also in the *De mundo*.[40] Friction as a cause of thunder and lightning we have found attributed to Democritus and to the early Stoics[41]; the latter were presumably Posidonius' source for this idea. I know no close parallel to the suggestion that air itself is rarefied, dried and heated; this looks like an idea original to Posidonius.

There is little that can be added about Posidonius from *NQ* II.21–30. Seneca introduces the two-exhalation theory only in the last of those chapters, from which it follows that either that theory played only a small part in Posidonius' theory of thunder and lightning, or those chapters are not very closely based on Posidonius. One point worth mention is that II.22 seems to contain a hint of a formal logical argument about the cause of the fire of lightning – a kind of argument which, other evidence suggests, Posidonius would have been interested in formulating. I discuss this in Chapter 21 (see especially p. 203–4).

Seneca's account of Posidonius' theory of thunder and lightning is only partially confirmed by Cicero; further support for it can be found by observing that Posidonius dealt with similar meteorological problems in a similar way, offering generally conservative explanations, and showing respect for the views of Aristotle and earlier Stoics. That this is so will, I hope, become clear in the following chapters.

Leaving aside the question of Posidonius' own contribution, we can perhaps see some development in Greek philosophers' views, at least about lightning. The earliest philosophers seem to have offered no explanation of lightning-fire, and sometimes apparently denied that fire was involved; Empedocles and Anaxagoras explained lightning as the appearance of pre-existing fire, Democritus and Aristotle thought it was caused by something pre-existing which was almost fire – for Democritus, lightning was caused by fire-producing atoms, which (probably) had only to be collected to produce actual fire; for Aristotle the cause was a highly inflammable "dry exhalation", which the smallest impulse might set alight. Democritus added friction – a familiar means of producing domestic fires – as part of the cause, and after Aristotle, philosophers tended to agree that friction was the cause, or a cause, of the fire of lightning. We also find in Seneca, possibly going back to Posidonius, an attempt to argue for his theory of lightning, instead of just asserting it.

And yet to us it seems evident that this theory, and several theories of the cause of thunder, are most implausible, in that they involve attributing to clouds properties which are manifestly inconsistent with everyday observations of clouds and the cloud-like bodies with which we have contact. Some ancient writers on thunder and lightning do record objections of this kind to their theories: at *NQ* II.25 Seneca imagines an objector saying: "Dicis ... nubes adtritas edere ignem, cum sint umidae, immo udae; quomodo ergo possunt gignere ignem, quem non magis verisimile est ex nube <fieri> quam ex aqua, <quae> ex nube nascitur",[42] "You say that clouds produce fire by friction, although they are moist, or rather watery; how then can they generate fire? It is no more probable that fire is made from cloud than that it is made from water, which is produced from cloud". At II.27.4 Seneca suggests that some thunder is the sound of clouds beating against each other, similar to the clapping of hands; at II.28.1 he mentions an objection to this: "Videmus ... nubes inpingi montibus nec sonum fieri", "we see clouds driven against mountains and no sound is produced". The Syriac Meteorology (1.(25)–(28)) posits a similar objection: "How is it possible that noise arises from clouds since they are not solid like stones and earthenware but rarefied like wool. ... For noise does not arise if a man beats tufts of wool, one with the other".[43] In all three passages an answer is given to the objection; and yet one can surely imagine a sceptic concluding that the philosophers' explanations of thunder and lightning are scarcely more plausible than the mythological explanations they are intended to replace. Without knowledge of electricity, progress towards the true explanation was not possible; but it is perhaps better, rather than abandoning a problem as insoluble, to put forward the best explanation available, even if it is a poor one, and so give others the chance to criticise and possibly improve it.

Notes

1 *Clouds* 404–11.
2 See above, Chapter 4 (p. 20–1).
3 Aëtius III.3.1 (DK 12A23).
4 Aëtius III.3.2 (DK 13A17).
5 That these two thinkers regarded thunder, etc., as due to wind is confirmed by Seneca *NQ* II.18 (DK 12A23) and II.17, and by Hippolytus, *Haer.* I.6.7 (DK 12A11) and I.7.8 (DK 13A7). For a discussion see Hine (1981) 279–80.
6 The idea is attributed only to Cleidemus (5th or 4th century B.C.). See Aristotle *Mete.* 370a10–15 (DK 62.1), also Seneca *NQ* II.55.4. (On Cleidemus see Hine [1981] 426–8). Later texts which list multiple causes of lightning do not mention phosphorescence: Epicurus, *Letter to Pythocles* 101–2; Lucretius VI.160–218; Syriac Meteorology, section 2 (see Daiber [1992] 262).

7 *Mete.* 369b12–17 (DK 31A63, 59A84). Similar accounts are given for Empedocles by Aëtius III.3.7 (also in DK 31A63), and for Anaxagoras by Aëtius III.3.4 and Seneca *NQ* II.12.3 and probably II.19 (all in DK 59A84; note that in *NQ* II.19 "Anaxagoras" is a conjecture); Hippolytus, *Haer.* I.8.11 (DK 59A42) does not mention *aithēr* but otherwise seems compatible with Aristotle's report.

8 An analogy attributed to Anaxagoras' pupil Archelaus (Aëtius III.3.5=DK 60A16).

9 For discussion of this passage see below, Chapter 21 (p. 205–6).

10 Aëtius III.3.11 (DK 68A93). The accompanying explanation of thunder I find incomprehensible: something crucial has surely been omitted.

11 Aristotle *De Caelo* III, 303a3–14 (DK 67A15) and 306b32–5.

12 Mete. 360a13.

13 *Mete.* 369a20–24. After Aristotle, Epicurus mentions, as one explanation of lightning, the squeezing out (ἐκπιασμός) of suitable particles from clouds (*Letter to Pythocles* 101).

14 *Mete.* 369a30–5

15 *Mete.* 340b27–8, 360a25.

16 *GC* 330b4.

17 *Mete.* 341b19–20.

18 Wind is part of the cause for Diogenes of Apollonia (Aëtius III.3.8 and Seneca *NQ* II.20; DK 64A16) and the main cause for Metrodorus of Chios (Aëtius III.3.3 = DK 70A15).

19 For Theophrastus see Simplicius, *In Ph.* p. 1236 Diels (Fortenbaugh et al. [1992] 336, text 176); for Strato, Aëtius III.3.13.

20 *Letter to Pythocles* 100. Cf. in 102 τὸ πνεῦμα ἀνειλούμενον τὸν βόμβον ἀποτελεῖν, "the wind rolling produces the rumble".

21 Diogenes Laertius VII.153 (*SVF* I 117).

22 Stobaeus, *Eclogae*, I p. 233.9 W. (*SVF* II 703).

23 Aëtius III.3.12 (*SVF* II 705).

24 Posidonius F135 EK (= F325 in Theiler [1982]). *NQ* II.55, printed as part of F135 EK, cannot be assumed to continue the account of Posidonius' theory. (So Hine [1981] 424. Vimercati [2004] 576 considers it Posidonian, but Kidd [1988] 508 only says that it "may be based on Posidonius". Theiler omits it from his F325.)

25 I follow the text of Hine (1981). I omit the final sentence of II.54.3 as being, as Hine (1981) 423 suggests, probably Seneca's comment, not part of his report of Posidonius.

26 For detailed discussions of this see Gilbert (1907) 634–6; Hine (1981) 421–3; Kidd (1988) 503–9.

27 Cicero, *Div.* II.44 (*SVF* II.699).

28 *NQ* II.57.1, "in ignem aer extenuatis nubibus vertitur", "air is turned to fire by clouds being rarefied": cloud here must be a dense form of air (so Oltramare [1961] 102, note 1). The close dependence of *NQ* II.57 on II.54 is shown by the fact that in II.57 Seneca goes on to mention as causes both friction (see note 30) and a smoky exhalation (II.57.3, "calidi fumidique natura emissa terris", "matter of a hot and smoky nature emitted by earth"). In II.57, as in II.54, the relation of the smoky exhalation to the rarefied air is not explained. In II.57.2 the air is rarefied by movement ("motus"), but we are not told what causes the movement.

29 *De mundo* 395a11–13, thunder is produced by εἰληθὲν πνεῦμα ἐν νέφει παχεῖ ... βιαίως ῥηγνύον τὰ συνεχῆ πιλήματα τοῦ νέφους, "wind whirling in a thick cloud ... violently breaking the close-packed material of the cloud".

30 E.g., II.57.2, there are most thunderbolts in summer because "facilius ... adtritu calidorum ignis existit", "fire comes to be more easily by friction of hot matter".

31 Several scholars (see references in Hine [1981] 320; Kidd [1988] 503) have supposed that Seneca is referring back to *NQ* II.12.4, where Seneca says he is describing Aristotle's theory, and begins, as in II.54, with an account of the theory of dry and wet

exhalations. This seems most unlikely, because II.12 not only names a different thinker but also describes a different theory of thunder: the dry exhalation is expelled by the cloud in II.12 but forces its way out in II.54, and II.12 mentions neither the rarefaction and heating of air, nor friction (see further Hine [1981] 321; Kidd [1988] 503–5). Hine (1981) 321 suggests that Seneca in II.54 is referring back to II.15 and II.27: reference to II.27 accords with Kidd's view, that all II.21–30 is referred to; in II.15 Seneca says it is the view of some Stoics ("quidam ex nostris") that air within clouds turns to fire spontaneously ("ipse ... se movendo accendit", "it [i.e., air] itself sets itself alight by moving"); Seneca in II.54 could be referring to this, but in II.54 he does not *say* that the air gets hot spontaneously. Theiler (1982) II. 207 suggests that "Revertar" in II.54 refers back to the lost beginning of *NQ* IVb: with many scholars, he thinks that our IVb was Seneca's first book, and our II was Seneca's sixth book (see Theiler [1982] II. 217). Could Seneca possibly have used this wording to refer back from his sixth book to his first?

32   Kidd (1988) 505–6.

33   This must be what he means by saying "Dimissis nunc praeceptoribus nostris incipimus per nos moveri et a confessis transimus ad dubia", "Setting aside our instructors, we now begin to be moved by ourselves, and proceed from what is agreed to what is doubtful."

34   Kidd (1988) 505–6.

35   The only earlier place in *NQ* II where Seneca has mentioned dry and wet exhalations is his account of Aristotle's theory at II.12.4, but he does speak of such a theory as his own at least at V.5.1, where "aquarum terrarumque evaporationes", "evaporations from earth and water" cause wind; and in Seneca's original order of the books of *NQ* our book V probably preceded our Book II. (See, for instance, Hine [1981] 6–29).

36   *NQ* II.26.6 and 30.1.

37   For Leucippus see Aëtius III.3.10 (DK 67A25); for the others see above.

38   On exhalation theories before Aristotle see below, Chapter 10 (p. 89–91). Theophrastus did not entirely reject dry exhalation (see *De igne* 7 and *De lapidibus* 50, in non-meteorological contexts, and reports of Theophrastus' wind theory in Alexander, *In Mete.* p. 93.35ff Hayduck, and Olympiodorus, *In Mete.* p. 97.5ff Stüve), but it does not occur in his explanations of thunder, etc., cited above (p. 68), and *De ventis* nowhere clearly mentions dry exhalation (see Coutant and Eichenlaub [1975] xliv–xlviii). The Syriac Meteorology does not mention exhalations in its account of thunder and lightning, though a distinction between "thick" and "fine" vapour has a part in its account of thunderbolts (see Daiber [1992] 261–5). One other text, the *De mundo*, may be earlier than Posidonius (see above, Chapter 3 [p. 13–14]) and claims to base its meteorology on the theory of dry and wet exhalations (394a9–15), but its actual account of thunder and lightning is entirely in terms of πνεῦμα, "wind" (395a11–16).

39   Syriac Meteorology 1.(6)–(8) (Daiber [1992] 261). This idea of wind rotating within a cloud is, I think, the only part of the theory attributed to Posidonius in *NQ* II.54 which is found in the Syriac Meteorology, so is arguably from Theophrastus, and is found in no author earlier than Theophrastus.

40   See p. 70 above and note 29.

41   It is also found in the Syriac Meteorology (e.g. 1.(21)–(22), 2.(3)–(9) [(Daiber {1992}] 261–2), and in Epicurus' *Letter to Pythocles* 100–1.

42   I give the text as printed by Hine (1981); the sense of the objection is not in doubt.

43   Daiber (1992) 262.

9 Lights in the sky

Comets, the Milky Way and other phenomena

This chapter concerns phenomena which, like lightning, involve the production of light in the space above our heads. Aristotle deals with them in *Mete*. I, chapters 4 to 8, in which he discusses meteors, the aurora borealis, comets, and the Milky Way, and explains them, like lightning, as caused by dry exhalation.

It is likely that Posidonius was interested in all these phenomena. At *NQ* VII.20.2[1] Seneca contrasts with thunderbolts other fires, which do not immediately perish:

> Alii vero ignes diu manent nec ante discedunt quam consumptum est omne quo pascebantur alimentum. Hoc loco sunt illa a Posidonio scripta miracula, columnae clipeique flagrantes aliaeque insigni novitate flammae

> Indeed other fires last for a long time, nor do they disappear before all the nourishment by which they were fed has been consumed. In this category are those marvels that Posidonius wrote about, burning columns and shields and other flames of remarkable novelty.

Seneca goes on to imply that such phenomena occur "compresso aere et in ardorem coacto", "when air has been put under pressure and forced into burning". It is not clear whether this is Posidonius' explanation as well as Seneca's.

We have some clues as to what the columns and shields were. Pliny *Nat.* II.100 mentions among strange phenomena in the sky "Clipeus ardens ab occasu ad ortum scintillans transcucurrit solis occasu L. Valerio C. Mario consulibus", "In the consulship of Lucius Valerius and Caius Marius [100 B.C.] a burning shield scattering sparks ran across the sky at sunset from west to east".[2] This was evidently a short-lived phenomenon (though longer-lasting than a lightning-strike): presumably a large meteor.

The "column" is mentioned in Aëtius III, Chapter 2. This is headed Περὶ κομητῶν καὶ διαττόντων καὶ δοκίδων, "On comets and shooting stars and beams"[3]; and under this heading, in III.2.5, one of the phenomena which Heraclides Ponticus is said to have explained is κίων, a "pillar" or "column". Seneca *NQ* VI.26.3 reports Callisthenes as having said

DOI: 10.4324/9780429399930-9

Inter multa prodigia quibus denuntiata est duarum urbium, Helices et Buris,
eversio, fuere maxime notabilia columna ignis immensi et Delos agitata.

Among many prodigies with which the destruction of two cities, Helice
and Buris, was announced the most notable were a column of an immense fire and an earthquake at Delos.

Aristotle, writing about this same event, does not mention a "column", but
does speak of a "great comet" (μέγας κομήτης), the light of which stretched
across a third of the sky.[4] Presumably Callisthenes' "column" was a phenomenon associated with that comet.

The ancients knew of other strange "flames" in the sky. Aristotle mentions
not only meteors (διαθέοντες ἀστέρες, literally "running stars"), but also the
related phenomena of "torches" (δαλοί) and "goats" (αἶγες)[5]; talking about
what is evidently the aurora borealis, he mentions χάσματά τε καὶ βόθυνοι καὶ
αἱματώδη χρώματα, "chasms, trenches and blood-red colours".[6] We have notices
which name Posidonius for explanations only of comets and the Milky Way,
and I argue below that a "Stoic" explanation of meteors is probably his; but he
would certainly have been ready with naturalistic explanations of other phenomena similar to these. That, however, may not have been their only significance for him. Seneca decries fear of the columns, shields and other remarkable
flames, saying of them:

Quae non adverterent animos, si ex consuetudine et lege decurrerent; ad
haec stupent omnes quae repentinum ex alto ignem efferunt

These things would not attract attention, if they came down customarily
and by rule; everyone is amazed at things which without warning bring
fire from on high.

This sentence comes immediately after the one which names Posidonius as writing about columns, shields and so forth, so it looks as though Posidonius is
included in Seneca's attack. It seems likely that Posidonius believed (and Seneca
did not) that such phenomena were portents of coming events on the earth.[7] It
was orthodox Stoic doctrine that μαντικὴν ὑφεστάναι ... πᾶσαν, εἰ καὶ πρόνοιαν
εἶναι, "every kind of divination exists, if indeed Providence exists".[8] As I have
said above,[9] it would not have been difficult for a Stoic pantheist to believe both
that an event had a naturalistic explanation and that that event had significance
in the providential scheme of things. But, if Posidonius did believe that unusual
celestial events were portents, we have no details of his views.

Meteors

Diogenes Laertius VII.153, in his account of Stoic meteorology, says:

Σέλας δὲ πυρὸς ἀθρόου ἔξαψιν ἐν ἀέρι φερομένου ταχέως καὶ φαντασίαν
μήκους ἐμφαίνοντος

A *selas* is a kindling of fire gathered in one mass which is carried swiftly
in the air and displays an appearance of length.

This sentence is evidently describing a meteor ("a small body of matter from
outer space that becomes incandescent as a result of friction with the earth's
atmosphere and appears as a streak of light"[10]). The word σέλας can mean a
light or fire of various kinds. As a word for a meteor or related phenomenon,
it was not part of the vocabulary of the earlier Greek writers on meteorology.
The word is not used in Aristotle's *Meteorologica.*, nor in Epicurus' *Letter to
Pythocles*; it is not used in this sense in any pre-Socratic fragment,[11] nor is it
used in what little evidence we have for the meteorological views of Zeno and
Chrysippus. Stoics before Posidonius, with little interest in meteorology, would
probably have used already familiar terms, such as Aristotle's διαθέοντες
ἀστέρες, if they felt that their account of the world would be incomplete without some reference to the subject; Posidonius, seriously interested in meteorology, is much more likely to have adopted new vocabulary. It is surely probable
that this sentence of Diogenes is reporting the view of Posidonius, or one based
on his.[12]

Comets

In a modern definition, a comet[13] is a "small solar system body that, when
passing close to the sun … begins to release gases … This produces a visible
atmosphere or coma and sometimes also a tail." The coma appears as a tenu-
ous, bright patch around the nucleus of the comet.[14] Unlike stars and planets,
a comet can be seen only for a limited time, and (so far as ancient observations
could determine) it has no regular orbit: the same comet was not seen to return.
Aristotle mentions that some comets have only a coma, while others have
a tail.[15]

To find the cause of comets, or of possibly related phenomena such as
shooting stars, was not a major concern of the first Greek philosophers. The
earliest whose view is recorded is Xenophanes, who said, according to Aëtius,
that such phenomena are νεφῶν πεπυρωμένων συστήματα ἢ κινήματα, "formations or movements of clouds which have been set on fire"[16]; he seems to have
said almost exactly the same of sun and stars,[17] so this was clearly not a developed physical theory, but part of his argument that such phenomena are not
divine.[18] Some later pre-Socratics were more seriously concerned with comets.
Anaxagoras and Democritus, Aristotle tells us, thought that a comet is "a conjunction of planets, when they appear to touch each other" (σύμφασιν τῶν
πλανήτων ἀστέρων, ὅταν…δόξωσι θιγγάνειν ἀλλήλων).[19] Aristotle also says that
some "Pythagoreans", Hippocrates of Chios and his pupil Aeschylus, thought

that a comet is one of the planets, the latter two adding that the tail (or the coma? In Greek τὴν κόμην) is due to our sight being reflected to the sun by moisture drawn up by the planet.[20] Aëtius also says that "Diogenes" thought that comets are stars (ἀστέρας)[21]: this is variously attributed to the pre-Socratic Diogenes of Apollonia and the Stoic Diogenes the Babylonian.[22]

Aristotle's own view (*Mete.* 344a9ff) is that below the perpetually-revolving fifth element there is hot, dry exhalation (ἀναθυμίασιν ξηρὰν καὶ θερμήν)/ elemental fire[23] – which is carried round with the revolution of the heavens. When suitably constituted (εὔκρατος), this often catches alight.

> ὅταν οὖν εἰς τὴν τοιαύτην πύκνωσιν ἐμπέσῃ διὰ τὴν ἄνωθεν κίνησιν ἀρχὴ πυρώδης, μήτε οὕτω πολλὴ λίαν ὥστε ταχὺ καὶ ἐπὶ πολὺ ἐκκαίειν, μήθ' οὕτως ἀσθενὴς ὥστε ἀποσβεσθῆναι ταχύ, ἀλλὰ πλείων καὶ ἐπὶ πολύ, ἅμα δὲ κάτωθεν συμπίπτῃ ἀναβαίνειν εὔκρατον ἀναθυμίασιν, ἀστὴρ τοῦτο γίγνεται κομήτης.

> Now when as a result of the upper motion there impinges upon a suitable condensation [presumably meaning, a dense mass of exhalation suitably constituted to catch alight] a fiery principle which is neither so very strong as to cause a rapid and widespread conflagration, nor so feeble as to be quickly extinguished, but which is yet strong enough and wide-spread enough; and when besides there coincides with it an exhalation from below of suitable consistency; then a comet is produced.[24]

This is supported by an analogy (rather an obscure one) with the burning of a pile of chaff.[25]

The comet may be independent, or may be apparently attached to a fixed star or planet. In the latter case, ὑπὸ τῆς κινήσεως συνίσταται ἡ ἀναθυμίασις … οὐ γὰρ πρὸς αὐτοῖς ἡ κόμη γίγνεται τοῖς ἄστροις, ἀλλ' ὥσπερ αἱ ἅλῳ περὶ τὸν ἥλιον φαίνονται, "the exhalation is put together by the movement [sc. of the star or planet] … for the tail [or coma?] is not formed at the stars themselves, but in the way haloes appear around the sun", that is, the tail, like a solar halo, is an atmospheric phenomenon, caused by exhalation which has accumulated below the star or planet.[26] I am unsure to what phenomenon Aristotle is referring.

I have quoted, following Aristotle, ancient explanations of three types: a comet has no separate existence, but the appearance of a comet is produced by the conjunction of two planets; or, it is a planet or star; or, it has a separate existence but is only a temporary accumulation of burning material. Seneca, too, at *NQ* VII 19 gives these as the three known explanations of comets. He attributes the first of them to "Zeno noster", Zeno the Stoic:

> Congruere iudicat stellas et radios inter se committere; hac societate luminis existere imaginem stellae longioris

> He considers that stars come together and join their rays to each other's; by this association of light there is the image of a longer star.

Seneca then mentions the two other views:

Quidam aiunt [sc. cometas] … habere cursus suos et post certa lustra in conspectum mortalium exire; quidam esse quidem, sed non quibus siderum nomen imponas, quia dilabuntur nec diu durant et exigui temporis mora dissipantur.

Some say that comets have their own courses and after certain long periods come out into the sight of mortals; some say that they exist indeed [i.e., are not an optical illusion, as Zeno thought], but are not bodies to which one would give the name of stars, because they decay, do not endure long, and are dissipated after lasting a short time.

Ancient thinkers did not have to choose one of these alternatives. Aëtius reports that Strato regarded a comet as ἄστρου φῶς περιληφθὲν νέφει πυκνῷ, "the light of a star surrounded by a dense cloud"[27] – presumably he thought the star was a permanent body but the coma transient. Epicurus, as one would expect, thought that comets might occur either as temporary fires, or as stars (presumably permanent bodies) that at intervals come into our view.[28] Most Stoics, however, according to Seneca, regarded comets as real but transient phenomena. He says of this view (*NQ* VII 20.1) "in hac sententia sunt plerique nostrorum", "most members of our school are of this latter opinion".[29] This view is clearly, as Kidd says, that of Posidonius, as is shown by the Scholia to Aratus, quoted in the next paragraph, but there is doubt whether it was the view of any earlier Stoics.[30]

The *Scholia to Aratus* 1091 describe Posidonius' theory of comets as follows (adopting the text printed by Kidd):

ὁ δὲ Ποσειδώνιος ἀρχὴν γενέσεώς φησιν ἴσχειν τοὺς κομήτας ὅταν τι τοῦ ἀέρος παχυμερέστερον εἰς τὸν αἰθέρα ἐκθλιβὲν τῇ τοῦ αἰθέρος δίνῃ ἐνδεθῇ, εἶτα πρὸς πλείονα δῖνον ἐπιρρεούσης τῆς τροφῆς φέρωνται.

Posidonius says that comets have the beginning of their formation when a rather dense bit of air is squeezed into the *aithēr* and is fixed in the *aithēr's* rotation and then they [i.e., comets] are carried into further rotation as nourishment keeps on flowing.[31]

Αἰθέρα and αἰθέρος are conjectures for the MS. readings ἀέρα and ἀέρος[32]; but the MS. reading cannot be right: the dense air must surely be squeezed out into something other than air. The conjecture seems guaranteed by the Stoic view reported at Diogenes Laertius VII 152: κομήτας … πυρὰ εἶναι ὑφεστῶτα πάχους ἀέρος εἰς τὸν αἰθερώδη τόπον ἀνενεχθέντος, "comets … are fires that consist of dense air borne up to the region of aether".[33] The Stoics believed that the heavenly bodies are "nourished" by vapour from the earth (see Chapter 10); Posidonius evidently thought that the same is true of comets. Presumably his theory was that a comet is formed when some dense air becomes fixed in the revolving *aithēr* which forms the heavens, and this dense air catches fire and is carried round with the *aithēr*, sustained by a continuous flow of vapour from below.

If we compare Posidonius' theory with those of earlier thinkers, it is clearly most like Aristotle's[34]; in both of them comets are due to a dense mass of suitable material high above the earth catching fire, and this fire lasts as long as it is maintained by exhalation from the earth. Posidonius seems to have said nothing about *dry* exhalation; the reason for this I discuss below. In one respect Posidonius' theory seems more plausible than Aristotle's: Aristotle's system required him to say that comets, being irregular phenomena, are sublunary; Posidonius, with his Stoic cosmology, places them in *aithēr*, in the same region as other heavenly bodies.[35]

The *Scholia to Aratus* and Seneca give some more details of Posidonius' account of comets. The *Scholia* say that, for Posidonius, comets are seen to change in size ὡς ἄν ποτε μὲν πλεῖον ἐπιδιδούσης τῆς τροφῆς αὔξεσθαι, ποτὲ δὲ λειπούσης συστέλλεσθαι, "as sometimes they grow, the nutriment providing more material, but sometimes, as it ceases, they are reduced"[36]; comets mostly form in the north, ἔνθα παχυμερὴς καὶ πεπιλημένος ἐστὶν ὁ ἀήρ, "where the air is thick and compressed" (so that, it is implied, the material needed for comets is abundant)[37]; also, comets are weather-signs, predicting drought as they form and rain as they dissolve: an idea, based on a theory of Aristotle's, which I discuss in a later chapter.[38]

Posidonius' explanations were inevitably incorrect; but what about the statements of fact which they aim to explain? It is true that the apparent size of a comet can change,[39] and presumably that had been observed. The association of comets with the northern sky was not a new idea: it was apparently the theory of Hippocrates and Aeschylus that a comet is a planet visible only in the northern sky, and only at midsummer.[40] Aristotle cites examples of comets which disprove this.[41] We do not know what evidence Posidonius had for his statement that comets form mostly in the north,[42] which appears to be false.[43]

Seneca attributes to Posidonius a statement which appears to be true (*NQ* VII.20.4): "Multos cometas non videmus, quia obscurantur radiis solis; quo deficiente quondam cometen apparuisse, quem sol vicinus obtexerat ... Posidonius tradit", "Many comets we do not see, because they are obscured by the rays of the sun. Posidonius reports ... that a comet once appeared when the sun was eclipsed, which the sun, close to it, had been hiding". In modern times previously unseen comets have been discovered when the sun was eclipsed,[44] and presumably this also happened in antiquity.

Additional information about what Posidonius said on comets may be contained in the part of the *Scholia to Aratus* 1091 which precedes the mention of Posidonius: the description of Posidonius' own theory (which is not criticised) is preceded by accounts of the theory of the Pythagoreans, and of that of Anaxagoras and Democritus, both of which are criticised; which suggests the possibility that the accounts and criticisms of these earlier theories are also from Posidonius.[45] The accounts agree with those given by Aristotle and summarised above, though with the addition that in the conjunction theory, the planets are "like mirrors reflecting light on each other"[46] and so produce the appearance of the comet. The criticisms are also mostly the same as Aristotle's, with the additions that, if comets were all one planet (as the Pythagoreans thought), then "expert scientists would certainly have observed its revolutions",[47] and comets would not differ in appearance. This is possibly a rare illustration of how Posidonius dealt with pre-Socratic meteorology.

Enough is recorded about the comet theories of writers certainly or probably later than Posidonius to enable us to estimate his influence. His theory was perhaps widely held by later Stoics,[48] but apparently not by others. In the *De mundo* a comet occurs in air, and is a variety of σέλας, a fire in the sky, which may move rapidly (evidently a shooting star) or may be stationary but elongated; the latter πλατυνομένη κατὰ θάτερον κομήτης καλεῖται, "if it spreads out towards one end, it is called a comet".[49] The origin of the fire is not stated. Seneca describes the theories of Epigenes and Apollonius Myndius (probably contemporaries of his own[50]). Epigenes, he says, held that some comets are immobile (and evidently lack tails), and are formed "ex intemperie aeris turbidi multa secum arida umidaque terris exhalata versantis", "from a storm of turbulent air whirling with itself much dry and wet material exhaled from the earth"[51]; others, which do move, "ex isdem causis fieri . . hoc tamen interesse quod terrarum exhalationes multa secum arida ferentes celsiorem petant partem et in editiora caeli aquilone pellantur", "are formed from the same cause ... but differ in this, that the terrestrial exhalations carrying much dry matter with them seek a higher position and are driven by the north wind into the further parts of the sky".[52] This is similar to Posidonius' theory in that both involve exhalations from the earth, but differs in other respects.[53] Apollonius Myndius[54] and Seneca himself[55] held that comets are permanent heavenly bodies. The one later writer who we know shared Posidonius' view of comets is Arrian[56] (a Stoic, as his work on Epictetus shows) who says of comets (and of phenomena he thought similar)

ἐκεῖνος ἂν κρατοίη ὁ λόγος <ὁ> ἀποφαίνων ἀέρος πιλήματα, ἀποθλιβόμενα καὶ ἐμπίπτοντα εἰς τὰ κατωτέρω καὶ τῷ ἀέρι ξυναφῇ τοῦ αἰθέρος, ἐξαφθέντα, ἔστ' ἂν ὑπάρχῃ περὶ αὐτοὺς ἡ τροφή, ξυμμένειν τε καὶ ξυμπερινοστεῖν τῷ αἰθέρι

that argument should prevail which shows that compressed masses of air are squeezed out [presumably from the main mass of air], burst into the lower part of the *aithēr*, which adjoins the air, and are set alight; and these masses, while there is nourishment around them, keep together and revolve with the *aithēr*.

He has previously said that the nourishment is ἄνω ἀναφερομένην, "carried upwards", so presumably it is some sort of exhalation. We cannot tell whether the idea that comets occur in *the lower part* of the *aithēr* was part of Posidonius' theory.

The Milky Way

The Milky Way,[57] as it appears from the earth, is "a hazy band of light seen in the night sky, formed from stars that cannot be individually distinguished by the naked eye". It does not look like an assemblage of stars, so it is not surprising if many ancient thinkers did not believe that it is one. Also, "its visibility can be greatly reduced by background light, such as light pollution or moonlight".[58] Ancient observers did not have to concern themselves with the pollution caused by bright artificial lights, but they were affected by moonlight: it is understandable if changes in the visibility of the Milky Way caused early observers to fail to realise that it is a permanent and invariable feature of the sky.

Parmenides is the first philosopher recorded to have held a view about the Milky Way. In one fragment[59] he promises to explain the origin of phenomena including γάλα οὐράνιον, "the milk of the sky"; the only hints of his explanation are the statements of Aëtius that for him the sun and the Milky Way are πυρὸς ἀναπνοήν, "a breath of fire",[60] and that the Milky Way is a "mixture of the dense and the rare" (τοῦ πυκνοῦ καὶ τοῦ ἀραιοῦ μίγμα).[61] Undatable, but presumably later, is the view recorded by Aristotle of some "Pythagoreans", who said that the Milky Way is ὁδὸν ... τῶν ἐκπεσόντων τινὸς ἀστέρων κατὰ τὴν λεγομένην ἐπὶ Φαέθοντος φθοράν, "the path taken by one of the stars [literally "stars that fell", sc. from their proper position] at the time of the legendary fall of Phaethon",[62] or, as Aristotle says other "Pythagoreans" thought, the path of the sun when it followed a course different from the present one; the Milky Way is apparently supposed to be the remains of the burning that occurred in one of these paths.[63]

Aristotle goes on to say that according to Anaxagoras and Democritus the Milky Way is the light of stars which are in the earth's shadow when the sun is below the earth (since the sun's rays obscure stars which are not in the earth's shadow).[64] He also mentions a third theory, that some say that the Milky Way is "a reflection of our vision to the sun, just as a comet was supposed to be". This may have been a "Pythagorean" theory.[65]

It may be that Aristotle misreports Democritus. We have two other reports of his theory which say that the Milky Way is the light of numerous small stars without mentioning the earth's shadow or the sun's rays obscuring other stars[66]: his theory may have differed from that of Anaxagoras. The only other pre-Aristotelian theory we know of is attributed to Metrodorus of Chios, who apparently said that the Milky Way occurs διὰ τὴν πάροδον τοῦ ἡλίου· τοῦτον γὰρ εἶναι τὸν ἡλιακὸν κύκλον, "because of the passage of the sun; for this is the sun's circle"[67] – this is to me obscure, unless it means a former course of the sun, and Metrodorus adopted the theory which Aristotle attributes to Pythagoreans.

Aristotle discusses the Milky Way in *Mete.* I Chapter 8. He regarded it as comparable to a comet: τὸ ἔσχατον τοῦ λεγομένου ἀέρος δύναμιν ἔχει πυρός, ὥστε τῇ κινήσει διακρινομένου τοῦ ἀέρος ἀποκρίνεσθαι τοιαύτην σύστασιν οἵαν καὶ τοὺς κομήτας ἀστέρας εἶναί φαμεν, "The last part of what is called air [i.e., the part furthest from the earth] has the potentiality of fire, so that when the air is divided up by motion a mass is separated off of the kind which we maintain also forms comets".[68] So, just as (according to Aristotle) a comet is sometimes a fixed star or planet with a tail apparently attached to it, τοῦτο δεῖ λαβεῖν γιγνόμενον περὶ ὅλον τὸν οὐρανὸν καὶ τὴν ἄνω φορὰν ἅπασαν, "one must suppose that this happens with the whole sky and the entire upper motion".[69] Aristotle has previously said that a comet is produced when a suitable mass of dry exhalation/elemental fire is set alight (see above), so by "the last part of what is called air" he must mean dry exhalation/elemental fire, and the Milky Way is in his view caused by the burning of a mass of this. The circle of the Milky Way does not coincide with the circle of the zodiac because the motion of the sun and planets disperses any suitable material.[70] There are many very large bright stars in the region of the Milky Way, and these must be a cause of it.[71]

Aristotle has cogent criticisms of the earlier theories he has mentioned: for instance, that if the Milky Way is a former course of the sun, then there should be a similar band of light in the circle of the zodiac, in which both sun and planets now move[72]; and if the Milky Way were either the light of stars shaded from the sun by the earth, or the reflection of our sight to the sun, then the Milky Way would move among the stars as the sun moves, whereas in fact its position in relation to the stars is always the same.[73] He does not say why he thought the Milky Way is sublunary, implying it is liable to change; he obviously knew that it does not change its position, but perhaps thought, as suggested above, that its brightness varies. Nor does he explain how fixed stars, in the outermost sphere of his cosmos, can affect the sublunary sphere.

Our sources tell of other theories which may have originated before Posidonius' time. Macrobius alleges that "Theophrastus lacteum dixit esse compagem qua de duobus hemisphaeriis caeli sphaera solidata est". "Theophrastus said that the Milky Way is the seam at which the sphere of the sky is made one from two hemispheres"[74] – a theory which sounds like a myth. Other explanations compare the Milky Way to cloud. Geminus[75] says it is formed ἐκ βραχυμερείας νεφελοειδοῦς, "from a cloud-like aggregate of small parts"; Aëtius[76] (naming no authority) describes it as κύκλος νεφελοειδὴς ἐν τῷ ἀέρι, "a cloud-like circle in the air". Neither statement says what its material actually is; but Achilles suggests it is ἐκ νεφῶν ἢ πίλημά τι ἀέρος διαυγές, is "from clouds, or is a bright dense mass of air".[77] For the Stoics, Mansfeld and Runia print a sentence from pseudo-Galen, *Historia philosopha*, about the Milky Way, with persuasive arguments that it is from the original text of Aëtius: οἱ Στωϊκοὶ τοῦ αἰθερίου πυρὸς ἀραιότητα ἀνώτερον τῶν πλανητῶν, "the Stoics [sc. say it is] ethereal fire in a rare state, higher than the planets".[78]

We have two reports of Posidonius' theory.[79] According to Aëtius, Posidonius thought that the Milky Way is πυρὸς σύστασιν ἄστρου μὲν μανωτέραν, αὐγῆς δὲ πυκνοτέραν, "a mass of fire, less dense than a star but denser than a beam of light"[80]; Macrobius says that Posidonius thought it is "caloris ... siderei infusionem", "a pouring of sidereal heat".[81]

I would regard Posidonius' theory, as reported by Aëtius, as a refinement of the "Stoic" theory, defining the exact degree of "rarity" of the fire of the Milky Way.[82] The theory also resembles Aristotle's: to both Aristotle and Posidonius, the Milky Way was the burning of a mass of material which does not form stars. (The theories which mention cloud are possibly similar, but, as we have them, they do not mention burning, and they may be later than Posidonius.) To Aëtius and Achilles (but perhaps not to Geminus) the Milky Way was a sublunary phenomenon, as it was to Aristotle, but there is no evidence that it was sublunary for Posidonius. In Aëtius' report of his theory, it is a mass of fire, with no suggestion that it is out of its natural place; Macrobius calls it "sidereal"; "the Stoics" placed the Milky Way above the planets. In any case, if to Posidonius comets are a phenomenon of the region of *aithēr*, then the Milky Way must surely be one also. If he thought that the Milky Way is liable to change, he did not, as a Stoic, have to assume that it was therefore sublunary.

Macrobius does not contradict Aëtius, since it is natural that heat should come from fire, but his interest is in the heat, which, in his account of Posidonius' theory, served a purpose. The Milky Way is

caloris ... siderei infusionem, quam ideo adversa zodiaco curvitas obliquavit, ut, quoniam sol numquam zodiaci excedendo terminos expertem fervoris sui partem caeli reliquam deserebat, hic circus a via solis in obliquum recedens universitatem flexu calido temperaret

a pouring of sidereal heat, which a curvature opposite to the zodiac has twisted, for this reason, that, because the sun by never leaving the bounds of the zodiac leaves the remainder of the sky destitute of its heat, this circle [the Milky Way] by departing at an angle away from the path of the sun tempers the universe with its warm curve.

This is obscurely phrased, but must mean, if Macrobius is right,[83] that Posidonius thought that the reason why the Milky Way does not coincide with the zodiac circle is to warm the part of the universe which the sun does not pass over: such providential governance of the world was a doctrine of the Stoics but not of Aristotle. It is unparalleled among the explanations of meteorological phenomena attributed to Posidonius, but on his theory the Milky Way is effectively a celestial phenomenon, and he probably did regard the effects of the Sun and Moon, two celestial bodies, as part of the providential governance of the world.[84]

According to Macrobius the agreement of most people ("plurium consensus") was with Posidonius' definition. Macrobius does also mention one man whose theory of the Milky Way may have been influenced by Posidonius[85]: Diodorus, who believed that the Milky Way is "ignem densetae concretaeque naturae", "fire of a dense and compressed essence" (with further details[86]); this could be a development of the "mass of fire" theory which Aëtius attributes to Posidonius, and we know that this Diodorus was later than Posidonius, and possibly his pupil.[87] Apart from this it is not clear, as Kidd notes,[88] to whom

Macrobius is referring. The evidence suggests that most ancient writers after Aristotle did not regard the study of the Milky Way as part of the discrete subject which Aristotle had called *meteōrologia*.[89] The Milky Way is included in Book III of the *Placita* of Aëtius, but there is no explanation of it in Seneca's *Naturales quaestiones*, nor in *De Mundo* Chapter 4. nor in the account of Stoic meteorology in Diogenes Laertius VII.151–4: surveys which in other ways cover very nearly the same range of subjects as Aristotle's *Meteorologica*. Epicurus' *Letter to Pythocles*, though it includes comets, shooting stars and some definitely astronomical subjects, does not mention the Milky Way. After Aristotle, when writers on nature speak of the Milky Way, it is usually just a familiar and important feature of the night sky, with no speculations about its nature. In Aratus' *Phaenomena* it is mentioned as a major circle of the sky, along with the Equator, the Ecliptic and the two Tropics[90]; when Seneca mentions it in *Naturales quaestiones*, it is as a point of comparison with a great comet which, Seneca says, equalled it.[91]

We hear of no-one in antiquity after Democritus who adopted the correct theory he probably held, that the Milky Way is the light of numerous stars. Presumably the Milky Way looked too unlike a collection of stars for this theory to seem credible. With this in mind, one might say that Posidonius, having moved the Milky Way outside the sublunary region, had a theory that was as plausible as any.

Notes

1 Part of Posidonius F132 EK. In my discussion of this passage I owe much to notes in Oltramare (1961) 321, and to Kidd (1988) 494–6.
2 Tr. Rackham (1938).
3 I give Pseudo-Plutarch's version of the title, preferred by Mansfeld and Runia (2020). Stobaeus has τῶν τοιούτων, "the like", for δοκίδων.
4 *Mete.* 343b1 and 18–25, also 344b34–345a1.
5 *Mete.* 341b2–3.
6 *Mete.* 342a35, with the translation of Lee (1952).
7 Kidd (1988) 496, commenting on this passage, says that Seneca is attacking Posidonius for "pandering to the general love of the miraculous". He does not say why Posidonius should have done this, but possibly he had in his mind some such explanation I here put forward.
8 Diogenes Laertius VII.149 (Posidonius F7 EK. I give the text as printed by EK).
9 See above, Chapter 4 (p. 26–7).
10 So defined in *Concise Oxford English dictionary* (1999).
11 The index to DK records its occurrence only in Empedocles, DK 31B84, where it is the light of a lantern, and in Democritus, DK 68B152, where it seems to refer to a flash of lightning.
12 I do not here discuss other occurrences of σέλας in this sort of context, as I do not think there is sufficient evidence of Posidonius' involvement.
13 For discussions of Posidonius' view of comets, see Gilbert (1907) 650–3; Kidd (1988) 490–6; Vimercati (2004) 550–2.
14 See, in Wikipedia article 'Comet', beginning of text and images of comets (read 30 August 2021; judging from the images, the coma may be the most conspicuous feature.)

15 Aristotle *Mete*. 344a21–23, ἀστὴρ ... γίγνεται κομήτης, ὅπως ἂν τὸ ἀναθυμιώμενον τύχῃ ἐσχηματισμένον· ἐὰν μὲν γὰρ πάντῃ ὁμοίως, κομήτης, ἐὰν δ᾽ ἐπὶ μῆκος, καλεῖται πωγωνίας, "a comet is produced, its exact form depending on the form taken by the exhalation [which Aristotle believes is the material of the comet] – if it extends equally in all directions, it is called a comet [literally "long-haired star"], if it extends lengthwise only it is called a bearded star". (Translation from Lee [1952].) This was not universal Greek usage: the *Scholia in Aratum* 1091 (printed as Posidonius F131a EK; the relevant part does not name Posidonius but may be derived from him; see below) say that a κομήτης has its κόμη above, a πωγωνίας has it below, i.e., the distinction depends on the direction of the tail.

16 Aëtius III.2.11 (DK 21A44).

17 They are ἐκ νεφῶν πεπυρωμένων, "from clouds which have been set on fire" (Aëtius II.20.3 and II.13.14 = DK 21A40 and 38).

18 See above, Chapter 4 (p. 23).

19 *Mete*. 342b27-29 (see DK59A81. English translation from Lee [1952]). See also Diogenes Laertius II 9 (DK 59A1) and Aëtius III.2.2 (in DK 59A81).

20 *Mete*. 342b30ff (DK 42,5). Aristotle in *Mete*. regularly speaks of reflection as the reflection of our sight to the object, not of the object to our sight (see below, Chapter 16 [p. 162]). Lee (1952) and Webster (1931) translate κόμην as "tail", but Aristotle may be following the usage he describes at 344a21–3 (see note 15), where κομήτης is a tailless comet: such a comet would look much more like a reflection of the sun than one with a tail. But if the Milky Way could be regarded as a reflection of the sun (see p. 82 below), no doubt a comet's tail could be.

21 Aëtius III.2.8 (DK 64A15).

22 DK print this as a view of Diogenes of Apollonia, Kidd (1988) 495 suggests the Babylonian.

23 See *Mete*. 340b19ff.

24 *Mete*. 344a16–21, with the translation of Lee (1952).

25 *Mete*. 344a25–32. For the apparent point of the analogy see the note in Lee (1952).

26 *Mete*. 344a34–b8.

27 Aëtius III.2.4 (Strato fragment 86 Wehrli).

28 *Letter to Pythocles* 111: πυρὸς ... διὰ χρόνων τινῶν ... συντρεφομένου, "fire gathered together through certain periods of time", must be temporary; but comets may also be ἄστρα, "stars", that "at certain times become visible" (ἐν χρόνοις τισὶν ... ἐκφανῆ γενέσθαι): presumably permanent bodies. Also permanent must be stars that appear "when the sky has a particular motion" (ἰδίαν τινὰ κίνησιν ... τοῦ οὐρανοῦ ἴσχοντος).

29 In VII 20–1 Seneca enlarges on this as a Stoic theory, citing Posidonius for some details (VII 20.2 and 4) but not saying that the whole account is from Posidonius; it cannot all be, since VII 21.3 mentions comets in the reigns of Claudius and Nero.

30 Kidd (1988) 495–6 maintains that Posidonius was the first Stoic to hold this theory; but this seems to me uncertain. In the little evidence that survives for definitely pre-Posidonian Stoic views on comets, we have the view of Boëthus (Aëtius III.2.7; *SVF* III p. 267) that a comet is ἀέρος ἀνημμένου φαντασίαν, "an appearance [or "perception"] of air that has been set on fire", and the view of Panaetius, that a comet is "falsam sideris speciem", "a false appearance of a star" (Seneca *NQ* VII.30.2 = Panaetius fr. 75 Straaten). Both statements appear to leave it doubtful what the reality is that produces the appearance: it may be something real but transient.

31 *Scholia in Aratum* 1091 (Posidonius F131a EK).

32 *Anonymus Parisinus 2422*, fol. 143 (F131b EK), closely parallels *Scholia in Aratum* 1091; the version in this MS. of the sentence I quote is partly nonsensical, but it does have the reading εἰς τὸν αἰθέρα ἐκθλιβὲν. (Theiler [1982] F316 [his edition of this fragment] retains ἀέρα and ἀέρος δίνῃ; his commentary, vol. II p. 204, does not refer to these readings.)

33 Translation of Mensch and Miller (2018).
34 So Kidd (1988) 492.
35 See above, Chapter 5 (p. 37).
36 F131a EK, lines 29–31, and F131b, lines 16–18.
37 F131b EK, lines 18–19; F131a, lines 31–3 are emended, no doubt rightly, to agree with it.
38 F131a EK, lines 33–7, and F131b, lines 20-3. See below, Chapter 17 (p. 172).
39 See examples in Wikipedia article "Comets", section "Coma". (Read 30 August 2021.)
40 Aristotle, *Mete.* 343a5–20 and 343a35–b1.
41 *Mete.* 343b1–7.
42 Pliny *Nat.* II.91 says that some comets move but others are stationary, "omnes ferme sub ipso septentrione", "almost all under the northern part of the sky".
43 Online *Encyclopaedia Britannica* article "Comets" says "Comets can appear at random from any direction". (Read 30 August 2021.)
44 Wikipedia article "C/1948V1": "The Eclipse Comet of 1948, formally known as C/1948V1, was an especially bright comet discovered during a solar eclipse. ... There have been several comets that have been seen during solar eclipses". (Read 30 August 2021.)
45 The relevant passage is printed as part of F131a EK (though it is omitted from the corresponding F316 in Theiler [1982]). On the likelihood of its being derived from Posidonius see Kidd (1988) 493.
46 The translation of καθάπερ ἐσόπτρων ἀντιλαμπόντων ἀλλήλοις in Kidd (1999) 184.
47 The translation of πάντως ἄν ... τὰς τούτου περιόδους οἱ περὶ τὰ μαθήματα δεινοὶ παρετήρησαν in Kidd (1999) 184.
48 Kidd (1988) 494–6 cites as evidence for this Seneca *NQ* VII.20.1, "In hac sententia sunt plerique nostrorum", "Most of our people [i.e., Stoics] are of this opinion". But the opinion, stated in VII.19.2, is simply that comets are real but impermanent bodies, an idea of which many variants are possible.
49 *De mundo* 395a29–32 and b3–9, with Furley's translation of b9, from Forster and Furley (1955). (The text is inconsistent: at 392b4 comets occur in the fiery region.)
50 So Oltramare (1961) 298–9; cf. Kidd (1988) 494, arguing that Seneca names Epigenes and Apollonius because they were the latest exponents of their theories.
51 *NQ* VII.6.1. Compare Aëtius III.2.6, Ἐπιγένης πνεύματος ἀναφορὰν γεωμιγοῦς πεπυρωμένου, "Epigenes [sc. thought a comet is] a rising of wind mixed with earth that has been set on fire".
52 *NQ* VII.7.2.
53 Kidd (1988) 492 sees a similarity between Posidonius' theory and that of Epigenes in that both involve a whirling motion. But it seems to me that when Posidonius said that a comet is fixed αἰθέρος δίνῃ he meant that it is carried in the rotation of the *aithēr* round the earth, and that the mention of πλείονα δῖνον means that it is carried further in that rotation; whereas the whirling in Epigenes' theory is a local whirlwind in the air.
54 *NQ* VII.4.1 and VII.17.1.
55 *NQ* VII.22.1.
56 Fragmenta de rebus physicis 6, in Roos/Wirth (1968).
57 For discussion of ancient theories of the Milky Way see Gilbert (1907) 658–62.
58 Both quotations from Wikipedia, article "Milky Way". (Read 31 August 2021.)
59 DK 28B11, line 2.
60 Aëtius II.7.1 (DK 28A37).
61 Aëtius III.1.4 (DK 28A43a); cf. Aëtius II.20.8a (DK 28A43).
62 *Mete.* 345a14–16, with translation from Lee (1952). Compare the first part of Aëtius III.1.2 (DK 58B37c), which recounts the same idea and has ἐκπεσόντος ... ἀπὸ τῆς ἰδίας ἕδρας, "falling from its own place", confirming the meaning of ἐκπεσόντων in Aristotle.

63 See *Mete.* 345a14–18 (DK 41.10).
64 *Mete.* 345a25–32. Aëtius III.1.5 confirms that this was Anaxagoras' theory. (Both passages in DK 59A80.)
65 *Mete.* 345b10–12 (DK 42.6). I quote the translation in Lee (1952) of ἀνάκλασιν … τῆς ἡμετέρας ὄψεως πρὸς τὸν ἥλιον, ὥσπερ καὶ τὸν ἀστέρα τὸν κομήτην. On this comet theory see above, p. 77–8. The latter part of Aëtius III.1.2 (DK 58B37c) states what seems to be this theory of the Milky Way, perhaps attributing it to Pythagoreans.
66 Aëtius III.1.6 (in DK68A91); Macrobius, *Commentarii in Ciceronis Somnium Scipionis* I.15.3–7 (see Posidonius F130 EK).
67 Aëtius III.1.3 (DK70A13).
68 *Mete.* 345b33–5.
69 *Mete.* 346a7–8.
70 *Mete.* 346a12f.
71 *Mete.* 346a19–31.
72 *Mete.* 345a19–25.
73 *Mete.* 345a32–7 and b13–25.
74 Macrobius, *Commentarii in Ciceronis Somnium Scipionis* I.15.3–7 (see Posidonis F130 EK).
75 Geminus, *Isagoge* V.68.
76 Aëtius III.1.1. He has just said that his Book III deals with τὰ ἀπὸ τοῦ κύκλου τῆς σελήνης καθήκοντα μέχρι πρὸς τὴν θέσιν τῆς γῆς, "things reaching from the circle of the moon to the position of the earth". Gilbert (1907) 661–2 attributes the theory summarised in Aëtius III.1.1 to the early Stoics, but this, as Kidd (1988) 490 points out, is very uncertain.
77 *Isagoga* 24 (Maass (1898) p. 55 lines 28–9).
78 Mansfeld and Runia (2020) 1150 (text = Aëtius III.1.10), 1159 (commentary).
79 For discussion of Posidonius' theory see Kidd (1988) 486–90; also Theiler (1982) II p. 177; Vimercati (2004) 552–3.
80 Aëtius III.1.8 (Posidonius F129 EK), in Mansfeld and Runia (2020) III.1.11. They read μανώτερον … πυκνότερον, and translate (p. 2109) "a solid structure (consisting) of a fire that is rarer than a star but denser than the brightest light", so the change of text hardly affects the meaning.
81 Macrobius, *Commentarii in Ciceronis Somnium Scipionis*, I.15.3–7 (Posidonius F130 EK).
82 I disagree with Mansfeld and Runia (2020) 1159 when they say that Posidonius' theory is "opposed" to that of other Stoics. Anything composed of fire must surely be ἀραιός, "rare".
83 Kidd (1988) 489 is doubtful about this.
84 See below, Chapter 18 (p. 181–2).
85 Macrobius mentions Diodorus before Posidonius, and does not suggest that Posidonius influenced him. However, as he mentions earlier authors in the order Theophrastus, Diodorus, Democritus, Posidonius, the order is clearly not meant to be chronological.
86 Macrobius, *Commentarii in Ciceronis Somnium Scipionis*, I.15.3–7 (Posidonius F130 EK), with translation from Kidd (1999) 183.
87 On Diodorus see Kidd (1988) 483–5, 487–8; Hultsch (1903).
88 Kidd (1988) 490.
89 See above, Chapter 1 (p. 1–2).
90 *Phaenomena* 460ff; the Milky Way is mentioned at 476 and 511.
91 *NQ* VII.15.2.

10 Exhalations

The concept of evaporation, as we would nowadays call it, or of "exhalation", as we usually say when talking about ancient theories, was important in ancient meteorology from the very beginning. I quoted earlier,¹ as the first hint of Greek meteorological theory known to us, lines from Hesiod which speak of mist which is drawn up from rivers and produces rain or wind; and this notion of mist – a body lighter than water, since it is above it, and finer than water, since we can walk through it without resistance, but derived from water – is a primitive form of the concept which Aristotle called ἀτμίς or ἀναθυμίασις, and we call "exhalation" when speaking of ancient meteorology.

A similar idea is implicit in fragment 30 of Xenophanes (I quote him because we have some of his actual words). Line 1 reads:

Πηγὴ δ' ἐστὶ θάλασσ(α) ὕδατος, πηγὴ δ' ἀνέμοιο

The sea is source of water and source of wind

and lines 5–6:

ἀλλὰ μέγας πόντος γενέτωρ νεφέων ἀνέμων τε
καὶ ποταμῶν

but the great sea is begetter of clouds and winds and rivers

Lines 2–4 are corrupt in the MS., but evidently mentioned rain (ὄμβριον ὕδωρ) among things which would not happen without the sea.² The fragment must mean that "the moisture being drawn up from the sea" (ἀνελκομένου ἐκ τῆς θαλάττης τοῦ ὑγροῦ, as Aëtius puts it) is the source of the other phenomena.³ In later chapters I shall quote other early sources which, in explanations of wind or rain, use this sort of vocabulary to describe what later authors would usually have referred to by some such term as ἀναθυμίασις.

That early thinkers should have thought in this way is not surprising. We all know that pools and streams, and domestic vessels containing water, are liable to dry up in hot weather unless the water is replenished, and we all know that

DOI: 10.4324/9780429399930-10

rain falls from clouds. It is also obvious, in domestic contexts, that the "steam" from boiling water looks not unlike cloud, and is liable to condense to water if it meets a cold surface; it was a simple inference that rain is formed in a similar way, by water being drawn up from the ground by the sun's heat, forming into clouds and then falling as rain – it is not so obvious why a variant of this process should be thought to produce wind, and I shall discuss this in the next chapter.

In this chapter I discuss some supposed complications of this simple process of water evaporating and condensing. One is the suggestion that there is more than one kind of exhalation. According to Diogenes Laertius, Heraclitus held that "exhalations occur from both earth and sea, some of them bright and pure, some of them dark" (γίνεσθαι ἀναθυμιάσεις ἀπό τε γῆς καὶ θαλάττης, ἃς μὲν λαμπρὰς καὶ καθαράς, ἃς δὲ σκοτεινάς). The burning of the bright exhalation produces the light of sun and stars and causes day and summer, the prevalence of the dark one causes night and winter.[4] Some scholars have denied or doubted that Heraclitus believed that there are two exhalations,[5] and this is not the place to discuss the matter; but two points need to be made.

First, a theory resembling that attributed to Heraclitus certainly did exist in the 5th century B.C., since it is found in the Hippocratic treatise *Airs waters places*,[6] Chapter 8, which deals with rainwater:

Ἐπειδὰν ἁρπασθῇ καὶ μετεωρισθῇ περιφερόμενον καὶ καταμεμιγμένον ἐς τὸν ἠέρα, τὸ μὲν θολερὸν αὐτοῦ καὶ νυκτοειδὲς ἐκκρίνεται καὶ ἐξίσταται καὶ γίνεται ἠὴρ καὶ ὁμίχλη, τὸ δὲ λαμπρότατον καὶ κουφότατον αὐτοῦ λείπεται καὶ γλυκαίνεται ὑπὸ τοῦ ἡλίου καιόμενόν τε καὶ ἑψόμενον.

When it [water] has been carried away aloft, and has combined with the air as it circles round, the turbid, dark part of it separates out, changes and becomes mist and fog, while the brightest and lightest part of it remains, and is sweetened, being burned and cooked by the sun.

(From Hippocrates, *Airs, waters, places*, 8, trans. W.H.S. Jones [1923, Harvard University Press, Loeb ed.], modified.)

Consequently, when the brightest part falls as rain, such rain is "the best, as one would expect" (ἄριστα κατὰ τὸ εἰκός).

Second, this concept, both as found in *Airs waters places* and as attributed to Heraclitus, is very different from the two-exhalation theory propounded by Aristotle in the *Meteorologica*. In the earlier theory the kinds of exhalation are bright and dark, not dry and wet. In *Airs waters places* both kinds are derived from water, while in Aristotle the dry one is from earth. When Diogenes attributes to Heraclitus exhalations "from both earth and sea", the exhalations from earth may well be from water on the earth rather than from earth as a material, and the same is true when Aëtius says "Heraclitus and the Stoics [sc. say that] the stars are nourished from the terrestrial exhalation" (Ἡράκλειτος καὶ οἱ Στωϊκοὶ τρέφεσθαι τοὺς ἀστέρας ἐκ τῆς ἐπιγείου [or ἀπὸ γῆς] ἀναθυμιάσεως).[7] If Heraclitus did hold a two-exhalation theory, the evidence indicates that *both* exhalations are wet: speaking of the exhalations, Diogenes says "the bright exhalation, set on fire in the circle of the sun, causes day" (τὴν λαμπρὰν ἀναθυμίασιν φλογωθεῖσαν ἐν τῷ κύκλῳ τοῦ ἡλίου ἡμέραν ποιεῖν), and adds, of the opposite exhalation (τὴν ἐναντίαν), that "from its darkness, moisture, being in excess, produces winter" (ἐκ τοῦ σκοτεινοῦ τὸ ὑγρὸν πλεονάζον χειμῶνα ἀπεργάζεσθαι).[8] Aëtius says that, to Heraclitus, sun and moon "emit their light, receiving the brightness from the moist exhalation" (δεχομένους τὰς ἀπὸ τῆς ὑγρᾶς ἀναθυμιάσεως αὐγάς, φωτίζεσθαι):[9] the exhalation which feeds the sun's fire is both bright and wet.

So this theory of multiple exhalations was available to later thinkers, as well as Aristotle's different one. The idea that the fire of the heavenly bodies is fed by moisture from the earth was influential: Heraclitus cannot have been the only man before Aristotle to have believed this, judging from the length at which Aristotle, naming no originator, attacks the idea.[10] After Aristotle, it became orthodox Stoic doctrine, and was accepted by Posidonius; but Posidonius, as we have seen, also accepted a form of Aristotle's two exhalation theory, so we must examine Posidonius' position in relation to Aristotle's theory.

In Aristotle's meteorology, dry exhalation causes comets, shooting stars and the like; it causes thunder, lightning and related phenomena; and it causes wind, and consequently earthquakes, which, according to a widely held ancient theory, are due to wind.[11] Wet exhalation causes rain and the other kinds of precipitation; and so the two exhalations are the basis of Aristotle's meteorology and are introduced as such at the beginning of *Meteorologica*.[12]

This theory did not greatly influence those who wrote on meteorology immediately after Aristotle. The evidence suggests that it had no great part in the meteorology of Theophrastus.[13] The scanty meteorological fragments of Strato[14] do not mention it. Epicurus in the *Letter to Pythocles*, though he seems once or twice to suggest that there may be more than one sort of exhalation,[15] never mentions a dry one. There is no mention of dry exhalation in the meteorological remains of the early Stoics.[16] The *De mundo* does use the dry exhalation hypothesis, but I have argued above that, though that work may predate Posidonius, it is probably later.[17]

Posidonius, then, was the first writer on meteorology that we know of after Aristotle (and, in a limited way, Theophrastus) who used the hypothesis of dry exhalation. For Posidonius, dry exhalation was the cause, or a cause, of thunder and lightning, but not of comets; and, though the evidence is slight, the balance of probability is, as we shall see, that he did not regard dry exhalation as the main cause or constituent of wind, as Aristotle had done. It therefore seems unlikely that, as some scholars have believed, he made the dry and wet exhalations the starting point of his meteorology, as Aristotle did.[18]

The likely reason why Posidonius made so much less use of the dry exhalation is a difference between Aristotelian and Stoic ideas about the four elements. At *Mete.* 341b6ff Aristotle says that exhalation is διπλῆν, τὴν μὲν ἀτμιδωδεστέραν, τὴν δὲ πνευματωδεστέραν, τὴν μὲν τοῦ ἐν τῇ γῇ καὶ ἐπὶ τῇ γῇ ὑγροῦ ἀτμίδα, τὴν δ' αὐτῆς τῆς γῆς οὔσης ξηρᾶς καπνώδη, "two-fold, the one more like water-vapour, the other more like wind; the one from moisture in the earth and on the earth is water-vapour; the one from the earth itself, which is dry, is smoky". So the wet exhalation is derived from moisture, the dry exhalation from earth as an element. Aristotle never explains why he put forward this theory. Various phenomena have been or might be suggested as supporting it. Gilbert[19] argued that the basis of the theory must be the re-emission from the earth of the heat it absorbs from the sun: a phenomenon sometimes obvious, as when a rock heated by the sun emits heat after the sun has gone in.[20] In ordinary human experience, a solid body such as wood emits whitish smoke when burned, which resembles the whitish vapour which rises from water when boiled; therefore, Aristotle may have reasoned, just as water obviously evaporates in hot weather although we cannot see the vapour, so too there must be invisible vapour from solid bodies. But he had another reason for assuming a dry exhalation from earth, which is implicit in the theory of elemental change which he expounds in *De generatione et corruptione*.[21] In that theory, change is rapid and easy between any pair of elements which share a common quality (but slower if there is no common quality); therefore, earth (being cold and dry) can change to elemental fire (hot and dry) as readily as water (cold and wet) changes to air (hot and wet); and so, if water by evaporating (or "being drawn up") gives rise to the regular phenomena of clouds and rain, as Aristotle and many earlier writers believed,[22] there must on Aristotle's theory be an equally regular "evaporation" from earth.[23]

The Stoics disagreed with Aristotle about the properties of the four elements. In their view, fire is the hot element (τὸ θερμόν), water the wet element, air the cold element, earth the dry element.[24] There is doubt whether Posidonius accepted this, at least as regards the coldness of air. Plutarch implies that Posidonius regarded air as the cold element, but what he quotes from Posidonius hardly proves it.[25] Simplicius gives an account of Aristotle's element theory in which he indicates that to Aristotle fire and air are both light in weight and hot (κοῦφα καὶ θερμά), and adds that Posidonius "everywhere uses this" (πανταχοῦ χρῆται).[26] However, no-one attributes to Posidonius Aristotle's view that each of the four elements is characterised by a pair of the four qualities hot, cold, wet and dry, and if he did not hold that view he cannot have accepted Aristotle's theory about changes between elements which share a common quality, and so cannot have had the reason which that theory gave to Aristotle for postulating an exhalation from earth. In Posidonius' theory, as given us by Seneca, the dry exhalation is not specified to be an exhalation from earth: *both* exhalations are "e terra terrenisque omnibus", "from the earth and everything terrestrial"[27] – and that presumably includes the water in and on the earth, and does not just mean earth as an element.

The change of water to air and then fire, and the reverse process, and the idea that water- vapour sustains the fire of the sun, are important in Stoic cosmogony and cosmology. In their view, says Diogenes Laertius, after a cosmos has been destroyed by fire, a new one is formed ὅταν ἐκ πυρὸς ἡ οὐσία τραπῇ δι' ἀέρος εἰς ὑγρότητα, εἶτα τὸ παχυμερὲς αὐτοῦ συστὰν ἀποτελεσθῇ γῆ, τὸ δὲ λεπτομερὲς ἐξαερωθῇ [or ἐξαραιωθῇ], καὶ τοῦτ' ἐπὶ πλέον λεπτυνθὲν πῦρ ἀπογεννήσῃ, "when the substance is transformed from fire through air into moisture, and then the dense part of the moisture congeals and becomes earth, while the fine part becomes air [or 'is well rarefied'], and this, when further rarefied, generates fire".[28] Posidonius is the most recent of five authors cited almost immediately after this as authorities on the Stoic view of the generation and destruction of the cosmos; this quotation evidently represents the ortho-dox Stoic view, accepted by Posidonius at least in principle.[29]

The Stoics also held that the sun is fed by exhalation from the sea – an idea, which, as we have seen, evidently originated with Heraclitus, and was widely enough held for Aristotle to take some trouble to attack it. Cleanthes and Chrysippus thought that the sun is ἄναμμα νοερὸν τὸ ἐκ θαλάττης, "the intelli-gent fiery mass from the sea",[30] or τὸ ἀθροισθὲν ἔξαμμα νοερὸν ἐκ τοῦ τῆς θαλάσσης ἀναθυμιάματος, "the collected intelligent fiery mass from the exhala-tion from the sea".[31] Posidonius seems to have accepted this, the orthodox Stoic view: Macrobius implies that Posidonius believed the sun to be nourished by moisture, "humore nutriri",[32] and Diogenes Laertius attributes to him the related theory that the moon is nourished by exhalation from fresh water.[33]

Thus the conversion of water to air and fire, and the "nourishment" of heav-enly bodies by exhalations from water, were familiar and orthodox ideas to Posidonius: it is not surprising that, though following Aristotle in other respects, he did not use dry exhalation in explaining astronomical phenomena such as comets. But he did use it in explaining thunder and lightning. These latter phenomena are frequently accompanied by heavy rain: he perhaps found it difficult to imagine water-vapour turning violently to fire and to water simul-taneously in the same place.

But in one respect Posidonius, or some Stoic, did use something like dry exhalation – an exhalation from earth – in his astronomy. As just mentioned, Cleanthes and Chrysippus thought that the sun is fuelled by exhalation from the sea. Chrysippus added to this the idea that the moon is "the collected fiery mass from the exhalation from fresh water" (τὸ ἀθροισθὲν ἔξαμμα…ἐκ τοῦ ἀπὸ τῶν ποτίμων ὑδάτων ἀναθυμιάματος)[34] – as just mentioned, Posidonius accepted this. Some Stoic thinker elaborated this theory further:

τρέφεσθαι δὲ τὰ ἔμπυρα ταῦτα καὶ τὰ ἄλλα ἄστρα, τὸν μὲν ἥλιον ἐκ μεγάλης θαλάττης … τὴν δὲ σελήνην ἐκ ποτίμων ὑδάτων, ἀερομιγῆ τυγχάνουσαν καὶ πρόσγειον οὖσαν, <καθά φησι> Ποσειδώνιος … τὰ δ' ἄλλα ἀπὸ γῆς.

These fiery bodies [sun and moon] and the other stars are nourished. The sun is nourished from the great sea . … The moon is nourished from fresh water, it happening to be mixed with air and being near the earth, as Posidonius says . … But the other stars are nourished from earth.[35]

In many contexts, an exhalation "from earth" might be an exhalation from water on the earth. I quoted earlier two passages where this may well be so,[36] but it surely cannot be in my last quotation, because it has already used up all the water: the sea nourishes the sun, fresh water the moon. What exhalation can be left except one from earth as a material?

This was not Chrysippus' view, since Plutarch[37] quotes him as saying: οἱ δ' ἀστέρες ἐκ θαλάσσης μετὰ τοῦ ἡλίου ἀνάπτονται, "the stars with the sun get their fire from the sea"; and the Stoics between Chrysippus and Posidonius were not especially interested in astronomy. Posidonius must be a strong candidate as the originator of this theory[38] (especially as the failure to account for the nourishment of stars as well as sun is mentioned by Aristotle as an objection to the sun-nourished-by-exhalation theory at *Mete.* 355a18–20, in a work we know Posidonius read).

A passage of Cleomedes is worth quoting here. He says that the air round the earth must form a sphere, ἀπὸ ὅλης αὐτῆς τῶν ἀναθυμιάσεων αἰρομένων καὶ ἐπισυρρεουσῶν καὶ οὕτως καὶ τὸ τοῦ ἀέρος σχῆμα ὅμοιον ἀπεργαζομένων, "since the exhalations are raised and flow together from the whole of it [the earth] and so make the shape of the air, too, similar [sc. to the shape of the earth]".[39] This implies that exhalations rise from every part of the earth, that is, from both land and sea; and this was perhaps the view of Posidonius, who had a motive for maximising the amount of exhalation produced by the earth, since he combined the traditional Stoic belief that all the heavenly bodies are nourished by exhalations from the earth with the astronomical realisation that the sun, at least, is many times the size of the earth and an enormous distance away.[40] How, then, can the earth produce enough exhalation? Exhalations rising from every part of the earth, and from earth as an element as well as from water, would help to provide an answer.

Further investigation of Posidonius' astronomy is outside the scope of this book. For present purposes, putting together what is said by Seneca at *NQ* II.54 and what I have quoted from Cleomedes, it is sufficient to conclude that Posidonius probably believed that exhalations occur from most, if not all, "earthy" materials as well as from liquids, and are available to explain meteorological as well astronomical phenomena.

Notes

1 See above, Chapter 4 (p. 20).
2 DK 21B30, which see for proposed emendations of lines 2–4.
3 See Aëtius III.4.4 (DK 21A46), which ends by quoting line 1 of DK 21B30.
4 Diogenes Laertius IX.9–11 (DK 22A1).
5 See especially Kirk (1954) 270–6, also Kirk, Raven and Schofield (1983) 202n.
6 The early date of *Airs, waters, places* is confirmed by the fact that ἠήρ is used to mean "mist" as often as it means "air". Using the line numbering of Jones (1923), the word means "mist" at §5 line 16 (if the MS. text is correct); §6 line 7; §8 line 32; §15 line 17; and §19 line 17. It means "air" in the title; at §6 line 17; §8 line 30; §15 line 25; and §19 line 31. (The two instances in §8 are included in my quotation.)
7 Aëtius II.17.4 (DK 22A11, *SVF* II.690).
8 Diogenes Laertius IX.11 (DK 22A1).
9 Aëtius II.28.6 (DK 22A12).
10 *Mete.* 354b34–355a32.
11 See below, Chapters 11 and 12.

12 Mete. I.3–4 (340b25ff, 341b5ff).

13 See above, Chapter 8 (p. 74) n. 38.

14 Listed in Chapter 4 (p. 25).

15 See *Letter to Pythocles* 99, clouds form κατὰ ῥευμάτων συλλογὴν ἀπό τε γῆς καὶ ὑδάτων, "by a gathering of currents from earth and waters"; also 106, on which see Chapter 11 (pp. 99, and 110 n. 18).

16 See above, Chapter 4 (p. 26–9).

17 See above, Chapter 3 (p. 13–14).

18 Theiler (1982, II p. 207) suggests that Posidonius did make the two exhalations the starting point of his meteorology; but the only evidence appears to be that they are the starting point for the meteorological section of *De mundo* (394a9–19), and I have argued above (Chapter 3, p. 12–13] that *De mundo* should not be regarded as evidence for Posidonius' view on this point.

19 Gilbert (1907) 449.

20 Not mentioned by Gilbert (1907). Suggested to me verbally by F.H. Sandbach.

21 *GC* II 4 (331a7–b2).

22 See below, Chapter 14 (p. 142).

23 The relation of Aristotle's element theory to his exhalation theory is also discussed by Wilson (2013) 35ff. His emphasis is different, but I do not think he would disagree with what I have written here.

24 See Diogenes Laertius VII.137, and other texts listed by Kidd (1988) 378. (Vimercati [2004] 521 is surely wrong to suggest that the Stoics generally thought air to be hot and moist; *SVF* II.430, which he cites, does not say this)

25 *De primo frigido* 951F (Posidonius F94 EK): to Posidonius, the cold of damp regions is due τὸ πρόσφατον εἶναι τὸν ἕλειον ἀέρα καὶ νοτερόν, "the fact that air in marshy places is newly-formed and damp". Kidd (1988) 378–9 implies that, in his opinion, Posidonius accepted the Stoic view that air is the cold element, but says that F94 EK is "hardly decisive" evidence that he did.

26 Simplicius *In Cael.* IV.3 (Posidonius F93a EK). Boechat (2016) argues that Posidonius accepted Aristotle's view as described by Simplicius.

27 See above, Chapter 8 (p. 69).

28 Diogenes Laertius VII.142 (Posidonius F304 in Theiler [1982]), with the translation of Mensch and Miller (2018), slightly modified. (They translate ἐξαερωθῇ, but Theiler, and Dorandi [2013], prefer the alternative reading ἐξαραιωθῇ.) Compare Chrysippus' view as given by Stobaeus, *Eclogae*, I p. 184 W. (*SVF* II 527).

29 See Kidd (1988) 118–21.

30 Cleanthes, according to Aëtius II.20.4 (*SVF* I 501). Aëtius II.20.16 (DK 22A12) attributes an identical description of the sun to Heraclitus, but it is surely likely that this is reading back into Heraclitus what was in fact a Stoic idea.

31 Chrysippus, according to Stobaeus, *Eclogae*, I p. 214,1 W. (*SVF* II 652).

32 *Saturnalia*, I.23.2 (Posidonius F118 EK); discussed by Kidd (1988) 458–61.

33 Diogenes Laertius VII.145 (Posidonius F10 EK). On this see further below.

34 Stobaeus, *Eclogae*, I p. 219.29 W. (*SVF* II 677).

35 Diogenes Laertius VII.145, partly printed as Posidonius F10 EK, more fully as F262 in Theiler (1982); but he notes (Theiler [1982] II p. 139) that what is printed is only partly from Posidonius.

36 Page 90 above.

37 *De Stoicorum repugnantiis* 1053A (*SVF* II 579).

38 Kidd (1988) 113–14, argues that the words τὰ δ' ἄλλα ἀπὸ γῆς are not part of the citation from Posidonius' Φυσικὸς λόγος; but at p. 459 he says that Posidonius "probably" held the theory that sea water nourishes the sun, fresh water the moon, and "earth moisture" the stars.

39 Cleomedes I.5, lines 126–8 Todd.

40 Cleomedes II.1.79–80 (II.1 lines 269–86 Todd) = Posidonius F115 EK.

11 Winds

The nature and cause of wind in general

There is a major gap in our evidence for what Posidonius thought about wind. We have reports of what he said about some particular winds – about, for instance, the name or the character (or both) of winds blowing from particular directions, or in particular places – but none about his answers to general questions such as "What is wind?" or "What causes wind?" We can only try to infer how he is likely to have answered such questions from reports of the views of unnamed Stoics (which may be his, or else would have influenced him or been influenced by him) and from surviving works which we know influenced him, especially Aristotle's *Meteorologica*, or which we know he influenced: principally Seneca's *Naturales quaestiones*.

About the question "What is wind?" the problem is fairly simple. Evidence going back to the 4th century B.C. shows most ancient writers agreeing that wind should be defined as "air in motion", or "a flow of air" or some similar phrase. The earliest surviving instances seem to be [Hippocrates], *De flatibus* 3, ἄνεμος … ἐστιν ἠέρος ῥεῦμα καὶ χεῦμα, "wind … is a flow and stream of air", and Plato, *Cratylus* 410B, suggesting that ἀήρ, 'air', may be so called ὅτι πνεῦμα ἐξ αὐτοῦ γίγνεται ῥέοντος· οἱ γὰρ ποιηταί που τὰ πνεύματα ἀήτας καλοῦσιν, "because wind arises from its [i.e., air] flowing; for the poets call the winds *aētas*" (i.e., ἀήρ gets its name from its resemblance to ἀήτας, blasts of wind). By Aristotle's time ἀὴρ κινούμενος, "air being moved" and κίνησις ἀέρος, "movement of air", seem to have been standard definitions of "wind" (πνεῦμα or ἄνεμος), though ones which Aristotle himself criticises[1] (essentially, I think, on the ground that not every movement of air can be called 'wind'[2]). Definitions of wind resembling these were normal in later antiquity,[3] including a Stoic one recorded by Aëtius: οἱ Στωϊκοὶ πᾶν πνεῦμα ἀέρος εἶναι ῥύσιν, "the Stoics [sc. say that] all wind is a flow of air".[4] The chances are that Posidonius would have agreed with this definition, accepted as it was by the ancient world generally, evidently including the Stoics, though he might have added some qualification, as in *Mete.* IV 387a29f, where the author, probably Aristotle,[5] defines wind as ῥύσις συνεχὴς ἀέρος ἐπὶ μῆκος, "a continuous flow of air at length" ("over a long distance" or "for a long time"?).

DOI: 10.4324/9780429399930-11

Such a definition says nothing about the cause of wind. My quotation from Aëtius continues ταῖς τῶν τόπων δὲ παραλλαγαῖς τὰς ἐπωνυμίας παραλάττουσαν, "[a flow of air] changing its names by changes of place", and he gives as examples Zephyrus, Apeliotes, Boreas and Lips, as the names of winds from the west, east, north and south respectively. The absence of any mention of a cause suggests that this is the view of an early Stoic, who avoided speculation about causes.[6]

We have two other accounts of a theory of wind held by unnamed Stoics. One is from the account of Stoic meteorology in Diogenes Laertius VII.152. The opening words are missing. What survives reads: παρὰ τοὺς τόπους, ἀφ' ὧν ῥέουσι. Τῆς δὲ γενέσεως αὐτῶν αἴτιον γίνεσθαι τὸν ἥλιον ἐξατμίζοντα τὰ νέφη, "according to the places from which they flow. The cause of their coming-to-be is the sun evaporating the clouds".[7] The first six words must be the end of a sentence like that in Aëtius, saying that the names of the winds change according to the places whence they blow. The other account is from Cicero, *De divinatione* II.44: "Placet enim Stoicis eos anhelitus terrae, qui frigidi sint, cum fluere coeperint, ventos esse", "for it is the opinion of the Stoics that those exhalations from the earth which are cold, when they begin to flow, are winds".[8]

Both passages are likely, prima facie, to be giving the view of Posidonius: Diogenes' account of Stoic meteorology cites Posidonius three times; Cicero was Posidonius' pupil. As I pointed out when quoting this passage previously,[9] Cicero's context does not require him to be accurate in detail; however, his account of Stoic wind theory is not necessarily inconsistent with that in Diogenes: a few lines later, in Diogenes' account of rain, cloud (νέφος) is identified with, or closely associated with, ἡ ἐκ γῆς ἢ ἐκ θαλάττης ἀνενεχθεῖσα ὑγρασία, "moisture drawn up from earth or from sea".[10]

It would not be surprising if Posidonius thought that exhalation was involved in the causation of wind, since it was a widespread ancient view. Aristotle, who so much influenced him, regarded the dry exhalation as the ἀρχὴ καὶ φύσις, "origin and nature" of all winds.[11] Seneca, much influenced by him, says (*NQ* V.4.3):

Numquid ergo hoc verius est dicere multa ex omni parte terrarum et assidua ferri corpuscula? Quae cum coacervata sunt, deinde extenuari sole coeperunt, quia omne quod in angusto dilatatur spatium maius desiderat, ventus existit.

Is there anything to say that is truer than this, that many particles are constantly being carried [sc. up] from every part of the earth? And when these have been massed together and afterwards begin to be rarefied by the sun, then wind occurs, because everything needs more space which is in a narrow space and becomes swollen.

The "many particles" are evidently what is referred to in *NQ* V.5.1 as "aquarum terrarumque evaporationes", "evaporations from water and earth". We may compare V.12.1: the violent wind *ecnephias* occurs when "vapor terrenus", "vapour from earth", contains both dry and wet materials, which conflict with each other. In Seneca's view, unlike Aristotle's, the exhalations which cause wind may include dry material, but dry material has no predominant role. The massing together of variegated particles of exhalation could well form the clouds mentioned in Diogenes' account of Stoic wind theory; and Seneca's mention of "particles from every part of the earth" fits with what I argued, in the last chapter, is likely to have been Posidonius' theory of exhalations.

There is one other piece of evidence which suggests that Posidonius did not see a special connection between wind and *dry* exhalation. To Aristotle comets are a sign of drought *and wind* (since both are caused by a predominance of dry exhalation),[12] to Posidonius they are a sign of drought, with no mention of wind.[13] He presumably did not associate wind with a predominance of dry matter in the atmosphere. This supports the suggestion that he probably thought, as Seneca evidently did, that exhalations from both water and earth were involved in the production of wind.

We might call "intuitive" the belief that wind is air in motion; this is not true, to modern ideas, of the belief that wind is derived from, or largely derived from, exhalations or vapour from earth or water, or that it is derived from cloud. This was, however, a very old Greek view. I have already quoted Hesiod, *Works and Days* 547–53, where mist, drawn from rivers, appears to blow as wind,[14] and Xenophanes fragment 30, where the sea is "source of wind" and "begetter of winds".[15] This is not the place to discuss in detail the evidence, mostly found in much later authors, for pre-Aristotelian wind theories,[16] but I will quote some relatively early texts which show predecessors of Aristotle deriving wind from water or from cloud. First, two passages from the Hippocratic corpus which presumably predate Aristotle: *De natura pueri* 25 (VII.522 Littré):

Τὰ ... πνεύματα ἡμῖν ἐστι πάντα ἀφ᾽ ὕδατος ... ἀπὸ γὰρ τῶν ποταμῶν πάντων πνεύματα χωρέει ἑκάστοτε καὶ τῶν νεφέων, τὰ δὲ νέφεα ἐστὶν ὕδωρ ξυνεχὲς ἐν ἠέρι.

All the winds we know come from water ... for winds always blow from all rivers and clouds, and clouds are a continuous volume of water in air.

And *De victu* II.38:

φύσιν ... ἔχει τὰ πνεύματα πάντα ὑγραίνειν καὶ ψύχειν ... διὰ τάδε· ἀνάγκη τὰ πνεύματα ταῦτα πνεῖν ἀπὸ χιόνος καὶ κρυστάλλου καὶ πάγων ἰσχυρῶν καὶ ποταμῶν καὶ λιμνέων καὶ γῆς ὑγρανθείσης καὶ ψυχθείσης.

All winds naturally moisten and cool ... for the following reason: these winds [he presumably means "all winds"] must blow from snow and ice and severely frozen places and rivers and lakes and land that has been moistened and cooled.

In both these passages, as in Hesiod and Xenophanes, winds are derived from water, or something derived from water (mist, cloud, ice). There is no word which we might translate as "vapour" or "exhalation", but some such idea must have been in the author's mind.

Second, a passage from Aristotle about an earlier theory (*Mete.* 353a35–b9):

οἱ ἀρχαῖοι ... οἱ σοφώτεροι τὴν ἀνθρωπίνην σοφίαν ... [sc. ὑπέλαβον] εἶναι τὸ πρῶτον ὑγρὸν ἅπαντα τὸν περὶ τὴν γῆν τόπον, ὑπὸ δὲ τοῦ ἡλίου ξηραινόμενον τὸ μὲν διατμίσαν πνεύματα καὶ τροπὰς ἡλίου καὶ σελήνης ... ποιεῖν

the ancients ... who were wiser in human [as opposed to mythological] wisdom ... [supposed that] at first the whole area about the earth was wet, but being dried by the sun that which evaporated caused winds and turnings of sun and moon [i.e., solstices, and the moon's equivalent movements].

Presumably, as these phenomena and winds still occur, the evaporation which causes them was thought to continue, and was not confined to the formation of the cosmos. Alexander, citing the authority of Theophrastus, tells us that the theory Aristotle describes was that of Anaximander and Diogenes of Apollonia.[17]

Epicurus was another thinker who regarded exhalations as a cause of wind, writing in the *Letter to Pythocles* that winds occur ἀλλοφυλίας τινὸς ... παρεισδυομένης, καὶ καθ' ὕδατος ἀφθόνου συλλογήν, "when some foreign matter ... penetrates [presumably into the air], and by a gathering of abundant water".[18] The "foreign matter" must be something different from water. Either can cause wind, but surely must get into the air as some sort of vapour in order to do so.

What was the origin, and what caused the popularity of theories which derive winds from vapour or exhalations? Pre-philosophical tradition may have had a part in this. Hesiod, *Works and Days* 547–53, is not the only passage of early poetry that speaks of wind coming from water: for example, *Odyssey* IV.567f, in Elysium αἰεὶ Ζεφύροιο λιγὺ πνείοντος ἀήτας Ὠκεανὸς ἀνίησιν, "Ocean ever sends up breezes of clear-blowing(?)[19] Zephyrus": the poet imagines a pleasant, cooling sea-breeze. Strong winds are usually accompanied by clouds, and so we have at *Iliad* II.145f, Εὖρός τε Νότος τε ... ἐπαΐξας πατρὸς Διὸς ἐκ νεοελάων, "Eurus and Notus swooping from the clouds of father Zeus". However, in other passages winds blow from a place on land (Mount Ida, or Thrace),[20] and Boreas is called αἰθρηγενής or αἰθρηγενέτης, "born from clear skies".[21] I do not suggest that any meteorological theory underlies any of these Homeric passages; in each of them, the poet describes only what he imagines he would see or feel.

Nevertheless, what the poets had said would have been in the minds of the first philosophers when they tried to find naturalistic explanations of wind. Somehow, the idea that winds come from water seemed the likeliest to many of them. A possible reason is that, as we know from everyday experience, a small quantity of boiling water can produce a volume of visible "vapour" much larger than itself, and can go on doing so for a considerable time. We also know that wind can blow from the same direction for days on end, that is, matter is constantly flowing from the direction from which the wind blows. The first thinkers would naturally have asked themselves, "how is this flow maintained?"; being used to the idea that winds blow from the sea, or rivers, or clouds, they would naturally answer that the flow was maintained in a way analogous to the flow of vapour from boiling water: something denser than the wind – water, or cloud, or mist – was being gradually evaporated or rarefied, and this gradual production of rare matter, that is, wind, from denser matter made it possible for the wind to continue blowing.

This view must have been reinforced by the rather primitive assumptions, which Aristotle clearly made and which were presumably widespread, that the named winds blow from a fixed location – one can discuss where that location is, for example, does Notus (the south wind) blow from the south pole or from the torrid zone, but Aristotle seems clear that there must be such a location[22] – and that opposite winds cannot blow simultaneously.[23] This must have made it difficult to envisage winds (apart from very localised whirlwinds) blowing in a circle, as we now know that they do round depressions and tropical storms. The view that wind is derived from water-vapour did not conflict with the definition of wind as air in motion, for those thinkers who believed, as many did, that water as it evaporates *becomes* air.

Up to the mid-4th century B.C. we hear only of moist exhalation, though sometimes of two varieties of it, but Aristotle, for reasons discussed in the last chapter, decided that there is a dry exhalation as well as a wet one (air being a mixture of both[24]), and apportioned meteorological phenomena between them. Rain being obviously wet, he assigned wind, with its different nature, to the dry one.[25] After him, Posidonius probably, as I would argue, and Seneca explicitly, thought that there were exhalations from all the materials on the earth, and that these provided the material cause of wind.

The efficient cause, or one of the efficient causes, was surely the sun. If the heavenly bodies are nourished by exhalations from the earth, presumably they also cause the exhalations to form; and both Diogenes Laertius VII.152, giving the Stoic view of wind, and Seneca *NQ* V.4.3, say that the sun evaporates or rarefies the cloud or the mass of exhalation to produce wind. It was an old view that the sun is cause, or part of the cause, of wind. Aristotle writes of the sun (along with terrestrial heat) causing the exhalation which produces wind,[26] and the sun causing evaporation is also a feature of the earlier theory he cites, which Alexander attributes to Anaximander and Diogenes of Apollonia. Doxographic writers name, if with less authority, other pre-Socratics as giving the sun a part in the production of wind.[27]

Aristotle declares that the natural motion of dry exhalation is upwards, but that wind moves horizontally because the air "follows the motion" (συνέπεται τῇ φορᾷ), and that "the origin of the motion [of wind] is from above" (τῆς … κινήσεως ἡ ἀρχὴ ἄνωθεν), apparently meaning that the diurnal revolution of the heavens, and of the higher part of the regions of air and elemental fire, cause the winds' motion – leaving unexplained the various directions in which winds blow.[28] Theophrastus' theory that the horizontal movement of wind is the resultant of natural movements in opposite directions, vertically upwards and vertically downwards,[29] probably also derives from the Aristotelian conviction that the natural motion of sublunary bodies is always vertically up or down. However, Posidonius, as a Stoic, need not have shared this conviction; he may well have thought that such explanations of wind's horizontal motion are otiose, and that the sun's heat is an adequate efficient cause of wind, as it seems to be in Seneca.

There is some evidence of other efficient causes. Aëtius, in his chapter on the sea's tides, says: Ποσειδώνιος ὑπὸ μὲν τῆς σελήνης κινεῖσθαι τοὺς ἀνέμους, ὑπὸ δὲ τούτων τὰ πελάγη, ἐν οἷς τὰ προειρημένα γίνεσθαι πάθη, "Posidonius says the winds are moved by the moon, and by these the seas are moved, in which the aforementioned phenomena [ebbs and floods] occur."[30] Whether this is correct is very doubtful, since another source, Priscianus Lydus, says that to Posidonius tides are due to gentle heat from the moon causing the sea to swell.[31] If Aëtius is correct, I think we must conclude that what the moon causes are just particular winds which cause the tides, not winds in general. We should not dismiss the idea that, for Posidonius, the moon might play some part in causing wind. Aristotle and Theophrastus had suggested that it does play a part, but only a subordinate one, as at *Mete.* II.8, 367b20ff, where the sun's heat reflected from the moon evidently causes wind, an effect which ceases when the moon is eclipsed.[32] *De generatione animalium* IV.10, 777b33–5, is another passage where both sun and moon are involved in causing winds. Theophrastus in *De ventis* 15–17 describes the sun as sometimes rousing and sometimes halting winds, and then adds ποιεῖ δὲ καὶ ἡ σελήνη ταὐτὰ πλὴν οὐχ ὁμοίως· οἷον γὰρ ἀσθενὴς ἥλιός ἐστι, "the moon has this effect also but not to the same degree, being a kind of weak sun".[33] In view of precedents like these, while it is surely incredible that Posidonius regarded the moon as a main cause of winds, he may have thought it had a minor part in causing them. What part wind may have played in Posidonius' account of tides I shall discuss further in Chapter 13.

One other possible cause of wind's motion is suggested by Seneca. Following the cause of wind he describes in *NQ* V.4.3, he goes on to describe a cause which he calls "more substantial and truer" ("valentior veriorque"): "habere aera naturalem vim movendi se", "air has a natural power of moving itself". He cites in support of this a claim that water moves itself in the absence of wind and alleged instances of spontaneous generation from water and from fire, and argues that, this being so, air cannot be inert.[34] It is not impossible that Posidonius held such a view; in a pantheistic universe there must be in everything something of the divine, so (one might suppose) some power of initiating action, and this should be particularly true of air, in view of the importance of πνεῦμα, "breath", in Stoic thought. Posidonius is reported as saying that god is πνεῦμα νοερὸν διῆκον δι' ἁπάσης οὐσίας, "an intelligent breath pervading all that is".[35] However, on this view the self-motion of air might be a cause of most meteorological phenomena, since they mostly involve or take place in air, and there is no explicit evidence that Posidonius regarded the self-motion of air as causing any of them.

No ancient thinker had even an inkling of the true cause of thunder and lightning. With wind it was possible to approach a little nearer to modern meteorological theory. In Seneca *NQ* V.4.3 – which, as I have argued, is likely to be close to the view of Posidonius – the sun causes the motion of wind (a widespread view in antiquity, but also recognised as true today[36]), and it does so by causing a mass of dense matter in the atmosphere to expand; which must mean, if an ancient thinker thought about it, that it pushes away the less dense air around it (since it does not push away something denser and stronger than itself, e.g., a moderate wind does not blow down a well-built wall); this is not so very far from the modern concept of wind being due to the tendency of air to flow from where its pressure is greater to where its pressure is less.[37] Seneca *NQ* V.5.1, repeating the theory of V.4.3, has "quae densa steterant, ut est necesse, extenuata nituntur in ampliorem locum", "what had been tightly packed inevitably struggles to find a broader space as it expands".[38] That does imply an exertion of pressure, may have been influenced by Posidonius, and is perhaps as near as any ancient thinker got to the explanation of wind as due to atmospheric pressure.

The wind-rose and wind names

Direct evidence for Posidonius' wind-rose[39] is confined to a single passage of Strabo, I.2.21,[40] which speaks of a view that there are only two principal winds, and says that its proponents cite in support of it Thrasyalces (about whom nothing more is said here[41]), and "the poet", i.e., Homer, who (they suggest) assimilated the wind Argestes to the wind Notus by his use of the phrase ἀργεστᾶο Νότοιο, and assimilated Zephyrus to Boreas, since he says they both "blow from Thrace" (Θρῄκηθεν ἄητον). Strabo then goes on: φησὶ δὲ Ποσειδώνιος, μηδένα οὕτως παραδεδωκέναι τοὺς ἀνέμους τῶν γνωρίμων περὶ ταῦτα, οἷον Ἀριστοτέλη, Τιμοσθένη, Βίωνα τὸν ἀστρολόγον· ἀλλὰ τὸν μὲν ἀπὸ θερινῶν ἀνατολῶν Καικίαν, "but Posidonius says that none of those who are knowledgeable about these matters have passed down this account of winds, for example Aristotle, Timosthenes, and Bion the astronomer; but Caecias blows from the summer sunrise"; and, he goes on, Lips blows from the winter sunset; Eurus blows from the winter sunrise, with Argestes opposite to it (i.e., blowing from the summer sunset), and Apeliotes and Zephyrus are in the middle (i.e., respectively, between Caecias and Eurus, and Lips and Argestes). This correctly reports the sense of Aristotle's account of these winds, at *Mete.* 363a25–b26, and that of Timosthenes as reported by Agathemerus[42] (we lack confirmation for Bion[43]); and the implication is that Posidonius agreed with them.

Strabo continues, in indirect speech so still presumably reporting Posidonius, with an explanation of Homeric phrases about winds,[44] and in the course of it he mentions a wind Λευκόνοτος, which in the wind-rose of Timosthenes, and also in that of Seneca,[45] is a wind from a particular direction, blowing from between Lips and Notus, a direction from which, according to Aristotle, no wind blows,[46] so the mention of Leuconotus could be seen as Posidonius following Timosthenes[47] and correcting Aristotle. However, Aristotle mentions λευκόνοτοι not as winds from a particular direction, but as south winds of a particular character, blowing rather irregularly after the winter solstice, the counterpart of the northerly Etesians which follow the summer solstice,[48] and Posidonius, too, in Strabo's citation, is interested in their character, not their precise direction (on this see below), so this citation is not good evidence for the inclusion of Leuconotus in Posidonius' wind-rose.

Posidonius evidently regarded as authorities on the wind-rose Aristotle, who describes 12 directions from which winds might blow, though denying that there is a wind from all of them,[49] and Timosthenes, whose wind-rose had 12 winds; also, wind-roses with 12 winds are described by Seneca,[50] who was influenced by Posidonius, and by the author of *De mundo*,[51] who either was influenced by Posidonius, or perhaps influenced him. So we can be reasonably sure that Posidonius' wind-rose had 12 winds, with three northerly and three southerly winds added to the three easterly and three westerly winds mentioned above (Caecias, Apeliotes, Eurus, Lips, Zephyrus and Argestes). Even with the four comparable texts just mentioned, we cannot with confidence complete Posidonius' wind-rose, since, as can be seen from Figure 11.1, the four texts disagree, especially about the winds blowing on either side of due south.

Aristotle, Timosthenes, Seneca and the *De mundo* agree in describing the directions of the three easterly and three westerly winds in terms of the direction of sunrise and sunset at midsummer, the equinoxes and midwinter. To judge from Strabo, quoted above, Posidonius did the same. The same four authors suggest no such precise direction for the winds, such as Thrascias, which blow on either side of due south and due north, usually describing them as blowing next to, or in between, a wind or winds adjacent to them (for instance, Aristotle[52] says that Thrascias is μέσος ἀργέστου καὶ ἀπαρκτίου, "midway between Argestes and Aparctias")[53]; however, Aristotle does note that the line between the positions from which Thrascias and Meses blow nearly corresponds to the circle which is "ever-visible" (διὰ παντὸς … φαινόμενον),[54] that is, to a position dependent on astronomical observation. Seneca, in what might be a development of this idea, says that the reason why there are believed to be 12 winds is that there are 12 "divisions of the sky" ("caeli discrimina"), which he describes in terms of the arctic and antarctic circles, the two tropics, the equator, the horizon, and the meridian.[55] Posidonius, as a student of astronomy as well as meteorology, is a possible author of this approach to the study of winds. It is not clear to me, from Seneca's account, how this scheme works; and, whoever devised it, it seems not to have influenced the way in which ancient authors set out their wind-roses.

Arist, Tim, Sen,
Mund: Thrascias

Arist: Aparctias or Boreas
Tim, *Mund*: Aparctias
(Sen: Latin name only)

Arist: Mesēs
Tim, *Mund*: Boreas
(Sen: Latin name only)

Pos, Arist, Tim, Sen,
Mund: Argestēs

L

A

B

Pos, Arist, Tim, Sen,
Mund: Caecias

K

C

Pos, Arist, Tim, Sen,
Mund: Zephyrus

J

D   Pos, Arist, Tim, Sen,
Mund: Apēliōtēs

Pos, Arist, Tim, Sen,
Mund: Lips

I

E

Arist: no wind.
Tim: Leuconotus orLibonotus
Sen: Leuconotus.
Mund: Libonotus or
Libophoenix

H

F

Arist: Eurus (also Euronotus)
Pos, Tim, Sen, *Mund*: Eurus

G

Arist: Notus
(also Leuconotus)
Tim, Sen, *Mund*: Notus

Arist: Phoenicias
Tim: Phoenix or Euronotus
Sen, *Mund*: Euronotus

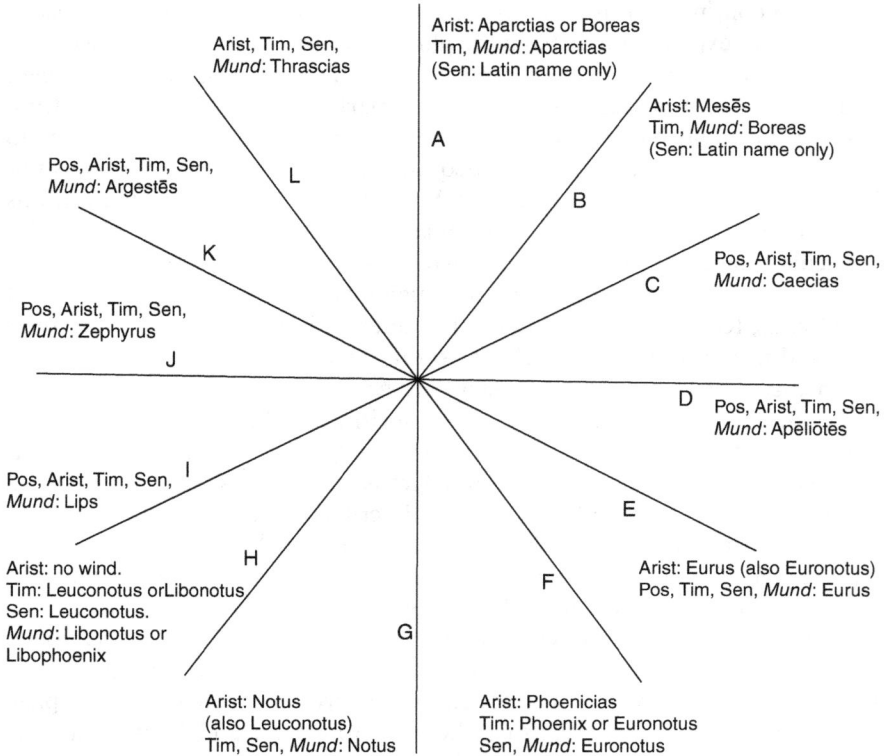

Figure 11.1 The wind-rose, in Posidonius and related authors.

A = due north; C = direction of midsummer sunrise; D = direction of equinoctial sunrise; E = direction of midwinter sunrise; G = due south; I = direction of midwinter sunset; J = direction of equinoctial sunset; K = direction of midsummer sunset. B, F, H and L are, respectively, the midpoints between A and C, E and G, G and I, and K and A. Pos = Posidonius (so far as his wind-rose is known). Arist = Aristotle, *Mete.* II.6, 363a21–364a4 (also II.5, 362a14–15). Tim = Timosthenes, as reported by Agathemerus; see Müller (1855–61) vol. 2, pp. 472–3. Sen = Seneca, *NQ* V.16.3–6. *Mund = De mundo* 394b19–35. Alternative wind-names are omitted, unless they are identical with or similar to a first-mentioned wind-name in one of the other four sources.

The *Iliad* and *Odyssey* name the four winds Boreas, Eurus, Notus and Zephyrus. Later pre-Aristotelian authors name in addition, at least, Caecias,[56] Apeliotes[57] and Lips.[58] There are very few mentions of named winds such as these in the fragments of the pre-Socratics,[59] and no evidence of a wind-rose (unless "Bion the astronomer" was indeed a pupil of Democritus; see note 43). However, the terminology of Aristotle's wind-rose clearly has been influenced by earlier usage: [Hippocrates] *Airs waters places* refers to the summer and winter risings and settings of the sun (and also to winds ἀπὸ τῶν ἄρκτων, "from the arctic") when describing wind-directions and how winds from different directions affect different locations.[60]

Ancient wind-roses could serve two purposes. They provided a way of describing from what direction the wind is blowing at a particular time and place: if the wind-rose is drawn as in Figure 11.1, then the observer is at the centre, and from the centre there radiate 12 lines, an equal distance from each other, each labelled with the name of a wind. Provided one direction is known, so as to align the wind-rose, then the user can readily name all the others, For this use, a wind-rose with a name attached to all 12 directions, like that of Timosthenes and presumably Posidonius, is clearly superior to Aristotle's, in which one direction has no name attached on the ground that no wind blows from it[61]; for obviously wind can blow, on occasion, from any direction.

The reason for this anomaly in Aristotle's wind-rose is that the ancient wind-roses were meant to be not just a way of naming the wind blowing at a particular time and place, but also a statement of what principal winds actually blow. They were thought to blow from the boundary of the inhabited world as known to the Greeks and Romans, the *oikoumenē*.[62] Thus Aristotle, after first implying that Notus blows from the south polar region, eventually decides that it blows from the Tropic of Cancer[63]; wind from the antarctic, he says, cannot reach us, since οὐδ' ὁ βορέας οὗτος εἰς τὴν ἐνταῦθα οἰκουμένην πᾶσαν [sc. διέχει], "our own north wind [Boreas] does not blow right across the region in which we live".[64] Clearly, Boreas originates in the arctic, and so is the chief wind in the northerly region where we live, but stops before it reaches the torrid zone.[65] Similarly, Timosthenes said that ἔθνη οἰκεῖν τὰ πέρατα, κατ' ἀπηλιώτην Βακτριανούς, κατ' εὖρον Ἰνδούς, "peoples inhabit the limits, by Apeliotes the Bactrians, by Eurus the Indians", and so on through the whole list of 12 winds.[66] These various peoples evidently are thought to live at the boundary of the *oikoumenē*, and it is implied that the winds named blow from, or from beyond, the places where they live. Seneca, too, after setting out his wind-rose, adds that there are local winds, about which he says "non est illis a latere universi mundi impetus", "their impulse is not from the edge of the whole world",[67] implying that the winds in the wind-rose do have their impulse from that edge.

Wind-roses like these are reasonable if it is believed that the *oikoumenē* is a fairly small area, roughly circular in shape, the centre of which is the home of the society using the wind-rose. Difficulties should have become apparent as the ancients' geographical knowledge grew wider, and as it came to be realised that the earth is spherical and that what has to be described is the system of winds in the northern temperate zone, whose shape is nothing like a circle, as Aristotle pointed out,[68] though he did not seriously attempt to solve the problems which this raised. For example, it was plausible enough to suppose that winds originated in the arctic or the torrid zone – parts of the world with a climate different from that of the Mediterranean region: but, as one travelled east or west along the temperate zone, the climate should remain the same until one got back to one's starting point: why then should east or west winds originate in any particular part of that zone?

The ancients were aware that there are local breezes which originate within the regions that they knew and only blow for a short distance. Theophrastus in the *De ventis* discusses several types of them, αὖραι, ἀπόγειαι, and τροπαί ("breezes" from lakes and rivers, "land-breezes", and "reverse breezes" from the sea), making clear his view that they are due to a locally formed exhalation or accumulation of material.[69] These, however, are minor phenomena. As geographical knowledge increased, one would have expected ancient thinkers to have become aware of strong and persistent winds which cannot be blowing from the point at the edge of the *oikoumenē* which marks the direction from which they blow: I shall come shortly to an instance in Posidonius' own experience. But the traditional wind-rose was not abandoned, as Seneca shows (and also the *De mundo*, if, as is likely, it postdates Posidonius). What is noticeable is that Seneca, unlike the others as we have them, immediately follows his wind-rose with a list of named winds which he says are local: *atabulus* in Apulia, *iapyx* in Calabria, *sciron* in Athens, *crageus* in Pamphylia and *circius* in Gaul, and he adds that there are many others, "nulla enim propemodum regio est quae non habeat aliquem flatum ex se nascentem et circa se cadentem", "for there is virtually no region which does not have some breeze born from itself and ceasing near itself".[70] *Iapyx*, *sciron* and *circius* are all alternative names of winds in at least one of the other wind-roses here discussed, so not there regarded as local.[71] Perhaps the inadequacy of the traditional wind-rose had begun to be recognised by Seneca, and possibly by others (including Posidonius?).

Features of particular winds

I quoted on p. 102 Strabo I.2.21 (Posidonius F137a EK), where he lists as from Posidonius' wind-rose the six winds Caecias, Lips, Eurus, Argestes, Apeliotes and Zephyrus. He then continues, in indirect speech so still presumably reporting Posidonius, with an explanation of Homeric phrases about winds: δυσαῆ Ζέφυρον, "ill-blowing Zephyrus" is "what we call Argestes"; λίγα πνέοντα Ζέφυρον, "clear-blowing(?) Zephyrus" (a pleasant wind, since in Homer it blows in Elysium)[72] is "what we call Zephyrus"; and ἀργέστην Νότον, is what we call Λευκόνοτον· οὗτος γὰρ ὀλίγα τὰ νέφη ποιεῖ, τοῦ λοιποῦ Νότου ὀλεροῦ πως ὄντος, "Leuconotus; for this wind makes few clouds, the rest of Notus being somehow turbid" – presumably meaning cloudy and misty.[73] He then quotes from Homer (*Iliad* XI.305–6):

ὡς ὁπότε Ζέφυρος νέφεα στυφελίξῃ
ἀργεστᾶο Νότοιο, βαθείῃ λαίλαπι τύπτων

as when Zephyrus drives away the clouds of *argestes* Notus, striking them with a deep storm

This, he explains, is the "ill-blowing" Zephyrus, which scatters the weak clouds gathered by Leuconotus, and says that ἀργέστης is applied to Notus "by way of an epithet" (ἐπιθέτως), evidently meaning "clearing" or "brightening" (compare ἀργής, "bright").[74] On this interpretation of Homer, "ill-blowing Zephyrus", that is, Argestes, seems not to be associated with cloud.

This tells us something of what Posidonius thought about the character of certain winds, and something also about his attitude to Homer. Homer, it is implied, is an author whom Posidonius would not dismiss, but who cannot be held to disprove the views of an authority such as Aristotle; instead, Homer must be interpreted so as to accord with the best authorities on the subject. (But we may suspect that Posidonius has modified the qualities of some winds to make them fit Homer, for example associating Leuconotus with a few weak clouds rather than none: see the next paragraph.)

Posidonius' characterisations of some winds in this passage gives an opportunity to compare what Theophrastus in *De ventis* (and sometimes Aristotle in *Mete.*) said of the same winds. On the Leuconoti, Theophrastus agrees with Aristotle that they are south winds corresponding to the northerly Etesians, and adds that they are αἴθριοι...καὶ ἀσυννεφεῖς ὡς ἐπίπαν. "winds of clear skies and generally cloudless",[75] which only roughly agrees with what Posidonius said of these winds as causing few clouds. Other south winds (νότοι), Posidonius says are turbid – associated, presumably, with mist and cloud. The nearest Theophrastus gets to this is to say that νότος is ἐπινεφὴς καὶ ὑέτιος "cloudy and rainy" for dwellers in the north,[76] which includes Greece.[77] On westerly winds Theophrastus, like Posidonius, quotes the Homeric epithet δυσαής, "ill-blowing", "stormy", but says it applies "in some places" (ἐνιαχοῦ) to ζέφυρος, while in other places ζέφυρος is μέτριος καὶ μαλακός, "moderate and gentle", presumably the same as Posidonius' λίγα πνέοντα Ζέφυρον.[78] On Argestes there is complete disagreement: Aristotle says, in different places, that it is associated with clear skies (αἴθριος), dryness (ξηρός), hail (χαλαζώδης), and tornadoes (ἐκνεφίαι)[79]; Theophrastus associates it with cloud (νεφέλη) and darkening of the sky (δασύνειν τὸν οὐρανόν)[80]; only Posidonius links it to Homer and says it is stormy (δυσαής). If one compares Posidonius with Theophrastus, one finds an almost complete divergence in vocabulary; there is some agreement in content, but this is weak evidence of the use of the earlier author by the later, since passages like these must be, to a large extent, a reporting of popular knowledge and beliefs. The comparison provides no evidence that Theophrastus *De ventis* was a significant source for Posidonius, and also shows that, however new his comments on Homer may have been, his observations on the characters of the winds were mostly not new.

Another presumably Hellenistic text on winds is Book XXVI of [Aristotle], *Problemata physica*. This book, too, has passages dealing with subjects on which we know something of Posidonius' views. The Leuconoti are nowhere mentioned, but XXVI.20 (941a34ff) asks why Notus ὅταν μὲν ἐλάττων ἦ αἴθριός ἐστιν, ὅταν δὲ μέγας, νεφώδης καὶ χρονιώτερος, "when it is less strong, [it brings] clear weather, but when it is strong, [it brings] clouds and lasts longer".[81] Similarly, XXVI.38 (944b25ff) asks why south winds μικρὰ μὲν πνέοντες οὐ ποιοῦσιν ἐπίνεψιν, μεγάλοι δὲ γενόμενοι ἐπινεφοῦσιν, "when they blow gently cause no overclouding, but when they become strong overcloud the sky". This has some resemblance to what seems to have been Posidonius' view, that most south winds bring cloud. XXVI.16 (942a5ff) says south winds are κυματοειδεῖς καὶ συνεστραμμένοι, "boisterous and whirling", which differs from the other two passages but would fit with a possible interpretation of ὀλεροῦ, in Strabo's report of Posidonius, as meaning "stormy". Like Posidonius, XXVI.31 (943b24) quotes the Homeric line about Zephyrus blowing in Elysium, but in a different form, omitting the word λίγα (or λιγύ) and saying just that breezes of Zephyrus διαπνείουσιν, "blow through", the Elysian plain, with no adjective or adverb to characterise them. As with Theophrastus, there is some resemblance in content between Posidonius and *Problemata* XXVI. All three drew on the same traditions, but the resemblance is not enough to suggest that Posidonius used *Problemata* XXVI.

Two other texts tell us what Posidonius said about winds in particular places. Strabo says[82]: Ἴδιον δέ τί φησι Ποσειδώνιος τηρῆσαι κατὰ τὸν ἀνάπλουν τὸν ἐκ τῆς Ἰβηρίας, ὅτι οἱ Εὖροι κατ' ἐκεῖνο τὸ πέλαγος ἕως τοῦ Σαρδῴου κόλπου πνέοιεν ἐτησίαι, "Posidonius says he observed a peculiarity in his voyage back from Iberia, that the east winds blow as etesians in that sea as far as the Sardinian Gulf", with the result that it took him three months to reach Italy. Posidonius spent a month at Cadiz at midsummer,[83] so this presumably refers to July to September. The etesian winds of Greece blow regularly from the north at this season, as we are told (for example) by Aristotle[84]; Posidonius had presumably enquired, and been told that the easterly winds which delayed him blow regularly at the same season in the western Mediterranean, and so were the equivalent of the etesian north winds of Greece. In modern times, among several winds which blow regularly enough in the western Mediterranean to have received a name, one is the Levante or Levanter, an "east to northeast wind that flows from the Alboran Channel [the western end of the Mediterranean] and is funnelled through the Straits of Gibraltar … most frequent from July to October and in March".[85] Presumably Posidonius encountered a Levanter.[86] In describing this east wind as etesian Posidonius probably thought that he was correcting Aristotle, who says that, for dwellers in the west, the etesian winds are north-westerly and westerly.[87] In fact, the pattern of winds in the western Mediterranean is complicated: besides the Levanter there is the Libeccio, a "westerly or south-westerly wind which predominates in northern Corsica all year round. … In summer it is most persistent".[88] Neither Aristotle nor Posidonius was entirely wrong.

Posidonius' experience should have affected his view of the wind-rose. He had encountered a persistent easterly wind near the western edge of the *oikoumenē*, and understood that it blew annually, at a time when a northerly wind is prevalent in Greece. But if this annual easterly wind was blowing from the eastern edge of the *oikoumenē*, it would have to blow through Greece to get to the west, and the prevalent summer wind in Greece would be easterly, not northerly.

Posidonius also had information about winds in India. Pliny (*Nat.* VI.57–8) says of him "eius venti adflatu iuvari Indiam salubremque fieri haud dubia ratione docuit", "he informed us convincingly that India benefited from the current of that wind to become a healthy country".[89] There is doubt about the text of the words that precede, but the wind mentioned is "Favonius", the west wind.[90] Presumably, Posidonius had some knowledge of the south-west monsoon and its importance for the supply of water and the growth of crops in India. Aristotle, too, may have heard of the monsoons: in support of his view that Boreas does not blow right across the *oikoumenē*, he says[91] περὶ τὴν ἔξω Λιβύης θάλατταν τὴν νοτίαν ... εὗροι καὶ ζέφυροι διαδεχόμενοι συνεχεῖς ἀεὶ πνέουσιν, "around the southern sea outside Africa ... continuous east and west winds always blow, succeeding each other". This might refer to the north-east and south-west monsoons, but Aristotle does not mention any effect on the climate of India. However, even if unknown to Aristotle, this effect was probably known to the Greeks before Posidonius' time. Strabo says (XV.1.13): ἐκ δὲ τῆς ἀναθυμιάσεως τῶν τοσούτων ποταμῶν καὶ ἐκ τῶν ἐτησίων, ὡς Ἐρατοσθένης φησί, βρέχεται τοῖς θερινοῖς ὄμβροις ἡ Ἰνδική, "from the evaporation from such great rivers [i.e., the rivers of India] and from the etesians, as Eratosthenes says, India is watered by summer rains". Assuming that "Etesians" means "annual winds", as in my last paragraph, but not "annual north winds", this must be another reference to the south-west monsoon.

I have here examined from their beginnings the development of Greek philosophical wind theories down to the time of Posidonius, and in several respects beyond him to Seneca. There are few innovations that our sources link to his name. He had experienced the "Levanter" – surely not the first Greek, but probably the first writer on meteorology, to do so. He differed from Aristotle in having a 12-wind wind-rose and better knowledge of the Indian monsoon; but these innovations were not original to him. What he said about winds mentioned by Homer may have been new, but is perhaps better regarded as literary criticism of Homer than as actual meteorological observation. More interesting and more likely to be new since Aristotle's time are some ideas found in Seneca: that wind is derived from exhalations from all or many kinds of terrestrial material, liquid and solid; that wind is the motion of expansion of dense material rarefied by the sun; that, alternatively, wind is due to air's self-motion; that there are astronomical reasons why a wind-rose should consist of 12 winds; that, even so, the wind-rose does not provide an adequate account of the winds that actually blow. I have tried to indicate which of these ideas seem to me likely to have been derived from Posidonius, and which might have led to a better understanding of wind. Even if they were not Posidonius' own ideas, it seems highly likely that he influenced the development of them.

Notes

1 See *Topics* 127a4f, 146b29, *Mete.* 349a16ff, 360a20–33.
2 In my opinion, Wilson (2013) 160 and 202 goes too far in suggesting that, for Aristotle, wind is not a flow of air at all.
3 I quote a few examples (selected from the list in Ideler [1832] 55n): Theophrastus (Aristotle's pupil), *De ventis* 29, μένειν... ὁ ἀὴρ οὐ δύναται... ἡ δὲ τούτου κίνησις ἄνεμος,

("the air cannot remain still ... and its movement is wind"; the Epicurean Lucretius, VI.685, "ventus enim fit, ubi est agitando percitus aer" ("for wind is produced when air is stirred by being set in motion"); the architect Vitruvius, I.6, "ventus est fluens aeris unda", ("wind is a flowing stream of air"); the encyclopedist Pliny, *Nat.* II.114, "ventus haut aliud intelligatur quam fluctus aeris", ("wind is understood to be nothing other than a stream of air").

4 Aëtius III.7.2 (*SVF* II.697). The context shows that this is a definition of "wind", not "breath".

5 In my opinion Lee (1952) ix and xiii–xxi demonstrates that *Mete.* IV is probably by Aristotle, though a separate treatise from *Mete.* I–III.

6 See above, Chapter 4 (p. 27–9).

7 *SVF* II.698.

8 *SVF* II.699.

9 Chapter 8, (p. 69–70).

10 Diogenes Laertius VII.153. See below, Chapter 14 (pp. 142, and 149 n. 8).

11 *Mete.* 360a11–13.

12 *Mete.* 344b19ff.

13 *Scholia in Aratum 1091* (Posidonius F131a EK). See below, Chapter 17 (pp. 172, and 179 n. 7).

14 See above, Chapter 4 (p. 21).

15 See above, Chapter 10 (p. 89).

16 I discuss this in my Ph.D. dissertation, Hall (1969).

17 *Mete.* 353b6–11 = DK 12A27; the same theory repeated, in rather different terms, at 355a21ff (DK 64A9); Alexander's comment: *In Mete.* p. 67.1ff Hayduck (DK 64A17, largely repeating 12A27).

18 *Letter to Pythocles* 106.

19 On the meaning of λιγὺ πνείοντος see below, p. 112 n. 72.

20 *Iliad* IX.5, XII.253.

21 *Iliad* XV.171, XIX.358; *Odyssey* V.296. (I do not claim that the Homeric passages cited in this paragraph are my own collection: see Mugler [1963] 55–9, 63f.)

22 *Mete.* 363a8–19.

23 *Mete.* 364a28–9. (However, at *Mete.* 344b36–345a1 Aristotle mentions a storm in which Boreas and Notus blew simultaneously in different places.)

24 *Mete.* 360a21–7.

25 *Mete.* 360a18ff.

26 *Mete.* 359b34–360a17.

27 Xenophanes: Aëtius III.4.4 (DK 21A46); Anaxagoras: Diogenes Laertius II.9 (DK 59A1) and Hippolytus, *Haer.* I.8.11 (DK 59A42); Metrodorus of Chios: Aëtius III.7.3 (DK 70A18). Aëtius also tells us more about the sun in Anaximander's wind theory: Aëtius III.7.1 (DK 12A24).

28 *Mete.* 361a22–34. Wilson (2013) 203–4 suggests that Aristotle means, not the diurnal revolution of the heavens, but the northward and southward motion of the sun as it moves along the ecliptic circle, though that, as Wilson says, involves other difficulties. This is not the place to discuss this problem.

29 See above, Chapter 3 (p. 16–17).

30 Aëtius III.17.4 (Posidonius F138 EK).

31 See below, Chapter 13 (p. 133–6).

32 Similar to this (but without mention of the sun) is pseudo-Aristotle, *Problems* XXVI.18, 942a22ff.

33 *De ventis* 17, with the translation of Coutant and Eichenlaub (1975).

34 *NQ* V.5–6.

35 *Scholia in Lucani Bellum civile, pars 1, Commenta Bernensia,* IX.578 (Posidonius F100 EK).

36 "Meteorology begins with the sun as the source of all atmospheric motion" (Sutton [1960] 30).

37 "In a fluid, motion is initiated … by differences of pressure, the flow, in the absence of other influences, being from regions of high to regions of low pressure" (Sutton [1960] 34).

38 Trans. Hine (2010).

39 By "wind-rose" I mean simply a diagram, or a verbal description, showing from what directions the named winds blow. (We have no diagrams from the authors I discuss, but they existed: Aristotle mentions his, *Mete*. 363a26.)

40 Posidonius F137a EK. For comments on this passage and discussion of Posidonius' wind-rose see Kidd (1988) 515–22: Theiler (1982) vol. II p. 16; Vimercati (2004) 576–8 (his conclusion seems to me unlikely: that Posidonius, regarding Aristotle and Timosthenes as authorities, yet had a wind-rose of only eight winds).

41 On Thrasyalces see also Chapter 15 (pp. 153 and 155–6).

42 See Müller (1855–61), vol. 2, p. 472–3. Timosthenes was active under Ptolemy II Philadelphus, c. 280–270 B.C. (Kidd [1988] 520).

43 Nothing is known for certain about this Bion beyond what Strabo says here. He has been identified with Bion of Abdera, follower of Democritus mentioned by Diogenes Laertius IV.58, but this seems quite uncertain. See Kidd (1988) 518.

44 See below, p. 106–7.

45 Seneca *NQ* V 16.6.

46 *Mete*. 363b27–364a1.

47 So Kidd (1988) 518–19.

48 *Mete*. 362a11–16.

49 *Mete*. 363a21–364a4.

50 *NQ* V.16.3–6. Seneca names Varro as his source, and gives priority to Latin wind-names, but he gives Greek names also, or only, for most of them.

51 *De mundo* 4, 394b19–35.

52 *Mete*. 363b29.

53 Seneca, weirdly, describes the position of some of these winds in terms of the positions on a couch at a Roman dinner: see *NQ* V.16.6 with the note of Oltramare (1961).

54 *Mete*. 363b32. On the "ever-visible" see above, Chapter 7 (pp. 55, 57).

55 *NQ* V.17.

56 Aristophanes, *Eq*. 437.

57 E.g., Thucydides 3.23.

58 Herodotus 2.25.

59 The fragments of Democritus' *parapēgma* mention several winds – Boreas, Notus, Lips, Zephyrus – as blowing on certain days (see DK 68B14; on the *parapēgma* see above, Chapter 4 [p. 22]), and he apparently had a theory about the effect of Notus on pregnant women (Aelian XII.17 [DK 68A152]). (There are also references to the northerly "etesian" winds attributed to Thales (see Chapter 4 [p. 21]), Democritus (Aëtius IV.1.4 = DK 68A99), and Metrodorus of Chios (Aëtius III.7.3 = DK 70A18), but "etesians" (ἐτησίαι) is never treated as a wind-name comparable to Boreas, etc.)

60 [Hippocrates] *Airs waters places* §3 lines 3–5 and 7, §4 lines 1–3, §5 lines 3–5 in Jones (1923). The origins and development of Aristotle's wind-rose are discussed by Wilson (2013) 207ff.

61 *Mete*. 363b33–364a4.

62 Aristotle holds that there is a similar system of winds in the southern hemisphere (*Mete*. 362b30–5). I do not think there is any evidence for Posidonius' view about this.

63 *Mete*. 361a5–22, 362a31–2, 363a8–18; discussed by Wilson (2013) 207–9.

64 *Mete*. 362b33–6, with the translation of Lee (1952) (and following Lee's text).

65 *Mete*. 363a3–8.

66 See Agathemerus II.7, in Müller (1855–61) vol. 2, p. 473.

67 *NQ* V.17.5.
68 *Mete.* 362a32–b30.
69 *De ventis* 23–6.
70 *NQ* V.17.5.
71 *Iapyx* is an alternative name for *argestēs* in *De mundo* 394b25–7, *scirōn* is an alternative for *argestēs* in Aristotle *Mete.* 363b25, *kirkios* or *kirkias* is an alternative for *thraskias* in Timosthenes and *De mundo* 394b30.
72 *Odyssey* IV.567. Translating this passage of Strabo, Jones (1917–32) vol. 1 p. 107 has "clear-blowing Zephyrus". Kidd (1999) p. 196 "loud-blowing Zephyrus". Λίγα and λιγύς are usually used of sound, meaning "loud", "clear-toned", "shrill" (see LSJ); but this seems inappropriate to wind.
73 Ὀλερός is a less usual form of θολερός, "muddy, foul, turbid" (see LSJ). With its use here compare Plato, *Timaeus* 58D, ἀήρ when θολερώτατος is ὀμίχλη τε καὶ σκότος, "mist and darkness".
74 On the relation of Leuconotus to other south winds, see also Galen *Opera* (in Kühn [1821–33] XVI pp. 409–10 and XVII [1] p. 655): some think Notus is always moist, but this is not so, for Leuconotus is dry.
75 Theophrastus, *De ventis*, 11.
76 Theophrastus, *De ventis*, 4.
77 Cf. Theophrastus, *De ventis*, 9, παρ' ἡμῖν ... καὶ ὅλως τοῖς ὑπὸ τὴν ἄρκτον οἰκοῦσιν, "to us ... and generally to those who live in the north".
78 Theophrastus, *De ventis*, 38, cf. 42, 43.
79 *Mete.* 364b7 and 29–30; 364b20; 364b23; 365a1–3.
80 Theophrastus, *De ventis*, 51, 61.
81 Translations in this paragraph are from Forster (1927).
82 III.2.5 (Posidonius T22 EK)
83 See Chapter 13 (p. 130).
84 *Mete.* 361b35–362a13.
85 Quoted from Weatheronline.co.uk/reports/wind/Levante (read 16 September 2021).
86 On this see also Kidd (1988) 18–19.
87 Θρασκίας καὶ ἀργέστας καὶ ζεφύρους. See *Mete.* 365a6–8.
88 See Wikipedia article Libeccio (read 16 September 2021).
89 Posidonius F212 EK, with translation from Kidd (1999) (omitting "west" before "wind").
90 On this passage see Kidd (1988) 756–9; Vimercati (2004) 607–8.
91 *Mete.* 363a5–8.

12   Earthquakes and volcanoes

Earthquake theories before Posidonius

Earthquakes, traditionally associated with Poseidon "the earthshaker", were, unsurprisingly, phenomena for which pre-Socratic philosophers sought naturalistic explanations. Like thunder and lightning, they were mysterious phenomena, though, as they involved places and materials more accessible than the clouds in the sky, it was easier to devise plausible explanations. The surviving evidence suggests that fewer pre-Socratics attempted to explain earthquakes than proposed explanations of thunder and lightning.

I have already quoted a line of Hesiod in which wind causes earthquakes (confirming Hine's suggestion[1] that this idea did not originate with philosophers), and Seneca's plausible but unconfirmable attribution to Thales of the idea that earthquakes are due to the earth being rocked by the water on which it floats.[2] Aristotle tells us about the earthquake theories of three other pre-Socratics.

According to Aristotle, Anaximenes said that

βρεχομένην τὴν γῆν καὶ ξηραινομένην ῥήγνυσθαι, καὶ ὑπὸ τούτων τῶν ἀπορρηγνυμένων κολωνῶν ἐμπιπτόντων σείεσθαι· διὸ καὶ γίγνεσθαι τοὺς σεισμοὺς ἔν τε τοῖς αὐχμοῖς καὶ πάλιν ἐν ταῖς ἐπομβρίαις

when the earth is drenched, or dried [sc. and cracks], it is broken up, and by these causes, as the hills are broken and collapse, it is shaken; for this reason, too, earthquakes occur in droughts, and again in times of heavy rain.[3]

Anaxagoras, Aristotle says, held that τὸν αἰθέρα πεφυκότα φέρεσθαι ἄνω, ἐμπίπτοντα δ' εἰς τὰ κάτω τῆς γῆς καὶ κοῖλα κινεῖν αὐτήν, "aithēr is naturally borne upwards, and when it strikes upon hollows in the lower parts of the earth, it moves it".[4] The idea evidently is that there is aithēr trapped in the earth, or beneath it (Anaxagoras' earth being flat), which shakes the earth as it tries to force its way to its natural place above. Democritus, in Aristotle's account, said that πλήρη τὴν γῆν ὕδατος οὖσαν, καὶ πολὺ δεχομένην ἕτερον

DOI: 10.4324/9780429399930-12

ὄμβριον ὕδωρ, ὑπὸ τούτου κινεῖσθαι, "the earth being full of water, and receiving much other water as rain, it is moved by this". There is an earthquake if there is too much water, the water "forcing its way out because the hollows [sc. in the earth] cannot contain it" (διὰ τὸ μὴ δύνασθαι δέχεσθαι τὰς κοιλίας ἀποβιαζόμενον), and also if the earth, "being dried up, draws [sc. water] into empty places from fuller ones, and [sc. the water] causes earthquakes by the impact of its passage" (ξηραινομένην ἕλκουσαν εἰς τοὺς κενοὺς τόπους ἐκ τῶν πληρεστέρων τὸ μεταβάλλον ἐμπῖπτον κινεῖν).[5] Democritus presumably accepted the theory attributed to Anaximenes, that earthquakes are associated with droughts and heavy rains.

Seneca, too, gives accounts of these three thinkers' theories,[6] consistent in principle with Aristotle's, but with additions and variations which I cannot here discuss, except to note that, according to Seneca, Anaximenes and Democritus both regarded wind (*spiritus*) as a cause of earthquakes, in addition to the causes attributed to these two thinkers by Aristotle.

Seneca ascribes to two pre-Socratics, Archelaus and Metrodorus of Chios, earthquake theories not mentioned by Aristotle. Archelaus, he says, thought that "venti in concava terrarum deferuntur; deinde, ubi iam omnia spatia plena sunt ... is qui supervenit spiritus priorem premit ... sic evenit ut terrae spiritu luctante et fugam quaerente moveantur", "winds are borne down into hollows of the earth; then, when all the spaces are full ... the wind which comes after presses on the former ... so it happens that the earth is moved, as wind struggles and seeks escape".[7] And so, Seneca adds, a calm in the air precedes earthquakes, because "the force of wind ... is held under the ground" ("vis spiritus ... in inferna sede retinetur"). This is the sort of elaboration which Seneca might have added without evidence to Archelaus' theory,[8] but it was the theory of Aristotle.[9]

Metrodorus' theory was, apparently, that

> quomodo cum in dolii cantant os, vox illa per totum cum quadam discussione percurrit ... sic speluncarum sub terrra pendentium vastitas habet aera suum, quem simul alius superne incidens percussit, agitat

> just as, when people sing in the mouth of a jar, the sound runs through the whole jar with a kind of vibration ... in the same way the immense space of caves beneath the earth contains its own air, which other air sets in motion when it falls from above and strikes it.[10]

Aristotle describes three pre-Socratic earthquake theories, none of which, in his account, involves wind or moving air. According to Seneca, wind or moving air was the only cause for Archelaus and Metrodorus, and one of a variety of causes for Anaximenes and Democritus. In view of the importance of wind in the earthquake theories of Aristotle and his successors, which I discuss below, I suspect that Seneca is at least partly right, and that Aristotle omits mention of wind in his account of earlier theories because, it being the theory he accepted, he had no reason to criticise it, or contrast it with his own theory.

Aristotle gives his own account of earthquakes in *Mete*. II.8. 365b21–366a5. He begins by saying that his theory of two exhalations is established as necessary, and that ἀνάγκη τούτων ὑπαρχόντων γίγνεσθαι τοὺς σεισμούς, "earthquakes are a necessary result of the existence of these exhalations".[11] He then asks ποῖον κινητικώτατον εἴη τῶν σωμάτων; "What sort of body is most able to cause motion?" and says that it must be τὸ ἐπὶ πλεῖστόν τε πεφυκὸς ἰέναι καὶ σφοδρότατον μάλιστα, "the substance whose natural motion is most prolonged and whose action is most violent", and that this is τὸ τάχιστα φερόμενον and τὸ λεπτότατον, "that which moves most quickly" and "the finest", and that this is πνεῦμα, "wind" (this being implicitly identified with dry exhalation). Therefore, αἴτιον … πνεῦμα τῆς κινήσεως, ὅταν εἴσω τύχῃ ῥυὲν τὸ ἔξω ἀναθυμιώμενον, "wind is the cause of the movement [sc. of the earth], when the external exhalation happens to flow inwards". [12] Aristotle here gives both a theory of earthquakes and an argument in support of it.

Seneca, after describing Archelaus' theory that wind causes earthquakes, goes on "in hac sententia licet ponas Aristotelem et discipulum eius Theophrastum", "as holding this opinion one may place Aristotle and his pupil Theophrastus", and describes a theory not precisely identical with Aristotle's but based on his theory of two exhalations.[13] In view of this, even though the Syriac Meteorology describes wind as only one of four causes, the others being based on the other elements, earth, water and fire,[14] I would conclude that wind and exhalation were at least the main causes of earthquakes to Theophrastus.

Strato, according to Seneca,[15] thought earthquakes due to a conflict between hot and cold matter, apparently without specifying what sort of matter it is – according to Aëtius, he did the same with thunder and lightning.[16] The idea seems to be that, when cold matter predominates within the earth, any hot matter gives way to it and shakes the earth in seeking to escape; and contrariwise, when hot matter predominates. I do not go into details, as this seems not to have influenced Posidonius.

Epicurus in *Letter to Pythocles* 105–6 describes several causes of earthquakes, beginning with different ways in which πνεῦμα, "wind", causes them: they may happen

καὶ κατὰ πνεύματος ἐν τῇ γῇ ἀπόληψιν καὶ παρὰ μικροὺς ὄγκους αὐτῆς παράθεσιν καὶ συνεχῆ κίνησιν, ὅταν κραδασμὸν τῇ γῇ παρασκευάζῃ. καὶ τὸ πνεῦμα τοῦτο ἢ ἔξωθεν ἐμπεριλαμβάνει <ἢ> ἐκ τοῦ πίπτειν ἐδάφη εἰς ἀντροειδεῖς τόπους τῆς γῆς ἐκπνευματοῦντα τὸν ἐπειλημένον ἀέρα

both by the imprisonment of wind in the earth and by the storing up and continuous motion [sc. of wind] beside small bodies of it [i.e., earth], when it causes shaking of the earth. And this wind it [i.e., the earth] takes within itself either from outside or by masses of soil falling into cavernous places in the earth and turning the trapped air into wind".

He then mentions a further cause (ἐκ τῶν πτώσεων ἐδαφῶν πολλῶν, "from the collapse of many pieces of ground") and says that there are others.

Seneca attributes to Epicurus a number of causes of earthquakes, including several which involve wind, and ends "nullam tamen illi placet causam motus esse maiorem quam spiritum", "however, he believes that no cause of the motion [sc. of the earth] is greater than wind".[17]

It is unclear what, if anything, the early Stoics had to say in explanation of earthquakes. If they kept to their principle of not proposing speculative causes, they would have said nothing. The one text which may give us an early Stoic view is Aëtius III.15.2:[18] οἱ δὲ Στωϊκοί φασι· σεισμός ἐστι τὸ ἐν τῇ γῇ ὑγρὸν εἰς ἀέρα διακρινόμενον καὶ ἐκπῖπτον, "the Stoics say: an earthquake is the moisture in the earth being dissolved into air and bursting out."

All these theories depend, to a greater or a lesser extent, on analogies with familiar events (Strato's is possibly an exception): Anaximenes was thinking of how water softens the ground and how the ground cracks as it dries, Anaxagoras of how smoke and flame rise up from burning fuel, Democritus of the effects of floodwater. Those who attributed earthquakes to wind were thinking of the effects of gales, and probably also, considering how often the enclosure of wind is mentioned, of air bursting or escaping from an inflated bladder. The analogy used by Metrodorus is explicit in Seneca's account. This is not to deny that the philosophers' general physical systems contributed to some theories: Anaxagoras may well have appealed to his principle "in everything a portion of everything" to explain how there is *aithēr* within or below the earth, and Aristotle starts his exposition from the exhalation theory on which he bases his meteorology; but when he tries to demonstrate the part that dry exhalation/wind plays in causing earthquakes and says it is the body most able to cause motion, whose action is most violent, and which moves most quickly, he is surely appealing to our physical experience of wind.

For all the thinkers so far discussed, except for Aristotle, we have only a basic theory, or theories, of how earthquakes occur, with sometimes a claim that weather associated with them accords with the theory. But Aristotle, after setting out his basic theory, goes on to a long discussion of earthquakes (*Mete.* 366a5–369a9). Much of it concerns the weather conditions under which earthquakes occur: they usually occur when there is no wind (because the exhalation is within the earth), so most often at night or at midday (because it is most windless then), but may also occur when there is a wind (because several winds may blow at once),[19] and so on (including the Anaximenean theory that they occur in times of drought and heavy rain – in droughts because there is then more dry exhalation than wet in the air, in heavy rains because the exhalation becomes strong by being compressed (πιληθῆναι) in a small space within the earth[20]). He also discusses other aspects of earthquakes, for instance, the sorts of place where they occur, with examples[21]; earthquakes which did not cease until the wind burst out of the earth, with examples of where this happened (these are, or include, volcanic eruptions, since he mentions one which sent up "embers and ashes", τὸν φέψαλον καὶ τὴν τέφραν)[22]; two different types of earthquake, with the shock running either horizontally or vertically[23]; and so on. Some of these I shall discuss in more detail in what follows, where Posidonius seems to be following Aristotle, or discussing the same phenomenon.

Most of the statements Aristotle makes in *Mete*. 366a5–369a9 are of things which the ancients could have observed, and some of them are certainly correct, as that major earthquakes may be followed by smaller aftershocks.[24] Though he says much about earthquakes and weather, he never claims that earthquakes can be *predicted* from any prevailing state of weather, as others have sometimes done. Modern scientific research has failed to discover any weather pattern that regularly precedes earthquakes[25]; and Aristotle's statements about earthquakes and weather are so broadly drawn that an earthquake under any weather conditions would be arguably compatible with his theory.

Posidonius on earthquakes and volcanoes

Posidonius' study of earthquakes[26] and kindred phenomena is praised by Strabo:[27]

τὸ δὲ ἐξαίρεσθαι τὴν γῆν ποτε καὶ ἱζήματα λαμβάνειν καὶ μεταβολὰς τὰς ἐκ τῶν σεισμῶν καὶ τῶν ἄλλων τῶν παραπλησίων, ὅσα διηριθμησάμεθα καὶ ἡμεῖς, ὀρθῶς κεῖται παρ' αὐτῷ

but his account is correct of the occasional rising and sinking the earth is subject to and of changes arising from earthquakes and the rest of such similar phenomena, all of which I too have enumerated.

He evidently adopted what had become the majority view, that earthquakes are due to wind, but defects in our evidence leave his exact theory obscure. The MSS. of Diogenes Laertius VII 154 have: εἰς τὰ κοιλώματα τῆς γῆς ἢ καθειρχθέντος πνεύματος ἐν τῇ γῇ, καθά φησι Ποσειδώνιος ἐν τῇ η, "into the hollows of the earth, or when wind is confined in the earth, as Posidonius says in his eighth book".[28] That this is about earthquakes is confirmed by a parallel passage (with no mention of Posidonius) in the *Suda*: Σεισμός· πνεύματος εἰς τὰ κοιλώματα τῆς γῆς ἐγκαθειρχθέντος, "Earthquake: wind being confined into the hollows of the earth".

Diogenes apparently mentioned two causes of earthquakes, of which he attributed both, or perhaps just the second (the confinement of wind in the earth), to Posidonius; there is clearly a lacuna before εἰς τὰ κοιλώματα, which editors have filled with, for instance, σεισμοὺς δὲ γίνεσθα εἰσδύοντος πνεύματος ("earthquakes occur when wind enters").[29] That Posidonius made πνεῦμα, wind, the main cause of earthquakes can hardly be doubted. Apart from Diogenes' evidence, it was, as we have seen, the view most widely held among his predecessors; Aristotle actually mentions the two causes apparently mentioned by Posidonius, the inward flow of exhalation/wind[30] and the confinement of it within the earth.[31] Also, Seneca argues for wind ("spiritus") as cause of earthquakes in a passage which twice cites Posidonius (though not for this point),[32] but there must be doubt about his exact theory: Kidd suggests that ἤ ("or") in the MSS. may be an error, and that the confinement of wind was perhaps the only cause of earthquakes mentioned in the text of Diogenes.[33] (This would be roughly consistent with the "Stoic" theory mentioned by Aëtius, where air formed within the earth and bursting out of it is the only cause of earthquakes.[34])

In view of the doubts about Posidonius' theory, and the number of earlier thinkers who made wind the cause, or a cause, of earthquakes, it would be rash to conclude that Aristotle was Posidonius' main or only source for his earth-quake theory. It is, however, worth noting that, in the chapter where Seneca declares his own belief that wind causes earthquakes, and goes on to cite Posidonius, he (Seneca) has an argument for his belief which recalls Aristotle's. He speaks of "spiritum ... quo nihil est in rerum natura potentius, nihil acrius, sine quo ne illa quidem quae vehementissima sunt valent", "wind ... than which nothing is more powerful, nothing fiercer; without it not even the most violent things have strength" (he instances winds arousing fires and causing waves in water).[35] This recalls Aristotle's argument that wind must cause earthquakes because it is the body "most able to cause motion ... and whose action is most violent",[36] though Seneca does not go on, as Aristotle does, to say that wind is the swiftest of bodies. When Aristotle and Seneca use nearly the same argument to support a view which Posidonius held, there must be a strong chance that Posidonius used the same argument.[37]

The chance is strengthened by the fact that Seneca immediately goes on to describe a theory about earthquakes which he attributes to Posidonius, and which Posidonius probably did derive from Aristotle. Aristotle says:

ὅταν μὲν οὖν ᾖ πολὺ τὸ πνεῦμα, κινεῖ τὴν γῆν, ὥσπερ δὲ ὁ τρόμος, ἐπὶ πλάτος· γίγνεται δ᾽ ὀλιγάκις καὶ κατά τινας τόπους, οἷον σφυγμός, ἄνω κάτωθεν.

So then, when the quantity of wind is large, it causes an earthquake shock which runs horizontally, like a shudder; occasionally in some places the shock runs up from below, like a throb."
(*Mete.* 368b23–5, trans. H.D.P. Lee [1952, Harvard University Press, Loeb ed.])

Seneca attributes a similar view to Posidonius:

Duo genera sunt, ut Posidonio placet, quibus movetur terra. Utrique nomen est proprium. Altera succussio est, cum terra quatitur et sursum ac deorsum movetur; altera inclinatio, qua in latera nutat alternis navigii more.

In the opinion of Posidonius, there are two ways in which the earth is moved. Each has its appropriate name. One is shaking-from-below, when the earth is shaken and moved up and down; the other is rocking, by which the earth sways from side to side alternately, like a ship.[38]

In both authors there are the same two sorts of earthquakes, shaking from side to side, and shaking up and down: surely Posidonius was using Aristotle.

A complication is that the passage from Diogenes Laertius VII 154 which I quoted above continues: εἶναι δ' αὐτῶν τοὺς μὲν σεισματίας, τοὺς δὲ χασματίας, τοὺς δὲ κλιματίας, τοὺς δὲ βρασματίας,[39] (in the translation of Mensch and Miller [2018], "they include jolts, fissures, tilts, and vertical shocks"). Diogenes does not actually *say* that this fourfold classification is that of Posidonius, rather than any other Stoic; but what other Stoic took such interest in earthquakes?[40] (Posidonius' pupil Asclepiodotus did,[41] but did Diogenes know his work?) EK print the sentence just quoted as part of Posidonius F12, as does Theiler (1982) in his F264 and Vimercati (2004) in his A149.

What the four types are is not immediately clear; Kidd's interpretation, which seems to me a good one, is as follows.[42] Four passages in ancient authors have comparable classifications of earthquakes: *De mundo* 396a1–16; Heraclitus Homericus, *Allegoriae* 38.6; Ammianus Marcellinus 17.7.13; and Lydus, *De ostentis* 53–4. From parallels in these works it is clear that, for the third type, the MS. reading καιματίας or καυματίας must be corrected to κλιματίας, and that these are shakings from side to side; that βρασματίαι are vertical shocks; and that χασματίαι are earthquakes which open up chasms or fissures in the earth. Κλιματίαι and βρασματίαι are, respectively, Seneca's "inclinatio" and "succussio". The nature of σεισματίαι is less clear, but Kidd suggests that probably, in contrast to χασματίαι, σεισματίαι are earthquakes which only shake the earth without opening fissures in it; and that "Diogenes may preserve a combination of two classifications"; that is, one may classify earthquakes as horizontal or vertical shocks, or one may classify them as earth-splitting shocks or merely earth-shaking shocks. However, no source states that these are two alternative classifications, and I suggest it might be better to say that the four classes need not be mutually exclusive, and that the same earthquake may be (say) both horizontal and merely earth-shaking.[43] I am reluctant to speculate how Seneca came to mention only the horizontal/vertical distinction as Posidonian. The distinction between horizontal and vertical shocks is Aristotelian, but a distinction between earthquakes which cause fissures and those which merely shake the earth looks like an innovation by Posidonius.

One might interpret χασματίαι differently. A χάσμα need not be a narrow opening. Herodotus IV.85 refers to the Aegean as χάσμα πελάγεος, "a *chasma* of sea". The word σεισμός is used of volcanic eruptions as well as of earthquakes. May not a χασματίας be, or include, an eruption which emerges through an opening in the earth in the sense of the crater of a volcano?

There are also records of Posidonius describing or discussing particular earthquakes and related phenomena (or supposed phenomena).[44] The submergence of Atlantis, recounted by Plato,[45] and the belief that the wanderings and banditry of the Cimbri were due to the submergence of their homeland are two of them: the first Posidonius thought possibly true; in the case of the Cimbri, if the MSS. of Strabo are correct, he suggested that the sea had gradually encroached on their land.[46] We have also an account of a tsunami which destroyed an army marching along the coast of Syria.[47] We have a detailed description of an occasion which occurred in Posidonius' own memory (κατὰ τὴν ἑαυτοῦ μνήμην) – but probably in his childhood – when a new island rose from the sea among the Lipari islands.[48] This was clearly a volcanic eruption: there is mention of heat, flame and smoke (θέρμη, φλόγας, καπνούς). Another passage describes a similar occasion in the Aegean[49] – mentions of smoke and fire ("fumus", "ignem") show that this too was volcanic. These three events are probably to be dated to, respectively, ca. 144/143 B.C.; 126 B.C.; and 197 B.C.[50] The following, also reportedly described or mentioned by Posidonius, are apparently undatable, or belong to no particular year: an earthquake which devastated Sidon and also affected the whole of Syria and the Cyclades and Euboea (a mention of πηλοῦ διαπύρου, "fiery mud", presumably lava, shows that what occurred in Euboea was a volcanic eruption)[51]; earthquakes which caused widespread devastation around Rhagae in Parthia[52]; and (apparently mentioned by Posidonius) the beneficial effects on Sicilian agriculture of volcanic ash from Mount Etna, and possibly other details of that volcano.[53] Posidonius was probably born around 135 B.C., so the datable earthquakes he is known to have described probably did not occur during his adult life: he must have depended on written sources or old memories, but evidently took trouble to obtain detailed accounts.

I have already mentioned analogies which must have been the basis of the theory that earthquakes are due to wind; Seneca mentions others (wind arousing fire and moving water); Posidonius was probably also influenced by the force of tradition. The basis of his classification of earthquakes cannot be determined when we are unsure what the classification was, though at least it would have been possible to observe whether or not an opening in the ground was involved. The details which Posidonius mentions of earthquakes and eruptions should generally have been observable in antiquity.

Assessment of Posidonius' contribution

That earthquakes are due, or most often due, to wind had become the majority view before Posidonius' time. The evidence suggests that he accepted this traditional majority view and tried to refine and improve it.

Aristotle had argued that wind is most powerful because it is swiftest.[54] This is not a strong argument: the ancients must have known in practice that a heavy weight moving relatively slowly (e.g., a battering ram) can exert more force than a light one moving fast (e.g., a flying arrow). Posidonius, or Seneca, or somebody tried to improve the argument, by stressing not the speed of wind, but its ability to set other things in motion.[55]

Aristotle devotes much space to arguing that his theory, that earthquakes are due to wind, is consistent with the weather conditions under which earthquakes usually occur, and (as a corollary) with the seasons of the year and times of day at which they usually occur, and with their occasional coincidence with lunar eclipses.[56] There is nothing of this in the fragments of Posidonius;

this might be due to the chance of what records survive, but it is suggestive that there is very little of it in Seneca *NQ* VI.[57] Posidonius may well have realised (it may not have been his original realisation) that coincidences of earthquakes with weather are too unreliable to be worth spending much time on.

Aristotle divided earthquakes into two types. Posidonius tried to improve on this, by describing four types, one of which (I have suggested) may have been, or have approximated to, the distinction of volcanoes from earthquakes.[58] There is no evidence that he proposed a special cause for volcanic eruptions, and, if he did make the distinction, other writers on this subject who wrote, or probably wrote, after him did not follow him. Seneca ignores the distinction, speaking of an emission of fire from a mountain as one of a number of unusual phenomena – such as rivers disappearing into the earth, or new ones appearing – which sometimes accompany earthquakes.[59] The author of the *De mundo* does distinguish between earthquakes and volcanoes, but not in the way that I suggest Posidonius may have done.[60] He says that the earth contains within itself πολλὰς ... πνεύματος καὶ πυρὸς πηγάς, "many sources of wind and fire", and that many of these have ἀναπνοὰς ... καὶ ἀναφυσήσεις, ὥσπερ Λιπάρα τε καὶ Αἴτνη καὶ τὰ ἐν Αἰόλου νήσοις· αἳ δὴ καὶ ῥέουσι πολλάκις ποταμοῦ δίκην, καὶ μύδρους ἀναρριπτοῦσι διαπύρους, "vents and blow-holes, like Lipara and Etna and the ones in the Aeolian islands. These often flow like rivers [presumably with lava] and throw up fiery, red-hot lumps". Having dealt with such fiery phenomena, the writer goes on: ὁμοίως δὲ καὶ τῶν πνευμάτων πολλὰ πολλαχοῦ γῆς στόμια ἀνέῳκται, "similarly, too, there are in many places on the earth's surface open vents for the winds", winds which may cause ecstasy, or wasting away, or prophecy, or death (ἐνθουσιᾶν, ἀτροφεῖν, χρησμῳδεῖν, ἀναιρεῖ), and (he continues) other winds within the earth cause earthquakes; in discussing the latter he describes the classification of earthquakes which resembles that attributed to Posidonius. So he does distinguish volcanoes from earthquakes, but not in the way that (I suggest) Posidonius may have done. The same is true of Pliny: in *Naturalis historia* II he deals with volcanoes in a section of the book which is about fiery phenomena[61] and is separate from his account of earthquakes.[62]

The list of four types of earthquakes attributed to Posidonius, whether or not it included a distinction of volcanoes from earthquakes, continued to influence writers on the subject up to the time of Lydus in the 6th century A.D., but it was not canonical: all the four similar classifications, in the *De mundo*, Heraclitus Homericus, Ammianus Marcellinus and Lydus, differ from it (and from each other) in the names or the number of the types, or both.[63] Seneca has only two of the four classes, and adds a third of his own, "tremor" (vibration).[64] Pliny describes several types of earthquake, but it is unclear how they relate to the other classifications.[65] Perhaps one might say that Posidonius' classification probably led later authors to think about the subject.

Aristotle in *Mete*. II.8 writes about various earthquakes and related phenomena that occur in particular places, or that occurred on a particular occasion, usually with an explanation of how the phenomena accord with his general theory of earthquakes.[66] It is evident that Posidonius took more trouble to record details of particular phenomena. Our evidence (but it may be misleading) suggests that he was less interested than Aristotle was in showing how the details he recorded support his general theories about earthquakes, though clearly some of them do. When the new island arose among the Lipari islands, the sea "remained with a sustained upward blast for some time" (συμμεῖναί τινα χρόνον ἀναφυσωμένην συνεχῶς),[67] which supports the theory that wind is the cause; the earthquake in Euboea ended when a fissure opened in the earth (χάσμα γῆς ἀνοιχθέν) and lava was emitted[68]: the earthquake was presumably a χασματίας.

There is no evidence that any pre-Socratic tried to describe or explain volcanoes, and Aristotle refers to them only briefly.[69] Mount Etna was known in mythology as a place of fire beneath which Zeus had imprisoned Typhon, and the work-place of Hephaestus;[70] and Euripides' Cyclops lives ὑπ' Αἴτνῃ, τῇ πυριστάκτῳ πέτρᾳ, "beneath Etna, the fire-streaming rock";[71] yet neither Aristotle nor any pre-Socratic is known to have referred to Mount Etna.[72] (Plato and Theophrastus presumably had that mountain in mind when they referred to lava [ῥύαξ] in Sicily.[73]) Posidonius, however, recorded details of several phenomena which clearly were volcanic or partly volcanic, and he seems to have given at least a partial account of Mount Etna. In this way, too, Posidonius, even if he himself gave no clear account of volcanoes as distinct from earthquakes, should have encouraged later writers to think about the subject; though the only surviving work which has probable signs of Posidonius' influence and a clear view of volcanoes is the *De mundo*. We do have one ancient work about a volcano: the pseudo-Virgilian poem *Aetna* (of unknown authorship, probably 1st century but before 79 A.D.[74]), which is of considerable interest to historians of science.[75] I can see no clear sign that the author used Posidonius: that wind causes the eruptions he could have taken from other authors, and he shows no interest in the one thing Posidonius is (probably) recorded as saying about Etna, that its ash benefits Sicilian agriculture.[76] But the poem shows an interest in volcanoes which Posidonius had probably helped to arouse.

The ancients, with their preference for regarding earthquakes as due to wind, were far indeed from the modern explanation of earthquakes in terms of slowly-moving tectonic plates and "faults" in rock formations. They were not so far astray when they regarded wind as the cause of volcanic eruptions. One type of eruption "occurs when magma [molten rock] heats ground water or surface water. The extreme temperature of the magma ... causes near instantaneous evaporation of water to steam, resulting in an explosion of steam, water, ash, rock [etc.]".[77] In the usual ancient view, water when it evaporates becomes air, and wind is air in motion; in these terms, it is not far from the truth to say that wind causes these eruptions (the "Stoic" theory of earthquakes recorded by Aëtius[78] seems especially close). No doubt the upward emission of steam and other matter during an eruption seemed to be evidence that wind is the cause.

The "Stony Plain"

This is a phenomenon which Aristotle, if not Posidonius, thought was the result of an earthquake. Posidonius is known to have travelled in the area concerned[79] and so is likely to have seen the phenomenon himself or heard the testimony of eyewitnesses; his view (as reported by Strabo) can be directly compared with Aristotle's. Aristotle in *Mete.* II.8 says that, when the type of earthquake occurs in which the shock runs up from below, ἐπιπολάζει πλῆθος λίθων, ὥσπερ τῶν ἐν τοῖς λίκνοις ἀναβραττομένων, "large quantities of stones come to the surface, like what is thrown up [i.e., chaff] in winnowing baskets", and he gives as one of three examples τὰ περὶ τὴν Λιγυστικὴν χώραν, "the situation in the district of Liguria".[80] This evidently refers to the plain of La Crau in south-east France. Strabo[81] describes this plain in some detail under the name of πεδίον Λιθῶδες, "Stony" plain, saying it is full of λίθων χειροπληθῶν, "fist-sized stones", and he quotes two explanations: Aristotle's, that the stones were brought to the surface by an earthquake (this must refer to the passage of *Mete.* just quoted), and that of Posidonius: λίμνην οὖσαν παγῆναι μετὰ κλυδασμοῦ, καὶ διὰ τοῦτο εἰς πλείονας μερισθῆναι λίθους, καθάπερ τοὺς ποταμίους κάχληκας καὶ τὰς ψήφους τὰς αἰγιαλίτιδας, ὁμοίως δὲ καὶ λείους καὶ ἰσομεγέθεις <πρὸς> τῇ ὁμοιότητι, "being a lake it was solidified as waves surged, and because of this it was divided into many stones, like river pebbles and stones on the seashore, and like them they are both smooth and of equal size in their similarity". This idea of a lake solidifying into stone would not have seemed strange to a Stoic (the earth is formed from moisture in Stoic cosmogony[82]), and Posidonius was at least right to compare the stones to those made smooth by a river or the sea: the modern theory is that the stones were carried to the plain of La Crau by the river Durance (which has since changed its course).[83] So Posidonius, who may well have seen the plain, shows more knowledge of it than Aristotle; especially if, as is likely,[84] the description of the plain, for which Strabo names no source, is from Posidonius.[85]

Notes

1 Hine (2002) 72–5.
2 Chapter 4 (p. 20–1).
3 *Mete.* 365b6–8 (DK 13A21).
4 *Mete.* 365a20–2 (DK 59A89).
5 *Mete.* 365b1–6 (DK 68A97).
6 For Anaximenes *NQ* VI.10.1–2 (printed in DK vol. 1 p. 488); for Anaxagoras, *NQ* VI.9.1 (DK 59A89); for Democritus *NQ* VI.20.1–4 (DK 68A98). There are also brief accounts, not worth discussing here, in Aëtius III.15.3 (for Anaximenes), III.15.4 (for Anaxagoras; DK 59A89) and III.15 1 and 7 (for Democritus; for III.15.7 see DK 28A44), and in Hippolytus, *Haer.* I.7.8 (for Anaximenes; DK 13A7) and I.8.12 (for Anaxagoras; DK 59A42). Ammianus XVII.7.12 (DK 12A28) attributes what seems Anaximenes' theory to Anaximander.
7 *NQ* VI.12.1–2 (DK 60A16a).
8 See above, Chapter 3 (p. 11).
9 *Mete.* 366a6–8.

10 *NQ* VI.19 (DK 70A21). The text of "in dolii – illa" is corrupt in the MSS. I give the text as printed by DK. There are other emendations, but they do not affect the general sense. Metrodorus is also included in Aëtius' account of earthquakes, but the process by which the earth is shaken is there lost in a lacuna (see Aëtius III.15.6, also part of DK 70A21).

11 *Mete.* 365b23, with translation from Lee (1952).

12 See *Mete.* 365b30–366a5. The translation of ἐπὶ πλεῖστον – μάλιστα is from Lee (1952).

13 *NQ* VI.13.1.

14 Syriac Meteorology section 15, in Daiber (1992) 270–1.

15 Seneca *NQ* VI.13.2–5.

16 See above, Chapter 8 (p. 68).

17 *NQ* VI.20.5–7.

18 *SVF* II.707.

19 For the supposed phenomena mentioned in this sentence, see *Mete.* 366a5–19.

20 *Mete.* 366b2–15.

21 *Mete.* 366a24–b1.

22 *Mete.* 366b30–367a11.

23 *Mete.* 368b23–32.

24 *Mete.* 367b33–368a13.

25 See Wikipedia article "Earthquake weather" (read 24 September 2021).

26 For discussions of Posidonius' earthquake theory see Gilbert (1907) 315–20; Grewe (2008) 150ff. Both are readier than seems to me justifiable to regard as evidence for Posidonius sources which do not name him, such as [Aristotle], *De mundo*, 395b18ff and (in Grewe's discussion, p. 159ff, of Posidonius' view of volcanoes) the pseudo-Virgilian *Aetna* (on which see p. 122 below). The same seems to me to be true of Theiler (1982), who takes *De mundo* 395b18ff as evidence for Posidonius (see his F341a and F342, and vol. II pp. 236–8). (On my view of the relation of *De mundo* to Posidonius see above, Chapter 3, p. 12–14).

27 Strabo II.3.6 = Posidonius F49.294–6 EK, with translation from Kidd (1999) 121.

28 Posidonius F12 EK.

29 So EK at F12 EK, Theiler (1982) at his F264, and Dorandi (2013). Hicks (1925), Long (1964) and Vimercati (2004) at his A149 have ῥυέντος, "flowing", for εἰσδύοντος. (I owe the *Suda* reference to Long's apparatus criticus.)

30 *Mete.* 366a4–5 (quoted above, p. 115).

31 *Mete.* 366b10: τὴν ἐντὸς ἀναθυμίασιν ... τῷ ἐναπολαμβάνεσθαι ἐν στενοτέροις τόποις, "the internal exhalation ... by being confined in narrow spaces".

32 Seneca, *NQ* VI 21–5. Posidonius is cited at VI.21.2 and VI.24.6.

33 See Kidd (1988) 116–18. He there suggests that the correct reading might be "something like" <σεισμοὺς δὲ γίνεσθαι πνεύματος> εἰς τὰ κοιλώματα τῆς γῆς ἐγκαθειρχθέντος, "earthquakes occur when wind has been confined in the hollows of the earth".

34 See above, p. 116 of this chapter.

35 *NQ* VI.21.1.

36 See above, p. 115.

37 I discuss this further in Chapter 21, pp. 203–5.

38 *NQ* VI.21.2 (Posidonius F230 EK).

39 In the text as printed by Hicks (1925), by Long (1964), by EK F12, by Vimercati (2004) A149, and by Dorandi (2013).

40 The suggestion of Vimercati (2004) 602 that the fourfold classification was a Stoic tradition seems to me unlikely.

41 See Seneca *NQ* VI.17.3 and VI.22.2

42 Kidd (1988) 817–19.

43 Other interpretations: Theiler (1982) in his F264 alters σεισματίας to ἰζηματίας, from *De mundo* 396a4, where ἰζηματίαι are earthquakes in which the earth sinks, as contrasted with those in which it is split open, the χασματίαι of Diogenes Laertius VII 154. The *De mundo*, with its different vocabulary and additional earthquake-types (see above, Chapter 3 [p. 12] seems to me a doubtful basis for amending Diogenes. Grewe (2008) 151 would presumably (the book prints no text) adopt from some MSS. the reading καυματίας for the third type of earthquake, and interprets this as "Erdbrände", i.e., volcanic eruptions. This conflicts with the parallels in other authors and leaves nothing to balance βρασματίας. Boechat (2016) 446–50, discussing the "earthquake" supposed to have formed the Stony Plain (see below, p. 123), thinks that in Aristotle's view it involved an eruption of lava, i.e., was volcanic; and Strabo IV.1.7 (Posidonius F229 EK) calls it a βράστης. If this is right, βράστης and βρασματίας are terms for a volcanic eruption. But this seems highly unlikely: there is no suggestion in Aristotle, or in Seneca's account of Posidonius, that the vertically moving earthquake involves fire or smoke or ash, and the analogy of the winnowing of chaff (*Mete.* 368b29) is inappropriate to liquid lava.

44 On ancient records of such phenomena by various authors, including Posidonius, see Hine (2002) 63–68.

45 *Timaeus* 25D.

46 On Atlantis and on the Cimbri see Strabo II.3.6 (= Posidonius F49.297–305 EK). According to the MSS. of this passage, Posidonius εἰκάζει...τὴν τῶν Κίμβρων ... ἐξανάστασιν ἐκ τῆς οἰκείας γενέσθαι κατὰ θαλάττης ἔφοδον οὐκ ἀθρόαν συμβᾶσαν, "conjectures ... that the migration of the Cimbri ... from their native land arose from an encroachment of the sea that occurred not all at once" (tr. Kidd [1988] 259). However, at VII.1.1–2 (= F272 EK) Strabo argues, evidently following Posidonius (see Kidd [1988] 929), that the Cimbri cannot have been driven from their homes by "a great tide" (μεγάλη πλημμυρίδι), because tidal phenomena are too regular to cause people familiar with them to leave their homes; Posidonius, Strabo goes on, thought the migrations of the Cimbri were due to their being ληστρικοὶ ... καὶ πλάνητες, "piratical and nomadic". Many scholars (e.g., Kidd [1988] 260–1; Vimercati [2004] 615–16) have seen this as inconsistent with Strabo II.3.6, and have therefore emended the text of II.3.6 into a denial that inundation caused the Cimbri to leave their homes. But would Strabo in that passage, where he is talking about Posidonius' knowledge of earthquakes and related phenomena, have included, without explaining why, an occasion on which Posidonius denied that any earthquake or inundation had occurred? Also, saying that the emigration was due to gradual (but permanent) encroachment by the sea is not inconsistent with denying that it was due to one high tide; and, while inundation might have driven the Cimbri from their homes, it could not, on its own, have caused their wanderings all the way to Italy that are detailed in Strabo VII.2.2. Theiler (1982) I p. 34, in his F13, keeps the MS. reading of Strabo II.3.6. (For a different account of Posidonius' view see Compatangelo-Soussignan [2016]: she concludes by suggesting that he had information about an exceptional storm surge causing the Cimbri to emigrate, but that "true to the teachings of the Aristotelian school, he ... sought, on second thoughts, to deny the catastrophic impact of the event". But an inundation would not be inconsistent with *Mete.* 352a28ff, on "great winters" and Deucalion's flood confined to Greece; and anyway, Posidonius was not an Aristotelian.)

47 Athenaeus VIII 333B–D (Posidonius F226 EK). For comment see Kidd (1988) 807–8.

48 Strabo VI.2.11 (= Posidonius F227 EK). For comments see Kidd (1988) 809–10; Theiler (1982) II p. 55; Vimercati (2004) 599.

49 Seneca *NQ* II.26.4–5 (Posidonius F228 EK). For comments see Kidd (1988) 810–12; Theiler (1982) II p. 207; Vimercati (2004) 599–600.

50 Kidd (1988) 808–9 and 811–12.
51 Strabo I.3.16 (= Posidonius F231 EK) and Seneca *NQ* VI.24.6 (F232 EK). For comments see Kidd (1988) 820–3; Theiler (1982) II p. 21. As Kidd (1988) 821 says, Strabo's wording makes it probable but not certain that the whole of EK's F231 is from Posidonius.
52 Strabo XI.9.1 (= Posidonius F233 EK). For comment see Kidd (1988) 823–4.
53 Strabo VI.2.3 (= Posidonius F234 EK). For comment see Kidd (1988) 824–6. The mention of Posidonius here is a widely accepted emendation. Mount Etna erupted violently in 122 B.C. (Orosius V.13.3; mentioned by Hülsen [1894]), which must have drawn the attention of Posidonius and his contemporaries to it.
54 See above, p. 115.
55 See above, p. 118.
56 *Mete.* 366a5–20, b2–15, 367a20–b32.
57 At *NQ* VI.1.1 Seneca says that the recent earthquake in Campania has happened in the days of winter, "quos vacare a tali periculo maiores nostri solebant promittere", "which our elders used to promise are free from such a danger"; at VI.12.2 he says, apparently as part of the theory of Archelaus, that calms in the air precede earthquakes, and adds "nunc quoque, cum hic motus in Campania fuit, quamvis hiberno tempore et inquieto, per superiores dies caelo aer stetit", "just now, too, when this earthquake occurred in Campania, although it was in a stormy winter season, in the preceding days the air in the sky was still". So Seneca does not reject this belief, though he immediately adds (VI.12.3) that wind and earthquake do sometimes coincide, if two winds blow at once, one above and one below the ground.
58 See above, p. 120. On the distinction, as made in antiquity, of volcanic eruptions from volcanoes, see Hine (2002) 58–60.
59 *NQ* VI.4.1.
60 *De mundo* 4, 395b17–396a1. Translations are from Furley in Forster and Furley (1955), slightly modified.
61 *Nat.* II.236–8 deals with volcanoes, as part of II.235–241 about fiery phenomena.
62 *Nat.* II.191–200.
63 See Kidd (1988) 817–19.
64 *NQ* VI.21.2 (partly quoted above, p. 118).
65 *Nat.* II.198.
66 *Mete.* 366a26–9, 367a1–9, 368b6–13 and 30–2.
67 See F227 EK. (I quote the translation from Kidd [1999].)
68 See F231 EK.
69 *Mete.* 366b30–367a11.
70 Aeschylus, *Prometheus vinctus* 365–74; cf. Pindar, *Pythians* I.20–7.
71 Euripides, *Cyclops* 298.
72 I have checked this from *Thesaurus linguae Graecae*. For the pre-Socratics, Mount Etna is mentioned only in connection with Empedocles, mostly in versions of the story of his death in its crater (Diogenes Laertius VIII.69 = DK 31A1; Strabo VI.2.8 and Horace, *Ars poetica* 465–7 – both = DK 31A16). Lucretius I.714ff (DK 31A21) speaks of Empedocles and Mount Etna in the same passage, without mentioning Empedocles' death, but also without suggesting that what is said of Mount Etna is taken from Empedocles.
73 Plato, *Phaedo* 111E; Theophrastus, *De lapidibus* 22.
74 The 79 A.D. eruption of Vesuvius is not mentioned.
75 See Taub (2008) 31–55.
76 See above, p. 120.
77 See Wikipedia article "Phreatic eruption" (read 27 Sept. 2021).
78 See above, p. 116.

79 Posidonius travelling in Liguria: Strabo III.4.17 (= Posidonius T23 EK); among the Celts in Transalpine Gaul: Strabo IV.4.5 (= Posidonius T19 and F274 EK). On Posidonius' travels in this area see Kidd (1988) 17–18 and 20.
80 *Mete.* 368b28–32.
81 Strabo IV.1.7 (= Posidonius F229 EK).
82 See Diogenes Laertius VII 142, quoted above, Chapter 10 (p. 92–3).
83 Kidd (1988) 815.
84 Kidd (1988) 813 says it "no doubt comes from Posidonius", giving no reason; it seems to be just a likely presumption from Posidonius' visit to the area and known interests.
85 In interpreting F229 EK I agree with Kidd that Strabo's language, especially ἐξ ὑγροῦ παγέντας μεταβαλεῖν, "undergo change from moisture, being solidified", must mean that, in Posidonius' view, water changes to stone. Not all scholars agree. Steinmetz (1962) suggests that Posidonius accepted Aristotle's account of the origin of the stones; Boechat (2016) 446–50 argues that Aristotle's and Posidonius' explanations are effectively the same. Neither view, it seems to me, is consistent with Strabo's account. More plausibly than these two, Vimercati (2004) 600–1 suggests that Strabo's account of Posidonius' view deals only with the shaping of the stones, ignoring anything Posidonius may have said about their origin. In my opinion, Kidd's view fits the evidence best. (For comments on this passage from a different point of view see Theiler [1982] II p. 44–45.)

13 The sea and its tides

The sea; its salt

The sea is the subject of Chapters 1–3 of the second book of Aristotle's *Meteorologica*, in which Aristotle discusses whether the sea has sources (πηγαί); whether it has come into being and will one day dry up, or whether (as Aristotle thinks) it lasts forever, with the water flowing into the sea being exactly balanced, over time, by evaporation; and especially, what the origin is of the sea's salt, which Aristotle attributes to his dry exhalation.[1] We have little evidence for Posidonius' views on these subjects: he cannot have thought that the sea would last forever, since in his view the cosmos will not; and we have what may be an account of his view about the salt of the sea.

Priscianus Lydus, in *Solutiones ad Chosroem* VI, gives an account of the sea's tides which is evidently taken from Posidonius (see below), and follows it with a discussion, which seems heavily indebted to Aristotle's *Meteorologica*, of why the influx from rivers does not cause the sea to rise, and of the origin of the sea's salt.[2] He then continues:

> natura autem aquarum est talis, et non solum in magno mari sed etiam in aliis pluribus locis. Declarat hoc et lacum talem narrando ...

> Well, the nature of water is like that, and not only in the great sea but in very many other places too. He maintains that there are tales of a lake like that ...[3]

Previously in this passage the unnamed subject of third-person singular verbs like "declarat" ("he maintains") is evidently Posidonius,[4] so one would expect that Priscianus here means "Posidonius maintains"; however, the account of an ultra-salty lake which follows exactly matches one in Aristotle.[5] If the account of salt lakes and springs in this passage is from Posidonius,[6] then we have his explanation of the salt, presumably applicable also to the sea:

> Horum igitur omnium causam dixerit quis connaturalem et unitam seu ingenitam naturam ignis: ardens enim terra eo magis et minus varias recipit formas fusionum.

DOI: 10.4324/9780429399930-13

So the cause of all this has been said to be a fiery element inherent or produced in them, because burnt earth produces different forms of effluence depending on the degree of combustion.[7]

If this was Posidonius' view, then it was related to Aristotle's, since Aristotle's dry exhalation has the qualities of elemental fire. It is also related to a pre-Aristotelian theory, that the sea's salt is due to an admixture of salty earth.[8]

The depth of the sea

Strabo I.3.4ff discusses the possibility of the sea, or part of it, becoming silted up and becoming dry land – he is much concerned with the Black Sea, called by him the Euxine (Εὔξεινος) or Pontus (Πόντος), which was believed by some to be the shallowest (βραχύτατα).[9] At I.3.9 Strabo concludes that it is possible for the whole sea (τὸ πέλαγος πᾶν) to become silted up, τοῦτο δ' ἂν συμβαίη κἂν τοῦ Σαρδονίου πελάγους βαθύτερον ὑποθώμεθα τὸν Πόντον, ὅπερ λέγεται τῶν ἀναμετρηθέντων βαθύτατόν που χιλίων ὀργυιῶν ὡς Ποσειδώνιός φησι, "this would happen even if we posit the Black Sea to be deeper than the Sea of Sardinia, which is said to be the deepest of those that have been measured, about a thousand fathoms, as Posidonius says."[10] This figure is consistent with other figures in ancient authors for the deepest sea depths,[11] though actually much too small even for the Mediterranean.[12] We do not know how this figure was arrived at, but Strabo's words imply that attempts at measurement were made.[13] Perhaps Posidonius' statement about maximum depth was made in a discussion of the possible silting up of the sea, but, if so, we do not know what view he took.[14] He did provide a measurement: an example of his occasional use of numbers in discussing meteorological subjects.

Tides

Posidonius' main contribution to the study of the sea concerned a subject of which Aristotle had no real knowledge: tides.[15] Soon after Aristotle's time, knowledge of oceanic tides began to spread among the Greeks. According to Aëtius III.17.3, Pytheas connected tides with the moon (unsurprisingly, as he had sailed in the Atlantic); his work (though possibly not what he said about tides) was known in old Greece soon after, if not before, 300 B.C., since it was known to Dicaearchus.[16] Knowledge of tides might also have reached Greece via news about the difficulties caused to Alexander the Great's ships by tides in the Indian Ocean.[17] That flood and ebb tides are related to the daily movements of the moon was known, in the 3rd century B.C., to Eratosthenes.[18] In the following century Seleucus (according to Aëtius III.17.9) believed that the earth rotates, and

ἀντικόπτειν αὐτῆς τῷ δίνῳ φησὶ τὴν περιστροφὴν τῆς σελήνης· τοῦ δὲ μεταξὺ ἀμφοτέρων τῶν σωμάτων ἀντιπερισπωμένου πνεύματος καὶ ἐμπίπτοντος εἰς τὸ Ἀτλαντικὸν πέλαγος κατὰ λόγον οὕτω συγκυμαίνεσθαι τὴν θάλασσαν[19]

he says that the revolution of the moon [sc. around the earth] resists its [the earth's] rotation; wind between the two bodies is drawn round in a contrary motion and falls on the Atlantic Ocean, and the sea is thus swollen in a wave in proportion [to the wind's strength, presumably].

The meaning of this is not clear (and it is unlikely that Seleucus, an easterner, wrote about the Atlantic); but evidently, in some way, the motions of earth and moon generate wind which causes the tides. I say more of Seleucus' theory in what follows.

Posidonius spent 30 days at midsummer at Gadeira (Cadiz) and himself observed the tides there[20]; he and a later writer, Athenodorus, are cited by Strabo as the authorities on tides.[21] Posidonius, says Strabo,[22]

φησὶ ... τὴν τοῦ ὠκεανοῦ κίνησιν ὑπέχειν ἀστροειδῆ περίοδον, τὴν μὲν ἡμερήσιον ἀποδιδοῦσαν, τὴν δὲ μηνιαίαν, τὴν δ᾽ ἐνιαυσιαίαν συμπαθῶς τῇ σελήνῃ

says that the movement of ocean undergoes a cycle of a type like a heavenly body, exhibiting diurnal, monthly and annual movement in joint affinity with the moon.

To summarise a lengthy passage of Strabo: each day when the moon is rising and is well above the horizon, and when it has set well below the horizon, the sea rises; when the moon is declining towards its setting, and when it is below the horizon but is approaching the point where it will rise, the sea falls; each month the flows are greatest at full moon and new moon, less at half moon. This Posidonius had presumably observed in his stay at Cadiz. Thirty days was too short a time for him to observe annual variation, but "he says he had ascertained from people in Cadiz" (παρὰ τῶν ἐν Γαδείροις πυθέσθαι φησί) that the tides were greater at the summer solstice, and he supposed (εἰκάζει) that the tides were greater at the winter solstice, less at the equinoxes.[23] (Here we should, with Kidd,[24] applaud the evident care with which Posidonius distinguished what he had seen from what he had been told and what was merely supposition.)

That tides are greater at midsummer and midwinter, less at the equinoxes, is false. Later ancient writers, Seneca, Pliny and Priscianus Lydus, speak of equinoctial tides as the greatest,[25] and, as Posidonius is known to have influenced them, some scholars have supposed that Strabo has misrepresented Posidonius on this point,[26] or that Posidonius himself adopted this view in a later work than the one which Strabo was using.[27] Strictly speaking, Seneca, Pliny and Priscianus were correct. A modern scientist comments on Posidonius' view:

spring-tides reach their *highest* levels around the equinoxes and their *lowest* levels at the solstices, as may be verified by a study of a modern tide table for Cadiz ... [but] one has to take rather careful averages to detect any systematic difference at all.[28]

One wonders how much evidence Seneca and the others had.

Strabo also tells us about Posidonius' comments on two earlier tidal theories. At III.3.3[29] he says:

τὸν Ἀριστοτέλη φησὶν ὁ Ποσειδώνιος οὐκ ὀρθῶς αἰτιᾶσθαι τὴν παραλίαν καὶ τὴν Μαυρουσίαν τῶν πλημμυρίδων καὶ τῶν ἀμπώτεων· παλιρροεῖν γὰρ φάναι τὴν θάλατταν διὰ τὸ τὰς ἀκτὰς ὑψηλάς τε καὶ τραχείας εἶναι, δεχομένας τε τὸ κῦμα σκληρῶς καὶ ἀνταποδιδούσας. τἀναντία γὰρ θινώδεις εἶναι καὶ ταπεινὰς τὰς πλείστας

Posidonius says that Aristotle was wrong to say that the seaboard [sc. of Iberia] and Morocco were the cause of flood-tides and ebb-tides. For he [i.e., Aristotle] says that the sea flows back because the headlands are high and rugged, catching the waves roughly and throwing them back. For, on the contrary, [sc. says Posidonius] most of the shore there is sandy and low-lying

The statement attributed to Aristotle occurs in no surviving work of his; he does say, at *Mete.* 354a5–9,

Ῥέουσα δ' ἡ θάλαττα φαίνεται κατά τε τὰς στενότητας, εἴ που διὰ τὴν περιέχουσαν γῆν εἰς μικρὸν ἐκ μεγάλου συνάγεται πελάγους, διὰ τὸ ταλαντεύεσθαι δεῦρο κἀκεῖσε πολλάκις. τοῦτο δ' ἐν μὲν πολλῷ πλήθει θαλάττης ἄδηλον

the sea obviously flows in narrow places, where from a wide ocean it is confined in a small space, because it oscillates frequently hither and thither. This is not noticeable in a wide expanse of sea.

This implies that straits make the sea's currents obvious, but not that it causes them; and it must refer, primarily at least, to straits within the Mediterranean; but perhaps Aristotle, or some follower of his, thought, and said in a work now lost, that a similar process explained, or made obvious, tides at the Straits of Gibraltar; or else Posidonius inferred from the *Mete.* passage that this is what he would have said[30]; or, Strabo's report is wrong.[31]

According to Strabo III.5.9,[32] Posidonius

Φησὶ ... Σέλευκον τὸν ἀπὸ τῆς Ἐρυθρᾶς θαλέττης καὶ ἀνωμαλίαν τινὰ ἐν τούτοις καὶ ὁμαλότητα λέγειν κατὰ τὰς τῶν ζῳδίων διαφοράς· ἐν μὲν γὰρ τοῖς ἰσημερινοῖς ζῳδίοις τῆς σελήνης οὔσης ὁμαλίζειν τὰ πάθη, ἐν δὲ τοῖς τροπικοῖς ἀνωμαλίαν εἶναι, καὶ πλήθει καὶ τάχει

says that Seleucus, the man from the Red Sea [our Persian Gulf], says that there is a certain unevenness in these phenomena [i.e., the tides] and evenness according to different signs of the zodiac; for when the moon is in the equinoctial signs [i.e., is over the equator], the phenomena are even, but when it is in tropical signs [i.e., is over or near the Tropic of Cancer or of Capricorn], there is unevenness both in amount and speed

As Kidd says,[33] the explanation of this by Sir George Darwin[34] is clearly correct, that Seleucus had observed the "diurnal inequality" of the tides, that is, "that the range of two successive tides is not the same"[35]: this happens most strongly when the moon is at its maximum distance north or south of the equator, not at all when it is over the equator. This inequality, though "almost evanescent" in the Atlantic, is "very great" in the Indian Ocean,[36] where presumably Seleucus observed it. He cannot have observed it, without being aware, before Posidonius, that the daily tidal changes, if not all the monthly ones, are dependent on the moon.

After this passage about Seleucus, Strabo goes on:

αὐτὸς δὲ κατὰ τὰς θερινὰς τροπὰς περὶ τὴν πανσέληνόν φησιν ἐν τῷ Ἡρακλείῳ γενόμενος τῷ ἐν Γαδείροις πλείους ἡμέρας μὴ δύνασθαι συνεῖναι τὰς ἐνιαυσίους διαφοράς

but he [i.e., Posidonius] says that he himself at full moon at the summer solstice was in the Heracleion at Cadiz for a good many days without being able to mark the annual differences.[37]

Something is surely wrong here. Posidonius cannot have hoped to observe *annual* differences in a stay at Cadiz of only 30 days. The moon's movements over the equator are monthly, not annual, and Posidonius did have time to observe differences in "diurnal inequality" occurring within a month; but none were perceptible. Surely Strabo has mistakenly thought that Posidonius was looking for annual differences, perhaps misled by references to zodiacal signs, which are usually mentioned in connection with annual changes.[38]

Strabo continues, clearly still reporting Posidonius:

περὶ μέντοι τὴν σύνοδον ἐκείνου τοῦ μηνὸς τηρῆσαι μεγάλην παραλλαγὴν ἐν Ἰλίπᾳ τῆς τοῦ Βαίτιος ἀνακοπῆς παρὰ τὰς ἔμπροσθεν, ἐν αἷς οὐδὲ ἕως ἡμίσους τὰς ὄχθας ἔβρεχε· τότε δ᾽ ὑπερχεῖσθαι τὸ ὕδωρ ὥσθ᾽ ὑδρεύεσθαι τοὺς στρατιώτας αὐτόθι (διέχει δ᾽ Ἰλίπα τῆς θαλάττης περὶ ἑπτακοσίους σταδίους)· τῶν δ᾽ ἐπὶ θαλάττῃ πεδίων καὶ ἐπὶ τριάκοντα σταδίους εἰς βάθος καλυπτομένων ὑπὸ τῆς πλημμυρίδος ... τὸ τῆς κρηπῖδος ὕψος τῆς τε τοῦ νεὼ τοῦ ἐν τῷ Ἡρακλείῳ καὶ τῆς τοῦ χώματος ὃ τοῦ λιμένος πρόκειται τοῦ ἐν Γαδείροις, οὐδ᾽ ἐπὶ δέκα πήχεις καλυπτόμενον ἀναμετρῆσαί φησι· κἂν προσθῇ δέ τις τὸ διπλάσιον τούτου κατὰ τὰς γενομένας ποτὲ παραυξήσεις, <οὐδ᾽> οὕτω παρασχεῖν ἂν τὴν ἔμφασιν, ἣν ἐν τοῖς πεδίοις παρέχεται τὸ μέγεθος τῆς πλημμυρίδος.

however at the conjunction [new moon] of that month he observed (?) at Ilipa [Alcalà del Rio] a great change in the wave recoil of the Baitis

[Guadalquivir] compared with earlier occurrences when the water level had not reached even half way up the banks; but at that time it flooded over so that soldiers drew their water where they were (Ilipa is about 700 stades from the sea); and although the flat land by the sea was covered actually to a distance inland of 30 stades by the flood ... the height of the base of the temple in the Heracleion and of the base of the mole that juts out in the harbour at Cadiz he says he measured as covered not even up to 10 cubits [presumably meaning that these bases were submerged to a depth of less than 10 cubits {between 4 and 5 metres}]. If one doubled that measure in line with rises that sometimes happened, even then one could not have presented the spectacle that the size of the flood presents in the plains.[39]

Strabo here appears to say something impossible, that Posidonius *observed the same tide* both at Alcalà del Rio and at Cadiz, over 100 kilometres away. Strabo implies that Posidonius stressed the contrast between a modest tidal rise at Cadiz and a much larger rise (presumably reported to him) in the estuary of the Guadalquivir[40]; perhaps Posidonius regarded this as a tidal inequality comparable to those described by Seleucus. But what caused the estuary to rise? Kidd suggests a flood in the Guadalquivir coinciding with a high tide,[41] Compatangelo-Soussignan posits a tsunami.[42] However, either a flood at midsummer or a tsunami would have been an exceptional event, and Posidonius was only in the area for 30 days. It must be at least as likely that either Posidonius misunderstood or misrepresented his informants, or Strabo misunderstood or misrepresented Posidonius, and that the exceptional tide on the Guadalquivir had actually occurred on some other occasion. Note that Posidonius provided some at least approximate numbers: the height of the rise at Cadiz, which he says was measured; the distance of Ilipa from the sea; and the distance inland to which flat land was flooded.

Strabo says nothing of any cause suggested by Posidonius of the tidal phenomena he described, unless there is a hint in the words, at III.5.8, συμπαθῶς τῇ σελήνῃ, "in joint affinity (or 'in sympathy') with the moon": in the Stoics' general theory, the co-ordination of regular changes, such as that of tides with the moon's movements, are due to, and proof of, "sympathy" between different parts of the cosmos[43] and demonstrate the providential government of the world.[44] But Posidonius, as we have seen, liked to discover the more immediate causes of phenomena. Two authors purport to tell how he explained the tides; unfortunately, they do not agree.

Aëtius III.17.4,[45] under the heading Πῶς ἄμπωτις καὶ πλήμμυραι γίνονται, "How ebb and floods occur", says Ποσειδώνιος ὑπὸ μὲν τῆς σελήνης κινεῖσθαι τοὺς ἀνέμους, ὑπὸ δὲ τούτων τὰ πελάγη, ἐν οἷς τὰ προειρημένα γίνεσθαι πάθη, "Posidonius says that winds are set in motion by the moon, and by these are moved the seas, in which the above-mentioned phenomena occur".

In contrast, Priscianus Lydus, in part of *Solutiones ad Chosroem* VI (Posidonius F219 EK), after describing tidal phenomena, says "Horum ... causas requirens Stoicus Posidonius ... discernit magis causam esse eius lunam et non solem", "Posidonius, the Stoic, seeking out the causes of these things ... noticed that the moon was the cause rather, not the sun"[46]; the sun's pure heat destroys water by making it evaporate; "lunae vero ignem non sincerum sed infirmiorem esse ... consumere autem quaecunque infert non potest, sed solummodo elevare umida et fluctificare", "the moon's heat is not pure but weaker. ... It cannot consume whatever it encounters, but only raises moisture and makes waves"[47] (a similar phenomenon occurs, it is suggested, when a kettle is moderately heated); "sic ... circuire cum luna undam maris, veluti ab ipsa exaltata et sic infirmata (?) redundare; respiciente autem in occasum coinclinare", "so thus ... the water of the sea follows the cycle of the moon, as when the moon is risen and thus weakened (?) there are flood tides, when it turns towards its setting the sea sinks with it".[48] ("Infirmata", "weakened", is surely an error by Priscianus or his translator: the moon's effect on the sea must be strongest when it is directly above it.) Spring tides are also explained:

> In plenilunio et coitu extollitur maxime unda, quoniam et lunae tunc magna adest virtus: in plenilunio enim totum eius in terram conversum a sole illustratur: in coitu autem illuminata desuper a sole aequalem in ea quae sunt in terra virtutem plenitudini praestat.

> The maximum rise of water occurs at full moon and conjunction [i.e., new moon], since at those points the potency of the moon is great; for at full moon the whole of it, turned towards the earth, is shining from the sun; at conjunction, illuminated from above by the sun, it supplies a potency equal to the full [i.e., equal to that at full moon] to things which are on the earth.[49]

Presumably the sun at conjunction, shining on the side of the moon away from the earth, arouses a potency in the moon which affects the earth, although we cannot see it.

The moon when below the earth also causes flood tides, "et causam esse ait confluxum fieri etiam circularem aquae naturam: itaque unda quasi in figura semicyclii elata sequitur lunam", "and he says that a reason for the flowing together of waters is also the circular nature of water; so waves follow the moon rising in a semicircular pattern".[50] Because the earth is spherical and (other things being equal) all water has an equal tendency to fall towards the earth's centre, it follows that all points on the surface of an area of still water are equidistant from the centre of the earth, so that its surface is a fraction of a sphere, as Aristotle argues at *De caelo* 287b5ff; to the Stoics, too, next to the earth, τὸ ὕδωρ σφαιροειδές, ἔχον τὸ αὐτὸ κέντρον τῇ γῇ, "the water has the shape of a sphere, having the same centre as the earth".[51] I suggest that, in the theory Priscianus is trying to describe, water has by nature a tendency to form into a sphere, so that, if the sea's surface is distorted at one point, as by the moon heating and raising it, it will tend to maintain its sphericity as far as possible by a corresponding rise at the opposite point of the earth, and a fall between those two points: in that way, a line on the surface of the water from the point where the moon is most distorting it to a point on the opposite side of the earth can still be a section of a circle – almost a semicircle – even though the complete circle is distorted where the moon's effect is greatest, and at the point opposite to that, and the centre of the circle of which each nearly complete semicircle is a section is not the centre of the earth. ("Eccentric circles", whose centre is not the centre of the earth, were familiar in ancient astronomy.[52]) This will explain how there can be a high tide when the moon is overhead and when it is over the opposite point on the sea's surface (see Figure 13.1).

Tides are highest at the equinoxes because "in utrisque enim sole existente in Ariete aut Brachiis et lunae in eadem hora magna est virtus coeuntis sole",

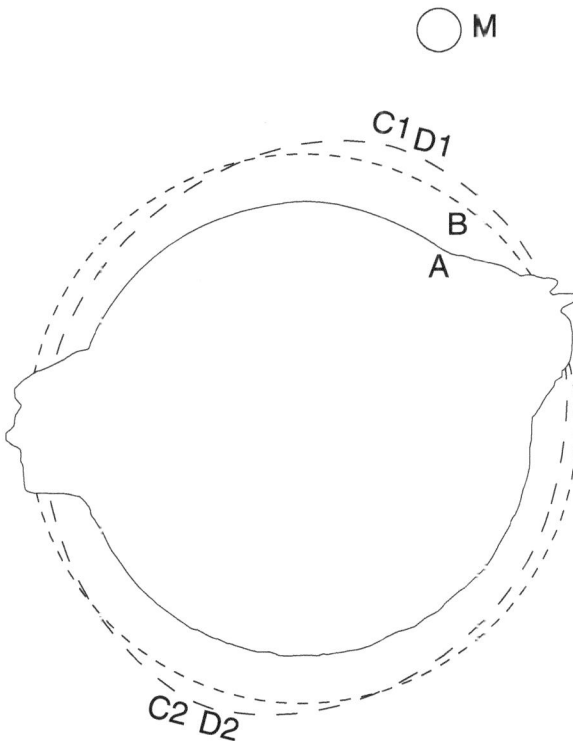

Figure 13.1 Suggested interpretation of Posidonius' explanation of the daily cycle of the tides, as described by Priscianus (Posidonius F219 lines 105–9 EK).

M = moon. A (continuous line) = surface of earth: approximately spherical, projecting in places above the surface of ocean to form land. B (line of short dashes) = surface of ocean if the moon did not affect it: a perfect sphere. C, D (lines of long dashes) = surface of ocean distorted by the moon so as to cause tides. At C1, D1 the moon overhead causes the ocean to swell: a flood tide. The ocean attempts to maintain its spherical shape as far as possible, swelling at C2, D2 (= flood tide at the point furthest from the moon) and subsiding between C1 and C2, and between D1 and D2 (= ebb tides), so that the curves C1 to C2 and D1 to D2 are perfectly formed but not quite complete semicircles, but with centres displaced from the centre of the earth.

"for in each of the equinoxes when the sun is in Aries and Libra [the equinoctial signs of the zodiac] the potency of the moon at the same hour [presumably meaning "being in the same direction in the sky"[53]] combining with the sun is great".[54] The idea is evidently that at these times the power of the moon is increased because it is combined with that of the sun (it is not clear why this combination should be particularly powerful at the equinoxes). An additional reason is given for high equinoctial tides: "posse autem hoc ipsum et horam anni ex natura lunae: calida enim est et umida, et hac virtute extollitur unda ... Ver autem et autumnus umidum mensurate et calidum", "this can also be from the nature of the moon at that time of year: for it is warm and moist, and it is from those qualities that the water rises. ... Spring and autumn ... are noticeably moist and warm."[55]

Priscianus gives the impression that all this is from Posidonius. He begins (F219.77–80 EK) by saying that Posidonius noticed the cause; the following lines 80 to 96 are mostly in indirect speech, implying that this is what Posidonius said; and in what follows there are words of "saying" (line 107 "ait", line 115 "confitetur") with no new subject introduced, so implying that it was Posidonius who said these things, and presumably what is in the context around them. It is hard to see why this comprehensive attempt to explain the tides should have been attributed to Posidonius unless at least the basis of it is his. Priscianus, or his immediate source, would naturally have corrected Posidonius' one definite error, the belief that tides are greatest at solstices. The idea that the moon causes tides by providing just the right, moderate amount of heat is reminiscent of Posidonius' theory that the earth is habitable at the equator because the sun provides an appropriate, moderated, degree of heat there.[56] (I admit that he was not the only thinker to provide such explanations.[57])

Pneuma, "breath" or "wind", was an important concept as a cause in Stoicism; Posidonius spoke of god as πνεῦμα νοερὸν διῆκον δι' ἀπάσης οὐσίας, "a rational breath extending through all that is".[58] It could be that Posidonius, regarding the connection of tides with the moon as evidence that Providence governs the world (see above), spoke of *pneuma* in this sense as involved in the moon's heating and raising of the sea, and that Aëtius' account of Posidonius' theory is a distortion of this.[59] Whether this is true or not, it is surely more likely that a highly abbreviated summary in a doxography, such as that of Aëtius, is gravely misleading, or misattributed,[60] than is a detailed account unaccompanied by alternatives, such as that in Priscianus.[61] If Priscianus is right, Posidonius did not simply describe the moon's effect on the sea as due to Stoic *pneuma*[62]; he said it was the effect of the moon's heat, analogous to the familiar process of heating water in a kettle.

How much of Posidonius' account of the tides was original to him? In describing the tides and their relation to the daily and monthly movements of the moon, he must have been at least partly anticipated by Seleucus, who cannot have discovered what he evidently did discover about the diurnal inequality of tides (see above) without realising that the daily tidal movements depend on the moon's movements – just possibly, Seleucus had not noticed, or had not recorded, the difference between spring and neap tides, and Posidonius was the first Greek thinker to do so. At the least, Posidonius gave a clear description of the tides, enough to make himself a recognised authority on the subject, as Strabo shows.

Posidonius' theories about the causes of tides were inevitably wrong, but how original were they? Aëtius III.17 names nine authors besides Posidonius as offering an explanation of ἀμπώτιδες καὶ πλήμμυραι, "ebb and floods". Four of the explanations have no relation to Posidonius' theory. Aëtius says that Plato referred tides to "the oscillatory motion" (τὴν αἰώραν) of the water; that Crates "the *grammatikos*" (Crates of Mallos?) thought the cause was a "convulsion" or "counter-movement" (ἀντισπασμόν) of the sea; that Apollodorus "the Corcyrean" thought it was "reverse currents" (παλιρροίας) from the ocean; and that the historian Timaeus thought that ebbs and floods are due to variations in the flow of rivers which discharge into the Atlantic.[63] Of these four, Plato at least was not really concerned with tides – his "explanation" is clearly derived from the myth at *Phaedo* 111E–112E.

Other views mentioned by Aëtius may have influenced Posidonius. Aëtius III.17.1 attributes to Aristotle and Heraclides a theory that the sun sets winds in motion and that these drive the Atlantic, causing it to flood. For Heraclides there is no other evidence,[64] but for Aristotle, some passages in *Mete.* suggest that there may be some truth in this,[65] insofar as Aristotle does involve the sun in causing wind,[66] and at *Mete.* 344b35 and 368a34ff he speaks of wind as causing a flood. This was a tsunami, but he may have thought the same about other rises and falls of the sea, so far as he knew about them. At *Mete.* 366a19f he compares the behaviour of earthquake-causing winds to ἄμπωτις and πλημμυρίς, "ebb" and "flood". This is not an explanation of tides, but suggests that he had similar ideas about rises and falls of the sea, and earthquakes. At *Mete.* 368a26–34, after speaking of wind as cause of both earthquakes and waves (κύματα) he says, in general terms, αἴτια ταῦτα μὲν ἄμφω ὡς ὕλη (πάσχει γάρ, ἀλλ' οὐ ποιεῖ), τὸ δὲ πνεῦμα ὡς ἀρχή, "both these [earth and water] are causes as being material (for they are passive, not active), but wind is cause as originator [of motion]". If the sea cannot move itself, but wind can move it, then it would be a natural assumption that wind causes tidal flows, as well as other movements of the sea: it would be natural for Aristotelians to conclude that the sun causes wind and that wind makes the sea rise and fall. Therefore, Posidonius might have taken from them the idea that heat causes tides. (There is no indication of this in the report, at Strabo III.3.3, of Posidonius' criticism of Aristotle's view of Atlantic tides; but, as mentioned above [p. 131], that report is doubtfully accurate.) Another theory that the sun causes tides is the one attributed by Aëtius III.17.2 to "the Messenian" (ὁ Μεσσήνιος, possibly Dicaearchus), that the sun is the cause πλημμύροντι τὰ πελάγη, "making the seas overflow".

Aristotle's surviving works show no real knowledge of oceanic tides; Dicaearchus may have known about them, since he knew the works of Pytheas.[67] Aëtius III.17.3 says of Pytheas τῇ πληρώσει τῆς σελήνης καὶ τῇ μειώσει τὰς ἑκατέρου τούτων αἰτίας ἀνατίθησι, "he attributes the causes of each of these [ebbs and floods] to the moon's becoming full and to its waning".[68] Having sailed in the Atlantic, Pytheas was obviously aware of the relation of the tides to the moon's movements, though it is not clear from this report how clearly he stated it, and Aëtius' report leaves it obscure how the moon acts as cause. I have already quoted Aëtius' final report, of the theory of Seleucus,[69] which is unclear in detail, but does say that in some way the motions of earth and moon generate wind which causes the tides. If Aëtius is right about Posidonius' theory, then Posidonius would seem to have derived it from Seleucus (while rejecting Seleucus' theory of the earth's rotation); if, as is probable, Priscianus is more nearly right about Posidonius, then the theory of "the Messenian" is closest to that of Posidonius, and that is not very close: in this case, Posidonius' explanation of tides would seem to have been largely original.

I should mention one other aspect of Posidonius' treatment of tides: he thought they had been mentioned by Homer. Strabo says:

Ποσειδώνιος δὲ καὶ ἐκ τοῦ σκοπέλους λέγειν τοτὲ μὲν καλυπτομένους, τοτὲ δὲ γυμνουμένους, καὶ ἐκ τοῦ ποταμὸν φάναι τὸν ὠκεανὸν εἰκάζει τὸ ῥοῶδες αὐτοῦ τὸ περὶ τὰς πλημμυρίδας ἐμφανίζεσθαι

Posidonius conjectures, when he [i.e., Homer] says headlands are sometimes covered, sometimes bared, and when he calls Oceanus 'river' [*Iliad* XIV.245], that its flow ... implies flow tides.[70]

The mention of headlands sometimes covered and sometime left bare apparently refers to the description, at *Odyssey* XII.235–43, of the effects of Charybdis sometimes spouting out water and sometimes sucking it in.[71] Posidonius cannot have regarded these passages as evidence that tides occur; rather, it is evidence of Homer's knowledge, and of how statements by him which appear fabulous can be rationally explained. This needs to be borne in mind when discussing other mentions by Posidonius of Homer's meteorology.

Notes

1 *Mete.* 358a15–27.
2 See Kidd (1988) 788–9.
3 Posidonius F219.134–6 with translation from Kidd (1999) 293.
4 F219 EK lines 19 "declarat", 107 "ait", 115 "confitetur". See p. 136 below.
5 Compare F219.136–40 EK with *Mete.* 359a18–23. See discussion in Kidd (1988) 789.
6 EK include it in their F219, as does Vimercati (2004) in his A139. Theiler (1982) omits it from his F313.
7 F219.147–50 EK with translation from Kidd (1999) 293.
8 Aristotle *Mete.* 357a9ff criticises this theory. It is attributed to Anaxagoras and Metrodorus of Chios (see Alexander, *In Mete.* 67.17ff Hayduck = DK 59A90; Aëtius III.16.5 = DK 70A19).
9 Strabo I.3.4.

10 Posidonius F221 EK, with the translation of Kidd (1999).
11 1,000 ὄργυιαι (fathoms) is about 10 stades. Plutarch (*Life of Aemilius Paulus* 15) mentions 10 stades as a figure given for the deepest sea-depths, Cleomedes (I.7 lines 123–4 Todd) and Fabianus (quoted by Pliny, *Nat.* II 224) say 15 stades. (See Capelle [1916] 21, 24).
12 1,000 ὄργυιαι is about 2,000 metres; in some parts of the Mediterranean the depth exceeds 3,000 metres (see *Times Comprehensive Atlas* [2014] plates 80, 81, 83, 84).
13 As does Aristotle *Mete.* 351a13, where he says of part of the Black Sea οὐδεὶς... πώποτε καθεὶς ἐδυνήθη πέρας εὑρεῖν, "no-one has ever yet been able by sounding to find the bottom".
14 It has been widely thought (e.g., by Theiler [1982] II 17–18) that the whole of Strabo I.3.9 is from Posidonius, but Kidd shows that there is no strong reason for attributing any of it to him, except for the detail about the sea depth. See Kidd (1988) 793–5 for detailed discussion.
15 This section is heavily dependent on Kidd's commentary on the relevant fragments, Posidonius F214–220 EK (Kidd [1988] 759–92). See also Theiler (1982) II 39–42; Vimercati (2004) 589–96. A recent study is Compatangelo-Soussignan (2015).
16 See Strabo II.4.2.
17 See Arrian, *Anabasis Alexandri* VI.19.1; Curtius Rufus IX.ix. 9ff. (Passages cited by Compatangelo-Soussignan [2013] 599 and 600 n. ix.)
18 See Strabo I.3.11.
19 I give the text as in Mansfeld and Runia (2020).
20 Strabo III.1.5 (Posidonius F119 EK) and III.5.9 (F218 EK).
21 Strabo I.1.9 and I.3.12 (Posidonius F214 and F215 EK).
22 Strabo III.5.8 (Posidonius F217.30ff EK). The quoted translation is from Kidd (1999) 283.
23 Strabo III.5.8 (Posidonius F217.49 ff EK).
24 Kidd (1988) 775; cf. Vimercati (2004) 593.
25 See Kidd (1988) 775–6; Seneca *NQ* III.28.6; Pliny *Nat.* II.215; Priscianus Lydus = Posidonius F219.52 ff and 110 ff EK. On Priscianus see below. On equinoctial tides see also Chapter 22 (p. 215).
26 This is "usually assumed" – Kidd (1988) 775.
27 That Posidonius changed his view is the opinion of Theiler (1982) II 41–42. He thinks that Strabo was using Posidonius Περὶ ὠκεανοῦ ("About ocean"), and that this was an early work (see Theiler [1982] II 6). Kidd (1988) 220 agrees that an early date is possible for this work.
28 Cartwright (1999) 8. I return to this point in Chapter 22 (p 215–6).
29 Posidonius F220 EK. I print τὴν παραλίαν καὶ τὴν Μαυρουσίαν, as in the MSS., which Kidd (1988) 792 says is "probably sufficiently clear", without emendations as printed in EK (and partly in Theiler [1982] F20).
30 So, in effect, Kidd (1988) 791.
31 As Kidd (1988) 791 and Theiler (1982) II p. 37 mention, Aëtius III.17.1 attributes to Aristotle what appears to be a quite different theory. (See below, p. 137.)
32 Posidonius F218 EK.
33 Kidd (1988) 778.
34 Darwin (1911) 88.
35 Darwin (1911) 162.
36 Darwin (1911) 88.
37 Translation from Kidd (1999) 286, slightly modified.
38 This seems to me likelier than the alternative which Kidd (1988) 778–9 prefers, that Posidonius *was* looking for annual differences, either because he had misunderstood Seleucus, or because Seleucus had misinterpreted his own observations.
39 Translation from Kidd (1999) 286–7, modified.

40 Nowadays the Guadalquivir is dammed at Alcalà del Rio, and is tidal up to that point. See, e.g., Diez Minguito et al. (2012).
41 Kidd (1988) 780.
42 Compatangelo-Soussignan (2013). She notes that a tidal rise of 10 cubits (taken to equal 4.62 metres), though not extraordinary, is greater than the normal tidal rise at Cadiz, and suggests that the phenomenon recorded by Posidonius, at Cadiz and the Guadalquivir, was a tsunami.
43 Sextus Empiricus *Adv. Math.* IX.79; Cicero *De divinatione* II.34 (passages printed by Theiler [1982] as Posidonius F354 and F379.)
44 Cicero *ND* II.19 (printed by Theiler [1982] as Posidonius F356): among proofs of providential government is the fact that "aestus maritimi ... ortu aut obitu lunae commoveri", "the tides of the sea are moved with the rising or setting of the moon". See Sambursky (1959) 42.
45 Posidonius F138 EK. On this see Kidd (1988) 522–5; Theiler (1982) II 204–5; Vimercati (2004) 589–90.
46 Posidonius F219.77–80 EK, with translation from Kidd (1999) 290–1. For discussion of the whole passage F219 EK, see Kidd (1988) 781–90; Theiler (1982) II p. 201–3 (the passages cited by Theiler do not seem to me to be close parallels); Vimercati (2004) 594–6.
47 Posidonius F219.82–6 EK, with translation from Kidd (1999) 291.
48 Posidonius F219.94–6 EK. In the translation, "so thus" to "when it turns" is from Kidd (1999), 291, slightly modified.
49 Posidonius F219.100–4 EK. In the translation, "the maximum" to "above by the sun" is from Kidd (1999) 291, slightly modified.
50 Posidonius F219.107–9 with translation from Kidd (1999), but with "circular nature" substituted for "cyclical nature".
51 Diogenes Laertius VII 155. Cf. Seneca *NQ* III.28.5.
52 See, e.g., Lloyd (1973) 61ff.
53 Cf. *Oxford Latin dictionary* (1982) "hora" 4: "The direction of the sun at a specified hour".
54 Posidonius F219.111–3 EK, with translation from Kidd (1999) 291–2, modified. For "Brachiis" meaning the zodiacal sign Libra see Kidd (1988) 788 and (on F126) 482.
55 Posidonius F219.119–24 EK, with translation from Kidd (1999) 292. See Kidd (1988) 788.
56 See above, Chapter 7 (p. 60).
57 Thus Aristotle *Mete.* 362a2ff says that winds blow not at midsummer (too dry), but just after it, when the sun's heat and the exhalation are in balance (σύμμετρος), and at *Mete.* 366b2ff maintains that earthquakes are most frequent in spring and autumn because then it is neither too hot nor too cold.
58 *Scholia in Lucani Bellum civile* ... IX 578 = Posidionius F100 EK. On the importance of πνεῦμα to the Stoics see (e.g.) Sandbach (1975) 71, 73, 75.
59 Something like this seems to be the conclusion of Theiler (1982) II 205. Kidd (1988) 525 suggests that Posidonius held both theories; but, as we have them, they seem incompatible. Kidd (1978a) 14 suggests that they are alternative hypotheses, which a philosopher must decide between; but nothing in our texts suggests this. Vimercati (2004) 589–90 rejects Aëtius' account.
60 Theiler (1982) II 204 suggests that the account in Aëtius may be misattributed to Posidonius.
61 Aëtius' account could be a conflation of two explanations of floods given by Posidonius, i.e., that some are caused by the moon and others by wind: in Strabo III.5.8–9 (Posidonius F217–8 EK) Posidonius' account of tides co-ordinated with lunar movements is immediately followed by his explanation of floods on the Ebro as caused by north winds.

62 As Compatangelo-Soussignan (2015) 92 seems to suggest.
63 Aëtius III.17.5–8.
64 There is none in Wehrli (1969b) or Schütrumpf (2008).
65 On this see Compatangelo-Soussignan (2015) 85–6.
66 *Mete.* 361b14ff, 362a2ff.
67 See above, p. 129.
68 This is the version of Stobaeus. Pseudo-Plutarch explicitly attributes floods to the moon becoming full and ebbs to its waning. Mansfeld and Runia [2020] 1331–2 prefer the text of Pseudo-Plutarch, which, as they show, is supported by parallels in Pseudo-Galen and Joannes Lydus. Neither version is true of the daily tidal movements,
69 See above, p. 129–30.
70 Strabo I.1.7 (Posidonius F216 EK), with translation from Kidd (1999), slightly modified.
71 See Kidd (1988) 765–7. Strabo I.1.7 quotes further Homeric passages as possible references to tides, and it has been thought that these too are from Posidonius; but Kidd shows that only the sentence I have quoted is certainly from him. Vimercati (2004) 591 says of the headlands sometimes covered and sometimes bare "come accade ai fiumi" but cites no Homeric text.

14 Rain, snow, hail and cloud

Rain

Posidonius,[1] like other ancient thinkers, was well aware that water on the earth evaporates, and that the vapour is formed into clouds, condenses there to water, and returns to the earth as rain. This is made clear, if evidence is needed, by Seneca's report which I quoted in discussing thunder and lightning, that in Posidonius' view "from the earth ... something moist is exhaled, and something dry and smoky. The latter is nourishment for thunderbolts, the former for rain".[2] I have argued in another chapter that this was an easily reached conclusion, which we find already in Hesiod.[3] That water evaporates (or, as early writers often put it, "is drawn up") and so gives rise to the regular phenomena of clouds and rain, is also attested as the view of many other early writers. We find it in the Hippocratic *Airs waters places* and in Herodotus, and among the pre-Socratics it is attributed to Anaximander, Xenophanes, Heraclitus, Democritus, and (implicitly) Metrodorus of Chios.[4]

Subsequently, it was the view of Aristotle,[5] and to Epicurus it was one (though not the only) way in which rain occurs.[6] For the Stoics, Chrysippus is reported as saying that "rain is movement of water from cloud", with no accompanying explanation of cloud – thereby showing his preference for avoiding mention of causes[7]; but in Diogenes Laertius' account of Stoic meteorology[8] (which Posidonius has influenced[9]) we find: ὑετὸν δὲ ἐκ νέφους μεταβολὴν εἰς ὕδωρ, ἐπειδὰν ἡ ἐκ γῆς ἢ ἐκ θαλάττης ἀνενεχθεῖσα ὑγρασία ὑφ᾽ ἡλίου μὴ τυγχάνῃ κατεργασίας, "rain is a change from cloud to water, when moisture drawn up by the sun from land or sea does not achieve completion" (presumably meaning: when the moisture is not completely changed to fire, as fuel for the sun or another heavenly body).

There was disagreement about the cause of the condensation to cloud and rain. To Aristotle it was cold: in the cool upper air, he says συνίσταται ... ἡ ἀτμὶς ψυχομένη διὰ ... τὴν ἀπόλειψιν τοῦ θερμοῦ, "the vapour condenses, being cooled owing to the loss of the heat"[10]; but Epicurus in the *Letter to Pythocles* speaks of compression as a cause of cloud and rain, with no mention of cold.[11] Posidonius would appear to have agreed with Aristotle: I have previously quoted

DOI: 10.4324/9780429399930-14

Cleomedes' report of Posidonius' arguments for the relative coolness of the equatorial region, with which is associated the occurrence of rain there.[12]

Snow

Diogenes Laertius says: χιόνα δὲ ὑγρὸν ἐκ νέφους πεπηγότος, ὡς Ποσειδώνιος ἐν τῷ η τοῦ Φυσικοῦ λόγου, "snow is moisture from a frozen cloud, as Posidonius says in the eighth book of his *Physikos logos*".[13] Aristotle in his *Meteorologica* says that snow is analogous to hoar-frost: ὅταν γὰρ παγῇ τὸ νέφος, χιών ἐστιν, ὅταν δ' ἡ ἀτμίς, πάχνη, "when cloud freezes, snow is produced; when vapour, hoar-frost".[14] Chrysippus held that snow is νέφος πεπηγὸς ἢ νέφους πῆξιν, "frozen cloud or freezing of cloud".[15] So Posidonius had both Aristotle's and Stoic authority for deriving snow from frozen cloud. Posidonius perhaps added something when he said that snow is "*moisture* from a frozen cloud" (my emphasis), which Kidd interprets as meaning that "Posidonius thought ... of a frozen cloud liquefying ... until the moisture consistency of snowflakes was reached"[16]; but the idea that snow is frozen matter partially thawed is surely a strange one, and I think there is a likelier interpretation.

Snowflakes look very different from the ice which is formed when a quantity of water freezes. Therefore, an explanation of snow gains plausibility if it states that snow is formed by the freezing of something other than water. Such an explanation we do find in the accounts of snow in *De mundo*, Seneca and Arrian. Arrian says that snow occurs because πρὶν παντελῶς ἐς ὕδωρ ξυστῆναι τὴν νεφέλην φθάναι παγῆναι ἐς χιόνα, "cloud, before it has completely condensed to water, is first frozen to snow".[17] Seneca says "hieme aer riget et ideo nondum in aquam vertitur, sed in nivem", "in winter air is stiff, and therefore it is not yet turned into water, but into snow".[18] *De mundo* says that snow is caused by the breaking up of dense clouds πρὸ τῆς εἰς ὕδωρ μεταβολῆς, "before they change into water", and that the cold of snow is due to ἡ σύμπηξις τοῦ ἐνόντος ὑγροῦ, "the solidification of the moisture that is in them".[19] So, to Arrian and Seneca cloud can freeze without turning to water first; in *De mundo* moisture, τὸ ὑγρόν, exists in clouds in some form,[20] and can become solid, before the change to water. Similarly, the Syriac *Meteorology* says that snow occurs "when coldness freezes the clouds before they turn into water".[21] Aristotle seems to have held the same view: he says that χιών...καὶ πάχνη ταὐτόν, "snow and hoar-frost are the same[22]; that when cloud freezes there is snow, but hoar-frost when vapour does (see quotation above); and that hoar-frost, παχνή, forms ὅταν ἡ ἀτμὶς παγῇ πρὶν εἰς ὕδωρ συγκριθῆναι, "when water-vapour freezes before it is condensed to water".[23] When such views were held by Aristotle (whom Posidonius often follows), by Arrian and Seneca (certainly influenced by Posidonius), and by the author of *De mundo* (who was influenced by Posidonius, or else influenced him), it is surely most likely that Posidonius' view of snow was similar to theirs; that, in the theory which Diogenes summarised, the cloud freezes, and moisture, ὑγρόν, in it freezes, and forms snow, without ever becoming water. Chrysippus, who also said that snow is frozen cloud, very likely held the same theory, which may go back to Anaximenes, if Hippolytus is correct: after speaking of Anaximenes' view of the formation of hail from clouds, he continues: χιόνα δὲ ὅταν αὐτὰ ταῦτα ἐνυγρότερα ὄντα πῆξιν λάβῃ, "but there is snow when these very things [i.e., clouds], being moister, undergo freezing".[24]

This was not the only explanation of snow. Plato at *Timaeus* 59E says that when water is "half-frozen" (ἡμιπαγὲς) above the earth, it is snow. Epicurus has multiple explanations of snow, with variants of the "snow is frozen cloud" theory, for example, that snow may form by "freezing with uniform rarity in clouds" (πῆξιν ἐν τοῖς νέφεσιν ὁμαλῇ ἀραιότητα ἔχουσαν), and also the idea that snow may be due to the freezing of fine rain (ὕδατος λεπτοῦ … πῆξιν).[25] But the theory described in the last paragraph seems to have been the commonest, and was surely that of Posidonius.

Hail

The fact that there are two forms of frozen precipitation from clouds presented problems to ancient thinkers, notably: why is there the difference between hail and snow, when both are, or appear to be, frozen water falling from clouds? And why does hail usually fall in warmer seasons of the year, although hail-stones appear to be water more deeply frozen than the snowflakes of cold seasons? Greek thinkers proposed answers to these questions, and there is evidence of how Posidonius answered them.

For Posidonius' theory of hail we have information in Seneca's *Naturales quaestiones* (IVb.3.1–2).[26] Unfortunately, all that precedes this passage is missing from this book of the *NQ*, and Seneca treats Posidonius' theory with some irony. The passage reads:

> Grandinem hoc modo fieri si tibi affirmavero quo apud nos glacies fit, gelata nube tota, nimis audacem rem fecero. Itaque ex his me testibus numero secundae notae qui vidisse quidem se negant. Aut, quod historici faciunt, et ipse faciam; illi, cum multa mentiti sunt ad arbitrium suum, unam aliquam rem nolunt spondere sed adiciunt: "Penes auctores fides erit". Ergo si mihi parum credis, Posidonius tibi auctoritatem promittit tam in illo quod praeteriit quam in hoc quod secuturum est; grandinem enim fieri ex nube aquosa iam et in umorem versa sic affirmabit tamquam interfuerit.

> If I tell you that hail is formed in the way in which ice is formed with us, the whole cloud having been frozen, I shall have done too bold a thing. And so I number myself among the second-category witnesses, who say that they have not seen for themselves. Or, I shall do as historians do. They, when they have told many lies on their own authority, are unwilling to guarantee some one thing, but add: "Credibility will depend on my authorities". Therefore, if you have not enough confidence in me, Posidonius puts

forward his authority, as much in what has passed as in this which will follow: he asserts, as though he had been there, that hail is formed from a watery cloud now even [Kidd {1999} "only just"] turned to liquid.

Seneca here begins with an explanation of hail in his own name, says that he is being too bold, and so appeals (but still with irony) to the authority of Posidonius. Presumably Seneca's explanation and that of Posidonius are the same or very nearly so. To decide what the explanation is, we must examine other ancient theories of hail.

The commonest ancient view was the obvious one, that hail is frozen water. Aristotle says ἔστι μὲν ... ἡ χάλαζα κρύσταλλος, πήγνυται δὲ τὸ ὕδωρ τοῦ χειμῶνος, "hail is ice, and water is frozen in winter",[27] implying that ice, and therefore hail, is frozen water. In what follows he is more explicit, saying, for instance, τὸ ψυχρὸν...ὕδωρ ποιῆσαν ἔπηξεν καὶ γίγνεται χάλαζα, "the cold ... having produced water [sc. in a cloud], freezes it, and hail is formed".[28] Before him, Plato at *Timaeus* 59D–E says that "water ... when, separated from fire and air, it is isolated ... and is compressed" (ὕδωρ ... ὅταν πυρὸς ἀποχωρισθὲν ἀέρος τε μονωθῇ ... ξυνέωσται δέ), then, if it freezes above the earth, it is hail. Earlier still, according to Hippolytus, Anaximenes said that hail occurs ὅταν ἀπὸ τῶν νεφῶν τὸ ὕδωρ καταφερόμενον παγῇ, "when the water that is being carried down from the clouds freezes",[29] presumably meaning that hail is frozen rain. For the sophist Antipho (as reported by Galen) hail is formed when "water is massed together and made dense" (συστρέφεται τὸ ὕδωρ καὶ πυκνοῦται).[30] The account of hail in Epicurus' *Letter to Pythocles* 106–7 is obscure and the text of the MSS. perhaps corrupt, but the formation of hail apparently may involve πῆξιν μετριωτέραν ὑδατοειδῶν τινων, "more moderate freezing of watery particles".[31]

However, we have reports of another theory apparently favoured by the Stoics, in which cloud freezes without water being mentioned, and is then broken up into hailstones – this, at least, is how it appears in our sources, though, as they are very brief, it is probably an over-simplification. Diogenes Laertius says the Stoic theory of hail is that it is νέφος πεπηγὸς ὑπὸ πνεύματος διαθρυφθέν, "frozen cloud broken into pieces by wind".[32] The *De mundo* has a variant of this type: hail occurs νιφετοῦ συστραφέντος, evidently meaning "when a mass of snowflakes is gathered together"; the size and speed of fall of the hailstones depends on τὰ μεγέθη τῶν ἀπορρηγνυμένων θραυσμάτων, "the size of the fragments broken off".[33] Anaxagoras may have originated this theory: if Aëtius is correct, he said that hail occurs ὅταν ἀπὸ τῶν παγέντων νεφῶν προωσθῇ τινα πρὸς τὴν γῆν, ἃ δὴ ταῖς καταφοραῖς ἀποψυχρούμενα στρογγυλοῦται, "when from the frozen clouds some bits are propelled towards the ground, which losing cold(?) by their descents become spherical".[34]

Chrysippus' view, as we have it, looks like a compromise: hail is ὑετοῦ πεπηγότος διάθρυψιν, "a breaking to pieces of frozen rain"[35] (this is surely over-simplified: rain consists of separate drops, with no need for breaking in pieces).

Given this background, what is Posidonius likely to have thought? In the normal ancient view, when a cloud changes to water, it is from a state intermediate between water and clear air, as is evident from the passages quoted above, in which cloud is frozen into snow before it has fully changed to water. When Posidonius said, "hail is formed from a watery cloud now even turned to liquid", he surely meant that hail is formed when a cloud freezes after its change to water has gone too far for it to become snow.[36] As often, he followed Aristotle and believed that hail is frozen water. As a Stoic, he may also have echoed the Stoic view reported by Diogenes, and said that hail is derived from frozen cloud, as the first sentence of *NQ* IVb.3.1 suggests; but he would have meant by "frozen cloud" a cloud converted to frozen water droplets, for there is no "breaking up" of a cloud in this passage of Seneca, who goes on to say (*NQ* IVb.3.3) that you can know why hailstones are round "cum adnotaveris stillicidium omne glomerari", "when you have noticed that every droplet is formed into a ball", that is, hailstones are round because water droplets are – this may well be Posidonius' view, though not said to be so.

We have no direct information about Posidonius' answers to other questions concerning hail, questions such as: What causes the freezing? How can a whole frozen cloud remain in the air? How can water-drops remain in the air for long enough to freeze before they reach the ground? Solutions had been proposed: Anaxagoras thought that the freezing occurs in the cold of the high atmosphere,[37] Aristotle that it is due to a process of ἀντιπερίστασις, in which cold is intensified so strongly by heat surrounding it that the water freezes before reaching the ground.[38] Seneca says that one reason why snowflakes, unlike hailstones, are not spherical, is that they do not fall from a great height,[39] suggesting that he thought, like Anaxagoras, that hailstones do fall from a height; which would seem to leave Posidonius equally likely to have adopted either view of what causes hailstones to freeze.

Seneca, in what survives of *NQ* IVb, only mentions Posidonius in the passage I have quoted, but he does twice describe, and reject, what he says are views of "our people", that is, the Stoics. There are reasons for thinking that these are Posidonius' views, or at least due to his influence.

At *NQ* IVb.5 Seneca describes a Stoic explanation of why there is hail in spring rather than in winter: the idea seems to be that in spring the thawing of the ice and frozen ground of Scythia, Pontus [the Black Sea] and the northern region releases cold air, and that consequently cold winds blow; their cold "binds and constricts" ("alligat et praestringit") the warm, moist air of the south, and so what would have been rain becomes hail. It is hard to believe that any Stoic before Posidonius was interested enough in meteorology to devise this theory, and Posidonius was also interested in geography[40]; the inventor of this theory, if not Posidonius himself, was surely a follower of his.

In *NQ* IVb.6–7 Seneca is scornful of Stoics who "declare that certain people are skilled in observing clouds and predict when hail is coming" ("quosdam peritos observandarum nubium esse affirmant et praedicere cum grando ventura sit") – they apparently claimed to recognise hail-clouds by their colour (a claim which seems to have been made in modern times also[41]). Seneca goes on to tell and call "incredible" the story of the "hail guards" ("chalazophylacas") of Cleonae, who bade the citizens perform propitiatory sacrifices when hail was threatened, and he is scornful of the rationalisation of such a practice that some had suggested, "that there is in blood itself some powerful force for turning away a cloud" ("esse in ipso sanguine vim quandam potentem avertendae nubis"). Here, too, we have a story that relates to two of Posidonius' interests, in meteorology and in divination. We have seen that Seneca appears to criticise Posidonius for his belief in astronomical portents.[42] If, as would be not unlikely, Posidonius took seriously claims that hail-clouds are recognisable, and the work of the hail-guards, and if Seneca had occasion to mention and criticise this in the lost part of *NQ* IVb, that would help to explain why Seneca speaks with such irony of Posidonius' theory of hail in *NQ* IVb.3.1–2.

The origin of ancient ideas about snow and hail

For the most part the ancients must have derived their ideas on snow and hail from experience, or analogies with phenomena they had experienced. Those who believed that the four elements, earth, water, air and fire, are transformable into each other, could point to the apparent change of air into cloud and cloud into water as evidence for that belief; but the formation of ice from water hardly accords with the belief, since ice only changes back into water; it was something known simply from experience. It was also from experience that the ancients knew that air at a height is colder than at sea level. Other ideas came from analogies: that hail must be frozen water depends on the analogy of ice forming at ground level. The idea of ἀντιπερίστασις (that a quality becomes more intense when isolated by a mass of its opposite) depends on an analogy with the apparent warmth of caves in winter. The origin is less obvious of the idea that cloud can freeze to snow without becoming water first, but it may owe at least something to the analogy of hoar-frost; we cannot know how carefully Aristotle had observed the formation of hoar-frost, but it seems that he was right to say it is formed by the freezing of water vapour.[43]

Cloud

No surviving account of cloud is attributed to Posidonius, or indeed to other Stoics, but its importance in the explanation of other phenomena makes some discussion necessary. For thinkers who believed that the different forms of matter can be transformed into each other, the obvious view – in the context of an explanation of rain – was that cloud is an intermediate stage, denser than air but less dense than water. This was already the view of Anaximenes.[44] Aristotle implies it when he writes εἰ δὴ γίγνεται ὕδωρ ἐξ ἀέρος καὶ ἀὴρ ἐξ ὕδατος, διὰ τίνα ποτ᾽ αἰτίαν οὐ συνίσταται νέφη κατὰ τὸν ἄνω τόπον; "if water comes to be from air and air from water, why do clouds not form in the upper region [sc. of the atmosphere]?"[45] He implies it again when he says ἔστι δ᾽ ἡ μὲν ἐξ ὕδατος ἀναθυμίασις ἀτμίς, ἡ δ᾽ ἐξ ἀέρος εἰς ὕδωρ νέφος, "the exhalation from water is water-vapour, and that from air into water is cloud"[46]; *De mundo* has νέφος δέ ἐστι πάχος ἀτμῶδες συνεστραμμένον, γόνιμον ὕδατος, "cloud is a dense, vaporous formation, productive of water".[47] Probably, Posidonius' view of cloud resembled one of these; his reported view of hail formed from "a watery cloud now even turned to liquid" definitely suggests it.

Such accounts of cloud must be derived partly from observation, and partly by analogy with the formation, in domestic contexts, of clouds of "steam" from boiling water and its condensation on cold surfaces. They do well in explanations of how clouds are a source of rain; they do a little less well in explaining snow and hail, for which ancient thinkers often envisaged cloud solidifying without becoming water first, and some (probably not including Posidonius) seem to have imagined cloud freezing to a solid mass suspended in the air. The idea that cloud is an intermediate stage between air and water seems quite incompatible with the analogies between cloud and inflated bladders and the like, which the ancients used to explain thunder and lightning,[48] and (as we shall see) the analogy between clouds and mirrors by which they explained rainbows and other optical phenomena. Epicurus has an atomist account which might serve these latter purposes: clouds may be formed παρὰ περιπλοκὰς ἀλληλούχων ἀτόμων καὶ ἐπιτηδείων εἰς τὸ τοῦτο τελέσαι, "because atoms which hold each other and are suitable to produce this result become mutually entangled"[49]; but it is so vague that it might be an account of almost any body which is at least temporarily cohesive. It may be that difficulties such as these deterred ancient thinkers from attempting to provide a full account of clouds.

Additional note on the interpretation of Seneca NQ IVb.3.1–2

This note gives my view on three problems discussed by Kidd (1988) 510–12.

1 What is the point of "apud nos" ("in the way in which ice is formed *with us*")? I agree with the argument of Kidd (1988) 510–11 that this must mean "on earth" (contrasting this with "in the air above us"). I suggest that the point of "with us" (sc. on earth) is that Seneca agreed with Aristotle[50] that actual ice, as opposed to snow and hoar-frost, is always formed from liquid water, as we know from experiencing ice at ground level; he says at *NQ* IVb.3.6 "grando nihil aliud est quam suspensa glacies", "hail is nothing else than ice hanging in mid air". This surely explains why Seneca quotes Posidonius' view that "hail is formed from a watery cloud ... turned to liquid" in support of his own statement that hail is formed "the whole cloud having been frozen": the cloud must be liquid water if it is to freeze completely, to real ice.

2 Why does Seneca call his statement about hail occurring when a cloud has been frozen "too bold"? As Kidd points out, and as we have seen, the concept of a frozen cloud was adopted by many writers on meteorology.

I suggest that, when Seneca says "I shall have done too bold a thing", he is echoing an objection by an imaginary interlocutor, that Seneca cannot have *seen* inside a hail cloud (cf. "witnesses, who say that they have not seen"). This would not be the only passage of *NQ* where such an interlocutor questions an improbable feature of clouds hypothesised by ancient meteorologists: compare, for instance, II.25, where an imagined interlocutor questions whether lightning can be produced from clouds by friction: an idea accepted in several ancient lightning theories (see Chapter 8, especially pp. 66–70).

3　What is the meaning of "quod praeteriit" (here translated "what has passed")? I agree with the argument of Kidd (1988) 512 that this phrase, contrasted with "hoc quod secuturum est", "this which will follow" immediately afterwards, must refer to something mentioned just previously, and that the translation in Oltramare (1961), "ce qu'il a passé sous silence" (supposed to refer to something in the lost part of the book), cannot be right here, although it is a more natural meaning of the phrase than "what has passed" with the sense of "what has just been mentioned". How could Seneca refer to Posidonius as an authority for something Posidonius had passed over in silence? But I doubt Kidd is right to say that the beginning of the fragment must be referred to. It seems to me more likely that Seneca is referring to something now lost which came just before that, in which Posidonius was named.

Notes

1　Aristotle in the relevant section of Mete., i.e. I.9–12, also deals with dew and hoarfrost; I ignore these here, as there is no evidence for Posidonius' view of them.
2　Seneca *NQ* II.54.1, quoted above, Chapter 8 (p. 69).
3　Hesiod *Op.* 547–53, quoted above, Chapter 4 (p. 20).
4　See [Hippocrates], *Airs waters places*, 8; Herodotus II 25. Among the pre-Socratics, see, for Anaximander, Hippolytus, *Haer.* I.6.7 (DK 12A11); for Xenophanes, fragment 30 and Aëtius III.4.4 = DK 21A46 (both partly quoted Chapter 10 [p. 89]), for Heraclitus, Diogenes Laertius IX 9 and 11 (DK 22A1), for Democritus, Aëtius IV.1.4 (DK 68A99); for Metrodorus, Aëtius III.4.3 (DK 70A16): clouds are formed "from watery matter rising" (ἀπὸ τῆς ὑδατώδους ἀναφορᾶς); clouds are obviously the source of rain.
5　*Mete.* I.9, especially 346b20–33.
6　*Letter to Pythocles* 99–100: one way in which clouds form is κατὰ ῥευμάτων συλλογὴν ἀπό τε γῆς καὶ ὑδάτων, "by a collection of currents from earth and waters"; and rain may be formed ἀπ᾽ αὐτῶν ἦ μὲν θλιβομένων, ἦ δὲ μεταβαλλόντων, "from them [i.e., clouds] being squeezed in one place, or changing in another".
7　See above, Chapter 4 (p. 28).
8　Diogenes Laertius VII.153.
9　See above, Chapter 3 (p. 11), and Chapter 4 (p. 27–8).
10　*Mete.* 346b26–31.

11 *Letter to Pythocles* 99–100, clouds may form παρὰ πιλήσεις ἀέρος, "by compression of air", and rain may form when clouds are "squeezed" (θλιβομένων).
12 See above, Chapter 7 (p. 59–60).
13 Diogenes Laertius VII 153 (Posidonius F11 EK).
14 *Mete.* 347b23, with translation from Lee (1952).
15 Stobaeus, *Eclogae* I p. 245.23 W. (*SVF* II 701).
16 Kidd (1988) 115. Vimercati (2004) 573 draws attention to the mention of moisture, but suggests no details of the process envisaged.
17 *Fragmenta de rebus physicis* 4 (see Roos/Wirth [1968] 192; Posidonius F336b in Theiler [1982]).
18 Seneca *NQ* IVb.4.2.
19 *De mundo* 394a34–36 (Posidonius F336a in Theiler [1982]).
20 The text has τοῦ ὑγροῦ...οὔπω χυθέντος οὐδὲ ἠραιωμένου, "the moisture not yet poured(?) or rarefied". The meaning and the point of this is unclear.
21 Daiber (1992) 267.
22 *Mete.* 347b17.
23 *Mete.* 347a17.
24 Hippolytus, *Haer.* I.7.7 (DK 13A7).
25 *Letter to Pythocles* 107–8.
26 Posidonius F136EK.
27 *Mete.* 347b35.
28 *Mete.* 348b17; cf. 348a4–6, 348a10–14. The Syriac Meteorology agrees with Aristotle: hail is formed when "drops of water are ... hardened by coldness" (Daiber [1992] 267).
29 Hippolytus, *Haer.* I.7.7 (DK 13A7). In the MSS. of Aëtius III.4.1 this process is described as producing snow, clearly because snow and hail have been confused: see DK 13A17 and apparatus criticus; Mansfeld and Runia (2020) 1208.
30 Galen *In Epid.* III.32 (DK 87B29).
31 Tr. Mensch and Miller (2018) 529. The more recent MSS. have πῆξιν, "freezing"; the oldest have τῆξιν, "melting", and Dorandi (2013) adopts this reading; but it surely makes no sense in an account of hail.
32 Diogenes Laertius VII.153.
33 *De mundo* 394b2–5.
34 Aëtius III.4.2 (DK 59A85).
35 Stobaeus, *Ecl.* I p. 245,23 W. = *SVF* II.701.
36 The view of Posidonius' hail theory taken by Vimercati (2004) 572 is close to this (but he is surely wrong to see no irony in what Seneca says of Posidonius). In view of the passages I have quoted concerning snow, the conclusion of Kidd (1988) 513 that Posidonius envisaged a frozen cloud partially melting seems to me most unlikely.
37 *Mete.* 348a15ff, b13f (DK 59A85).
38 *Mete.* 348b2ff.
39 *NQ* IVb.3.5.
40 What we know of his work Περὶ ὠκεανοῦ ("On ocean") is enough to prove this: see Posidonius F49 EK.
41 "Many observers have remarked that hail-bearing clouds have a greenish colour", according to Lane (1968) vol. 2, p. 36.
42 See above, Chapter 9 (p. 76).
43 *Mete.* 347a17, quoted above. Compare https://www.britannica.com/science/hoarfrost: "Hoarfrost ... is formed by direct condensation of water vapour to ice" (Read 23 November 2021).
44 Simplicius, *Phys.* 24.26ff (DK 13A5); Hippolytus, *Haer.* I.7.3 and 7 (DK 13A7); Aëtius III.4.1 (DK 13A17).
45 *Mete.* 340a24–5.

46 *Mete.* 346b32–3.
47 *De mundo* 394a27, with tr. of Furley in Forster and Furley (1955).
48 See above, Chapter 8 (especially p. 65–6).
49 *Letter to Pythocles* 99.
50 *Mete.* 348a6, οὔτε γὰρ παγῆναι δυνατὸν πρὶν γενέσθαι ὕδωρ, "for it cannot be frozen [sc. to hail, i.e., ice] before it becomes water".

15 Rivers

The Nile floods

Aristotle discusses rivers in Chapter 13 of the first book of his *Meteorologica*, arguing that they must be fed, at least in part, by condensation of vapour within the earth, and showing that all the greatest rivers have their sources in mountains. There is no evidence that Posidonius discussed these matters, but he did comment on one related topic: the cause of the Nile's summer flood. This was a subject which had been much discussed by ancient authors, certainly from the 5th century B.C. onwards, and very likely from the 6th century. We still have Herodotus' review of the problem,[1] and the evidence is quite good that it had been a concern of Thales, since it includes the Latin *De inundatione Nili*, which may well be a translation of a genuine work by Aristotle.[2]

Early thinkers proposed a variety of solutions. Thales is said to have held that the summer flood occurs because the northerly Etesian winds blow at this season and hold the Nile back in its course; Anaxagoras thought the flood due to the melting of snow on African mountains[3]; Herodotus thought that in winter the sun, being in the south, partially dries up the Nile, but in summer the sun has moved north and affects the Nile less, so that the Nile then rises[4]; Eudoxus apparently reported "the priests" as saying that during our summer it is winter in the Antarctic, and the Nile flows from there and so floods in our summer.[5] I need not examine these and other theories here, since increased geographical knowledge provided important information on the subject: Greeks and Romans came to learn about heavy summer rains in the mountains of Ethiopia, and realised that these caused the Nile's summer flood – this is not true of every writer about the Nile, but it is true of Posidonius. As he was also interested in the history of this realisation, it will be convenient to look first at the evidence for his views, and to consider afterwards which other Greek writers had held similar views before him.

Strabo reports Posidonius as saying, in opposition to Polybius, that the land at the equator is a plain at sea-level (see above, Chapter 7 [p. 59]), τοὺς δὲ πληροῦντας τὸν Νεῖλον ὄμβρους ἐκ τῶν Αἰθιοπικῶν ὀρῶν συμβαίνειν, "but the rains which fill the Nile come from the Ethiopian mountains".[6] The implication is that yes, there are mountains where heavy rains fall and cause the Nile to flood, but they are not on the equator (which is correct). Similarly, Cleomedes reports Posidonius as saying, in support of his belief in a

DOI: 10.4324/9780429399930-15

temperate equatorial zone, περὶ τὴν Αἰθιοπίαν ὄμβροι συνεχεῖς καταφέρεσθαι ἱστοροῦνται περὶ θέρος καὶ μάλιστα τὴν ἀκμὴν αὐτοῦ· ἀφ' ὧν καὶ ὁ Νεῖλος πληθύειν τοῦ θέρους ὑπονοεῖται, "in Ethiopia continuous rainfall is reported in summer, and particularly at the height of summer; it is from this actually that the flooding of the Nile in summer is conjectured".[7]

This was, as Posidonius himself said, not a new view. Strabo, on evidence which I shall discuss below, regards it as an established fact that the Nile floods are due to summer rains in Africa,

μηδὲ τοιούτων δεῖσθαι μαρτύρων, οἵους Ποσειδώνιος εἴρηκε. Φησὶ γὰρ Καλλισθένη λέγειν τὴν ἐκ τῶν ὄμβρων αἰτίαν τῶν θερινῶν. παρὰ Ἀριστοτέλους λαβόντα, ἐκεῖνον δὲ παρὰ Θρασυάλκου τοῦ Θασίου (τῶν ἀρχαίων δὲ φυσικῶν εἷς οὗτος), ἐκεῖνον δὲ παρ' ἄλλου, τὸν δὲ παρ' Ὁμήρου διπετέα φάσκοντος τὸν Νεῖλον

did not need ... witnesses of the kind that Posidonius has given. His account is that Callisthenes speaks of the cause from summer rains, taking it from Aristotle, and Aristotle from Thrasyalces from Thasos (he was one of the early physicists), and he took it from another, and he from Homer, who says the Nile is 'fallen from Zeus' [i.e., fed by rain]".

(Part of Strabo XVII.1.5, trans. I.G. Kidd [1999, Cambridge University Press, p. 296] modified)[8]

There is a further indication of Posidonius' view in a papyrus fragment, P. Oxy. 4458,[9] of the 3rd century A.D., which, R.L. Fowler has argued,[10] is a fragment of Posidonius' own work. It is clear that it is about the summer flood of the Nile, because the first column of the fragment includes the Greek text of the account of Herodotus' Nile theory which is found in Latin in the Aristotelian (or Pseudo-Aristotelian) *De inundatione Nili*.[11] There is a parallel with Strabo's account of Posidonius' view in lines 20–7 of column 2 of the fragment, which read as follows:

20 .. τὴν ἱστορίαν[
   .. νους ἐβουλομ[
   ... Ἀριστοτέλην οτ[
   [...] αὐτοῦ Θρασυκλ[
   [.. τ]ῶν πάλαι σοφῶ[ν
25 .. [τ]ὴν γνώμην .[
   .. ὡμολόγηκεν[
   [.].ρω δ' ἀνατιθεις[

This is too fragmentary for it to be worth attempting a continuous English translation, but clearly we have here a mention of Aristotle followed by "Thrasycl ... of the wise men of old", who is surely meant to be the man Strabo calls "Thrasyalces ... one of the early physicists". In line 21 νους could be the final letters of Καλλισθένους, and in line 27 ρω could be part of Ὁμήρῳ, giving us mentions of Callisthenes and Homer; there is room for brief mention of "another" at the end of line 26. This, Fowler's reconstruction of the text, seems highly probable, as it gives the same sequence of names that Strabo XVII.1.5 says were cited as witnesses by Posidonius; which leaves little doubt that this text is, or is following, the text of Posidonius there followed by Strabo. The second word of line 21 must be ἐβουλόμην, "I wished"; this, Fowler comments, is "a first-person authorial statement ... which can hardly be imagined as the utterance of someone else using Poseidonios; it is therefore a safe conclusion that the papyrus is Poseidonios himself". This conclusion seems to me less than certain: it is surely possible that a later writer took information from Posidonius and incorporated in it a first-person statement.[12]

As it seems clear that these lines are by or derived from Posidonius, there must be a strong chance that what immediately precedes is also by or derived from him. Lines 12 to 19 are fragmentary but evidently mentioned men sailing (πλεόντων) on a stretch of water with a name probably beginning Αρ; the cinnamon-producing country; elephant hunting; and (Ptolemy) Philadelphus. As D. Hughes, in publishing the fragment, and Fowler point out, Strabo XVII.1.5 explains this: he says that summer rains, falling especially in the mountains of Ethiopia, had been observed by men sailing τὸν Ἀράβιον κόλπον, "the Arabian Gulf", our Red Sea, as far as the cinnamon-producing country, and by men sent to hunt elephants by the Ptolemaic kings, especially Philadelphus (King of Egypt 282–246 B.C.). So, it is highly probable that this evidence was also cited by Posidonius.

After the mention of Philadelphus, Strabo goes on to speak of things not in the papyrus and presumably not mentioned by Posidonius: of expeditions by earlier kings into lands south of Egypt and the records kept by Egyptian priests; and Strabo comments:

> θαυμαστὸν οὖν πῶς ... οὐ τελέως ἐναργὴς ἦν ἡ περὶ τῶν ὄμβρων ἱστορία τοῖς τότε, καὶ ταῦτα τῶν ἱερέων ... ἀναφερόντων εἰς τὰ ἱερὰ γράμματα ... ὅσα μάθησιν περιττὴν ἐπιφαίνει

> it is surprising therefore that ... the story about the summer rains was not perfectly clear to the men of that time, especially when their priests ... report ... in their sacred writings[13] anything that displays curious learning.
> (Part of Strabo XVII.1.5, trans. I.G. Kidd [1999, Cambridge University Press, p. 295–6], slightly modified)[14]

In the context this must mean "it is incredible that the story was not perfectly clear"; for this is evidently the reason why Strabo goes on to say that there was no need to cite, as Posidonius did, Callisthenes and other Greeks as evidence for the occurrence of the rains: this has surely been introduced by Strabo to give himself the chance to do what he likes to do and criticise Posidonius. Strabo does not claim definite knowledge that old Egyptian records mentioned summer rains in Ethiopia as the cause of Nile floods, and very likely there were no such records: "the cause of the Nile flood was discussed in religious categories in Pharaonic culture".[15] On the other hand, sailors to the cinnamon-producing country, apparently Somalia,[16] would have been much nearer to the Ethiopian mountains than southern Egypt is, and if they did not travel to the mountains themselves, would be likely to have met people from there, and the same applies to men sent south from Egypt to hunt elephants. Such testimony, almost certainly cited by Posidonius, would have had some authority; we may admit, with Strabo, that the views of Callisthenes and the others, though historically interesting, are not authoritative, in the absence of information about the evidence on which their views are based. It is also likely, from the first column of P. Oxy. 4458, that Posidonius quoted the account of Herodotus' theory from the Greek original of *De inundatione Nili*, though here it seems to me possible, if the fragment is derived from Posidonius but not actually by him, that the author drew on the Aristotelian treatise also, and that there was some indication of a change from Aristotle to Posidonius in the text that is missing between the two columns of the fragment.

Of the authors cited by Posidonius, the view of Callisthenes is reported by Johannes Lydus, summarising a lost part of Seneca's *Naturales quaestiones*: Callisthenes said that he campaigned with Alexander, καὶ γενόμενον ἐπὶ τῆς Αἰθιοπίας εὑρεῖν τὸν Νεῖλον ἐξ ἀπείρων ὄμβρων κατ' ἐκείνην γενομένων καταφερόμενον., "and being in Ethiopia he found that the Nile was borne down [sc. in flood] as a result of unlimited rains occurring there".[17] Callisthenes, accompanying Alexander, cannot have himself experienced the summer rains of Ethiopia,[18] but presumably he was told and recorded that there were such rains. (Another text, the "Anonymus Florentinus" on the rise of the Nile[19] ascribes this view to Callisthenes, adding that the cause is the etesian winds carrying clouds to the mountains of Ethiopia; but he does not suggest that Callisthenes had been there.)

Aristotle's view of the Nile floods is not recorded in any work of his preserved in Greek, but he does say at *Mete.* 349a5–7: γίγνεται δὲ καὶ περὶ τὴν Ἀραβίαν καὶ τὴν Αἰθιοπίαν τοῦ θέρους τὰ ὕδατα καὶ οὐ τοῦ χειμῶνος, καὶ ταῦτα ῥαγδαῖα, καὶ τῆς αὐτῆς ἡμέρας πολλάκις, "in Arabia and Aethiopia rain falls in the summer and not in the winter, and falls with violence and many times on the same day".[20] Later writers attribute to Aristotle the theory that the etesian winds blow clouds against mountains in the south, where they condense into rain and so cause the Nile floods,[21] and this theory is accepted in the possibly Aristotelian *De inundatione Nili*; we can hardly doubt, in view of the *Meteorologica* passage, that Aristotle accepted this theory or something very like it.

Of Thrasyalces we know hardly anything more than Strabo tells us[22]; the only other author to mention his Nile theory is Lydus, summarising Seneca,[23] who says he held that τῆς ... Αἰθιοπίας ὑψηλοῖς ... ὄρεσι διεζωσμένης ὑποδεχομένης τε τὰς νεφέλας πρὸς τῶν ἐτησίων ὠθουμένας ἐκδιδόναι τὸν Νεῖλον, "because Ethiopia is surrounded by high mountains and receives the clouds which are driven by the etesians, the Nile floods". This is the same theory that is attributed to Aristotle. We do not know who the "other" was, who was Thrasyalces' source,[24] but we do find a similar theory attributed to Democritus.[25] It seems odd that Democritus is not mentioned: the attribution of this Nile theory to him, even if incorrect, must surely have been made before Posidonius' day.

It is far-fetched to read this theory into Homer's single word διπετέα.[26] However, it is clear, from reports of what Posidonius said about winds and tides in Homer[27] that in Posidonius' view what Homer says on meteorological subjects has to be interpreted so as to mean what Posidonius knew from other evidence to be the case. For the Nile, Posidonius was probably pleased to find that there was a plausible reason why Homer should have called the Nile "fallen from Zeus", that is, from the sky, when the Nile is an Egyptian river and Egypt itself is notoriously rainless.

Several ancient writers refer to eye-witness reports of summer rains in the mountains of Ethiopia. Disregarding what Lydus says of Callisthenes, eye-witness reports are mentioned by Eratosthenes (as reported by Proclus),[28] by the *De inundatione Nili*,[29] and by Strabo[30]; Diodorus, apparently following Agatharchides, refers to testimony given ὑπὸ τῶν περὶ τοὺς τόπους οἰκούντων βαρβάρων, "by the barbarians who live about the area".[31] Only Strabo gives any details of the witnesses; I have summarised what he says on page 154–5 above, and argued that it includes matter of value as an authority (which is almost certainly taken from Posidonius). Though Kidd is sceptical about Strabo's witnesses,[32] the reports of summer rains are true[33]; it is surely unreasonable to doubt that there were eye-witness accounts of summer rains in Ethiopian mountains made by, or to, some Greek travellers.

As Strabo remarks,[34] there remains the question of what causes these summer rains. To Aristotle, the cause is ἀντιπερίστασις: the tropical heat surrounds and somehow concentrates the cold of the cloud and so causes condensation.[35] (This need not be the sole cause: the cloud must come from somewhere, perhaps driven by the etesians, as in other accounts.) Cleomedes indicates that, to Posidonius, the relatively temperate climate of the equatorial zone is part of the cause; this is surely just a necessary precondition, but it makes Aristotelian ἀντιπερίστασις unlikely as the main cause. Thrasyalces (as reported by Joannes Lydus), Democritus (as reported by Aëtius, Diodorus and the "Anonymus Florentinus"), Callisthenes (as reported by the same Anonymus), and the *De inundatione Nili*[36] all say that the cause is that the etesian winds blow clouds and moisture from the north against the mountains of Ethiopia and so cause rain there (if not by ἀντιπερίστασις then perhaps by the compression or the shattering of the clouds[37]). As we have no hint to the contrary, the likelihood is that Posidonius accepted this, the standard theory. At II.3.2 Strabo says that, according to Polybius, rain falls on the high land at the equator τῶν βορείων νεφῶν κατὰ τοὺς ἐτησίας ἐκεῖ τοῖς ἀναστήμασι προσπιπτόντων πλείστων, "since the clouds from the north are driven in great number by the etesian winds against the rising ground there".[38] Strabo is interested in disagreements between Polybius and Posidonius, but does not suggest that Posidonius criticised the idea that etesian winds cause the rains by blowing clouds from the north.

However, Agatharchides' view, as reported by Diodorus, was different: the Nile floods are caused by summer rain in the Ethiopian mountains, but the cause of the rains is undiscoverable.[39] It is possible that Posidonius agreed with this.

On the available evidence, Posidonius contributed little that was original to the explanation of the Nile floods: at most, he clarified the position of the Ethiopian mountains and made it clear that equatorial lands are more temperate than traditionally supposed. However, he did adopt the best available theory, backed by the evidence of eye-witness reports of summer rains; and, as I have argued, he took the trouble to state where the eye-witness reports had come from. This was not enough to convince later ancient writers of their truth. The theory that the etesian winds cause the rains is incorrect, and was argued against in antiquity: Diodorus, probably following Agatharchides, has several arguments against the theory, for instance that the Nile starts to flood before the etesians begin to blow[40]; Aelius Aristides, in the 2nd century A.D., put forward the strong argument of eye-witness evidence that clouds are not blown southwards in summer over upper Egypt towards Ethiopia, as the theory about the etesians requires.[41] In the conditions of his time Agatharchides was right: the cause of the rains was undiscoverable. The summer rains are actually due to climatic conditions in tropical latitudes, of which the ancients could have had no knowledge.[42]

Very few, if any, educated Greeks or Romans can have personally experienced the summer rains of the mountains of Ethiopia, or even met people familiar with them. It would be easy to suspect that reports of summer rains there might be no more than travellers' tales. Epicurean thinkers can have seen no reason to forgo their usual assumption of multiple possible causes when explaining the Nile floods. Lucretius (VI.712–37) suggests four causes, three of which have been mentioned already: the northerly etesian winds may hold the river back in its course; or, winds may build up sandbanks which hold the river back; or, the etesian winds may blow clouds onto mountains in the south and cause rain there; or, the floods may be due to the sun melting snow on African mountains. The explanation which had some support from eye-witness evidence is mentioned third. And Lucretius was not alone. In the centuries that followed his, there were writers – not Epicureans: the elder Pliny, Lucan, Aristides – who gave no preference to the theory that the Nile floods are due to summer rains in Ethiopian mountains, but gave equal weight to older theories, in which those summer rains play no part[43]; though this is not true of every writer.[44] So difficult was it, in antiquity, to convince the literate public of the truth of a new discovery.

Notes

1 Herodotus II.19–26.
2 On Thales, and on the *De inundatione Nili*, see above, Chapter 4 (p. 21).
3 See *De inundatione Nili* (p. 749, col. 1,9ff in Gigon [1987]); Diodorus I.38.4; Seneca *NQ* IVa.2.17; Aëtius IV.1.3. Herodotus II.22 describes this theory but does not name its author (DK 59A91).

4   Herodotus II.24–6.
5   Aëtius IV.1.7. *De inundatione Nili* (p. 750, col. 1,3ff in Gigon [1987]) attributes this theory to "Nicagoras".
6   Strabo II.3.3 (Posidonius F49.130–5 EK).
7   Cleomedes I.6.32 (I.4 lines 107–9 Todd) = Posidonius F210.20–23 EK with translation from Kidd (1999) 276.
8   Posidonius F222.8–15 EK. On this passage see Kidd (1988) 795–9; Theiler (1982) II p. 68; Vimercati (2004) 596–7.
9   Published by Hughes (1998) 66–71 and plate VII.
10 Fowler (2000).
11 Pointed out by Jakobi and Luppe (2000).
12 It is not clear what the author "wished". Fowler's conclusion, that probably "Poseidonios wanted to cite all these people but ... could not, either because the works were not available or because they did not contain the views attributed to them", seems to me incompatible with Strabo's statement that Posidonius cited them as witnesses.
13 Presumably meaning "hieroglyphics", cf. Herodotus II.36.
14 Posidonius F222.1–4 EK. I modify Kidd's translation by altering his "secret" to "sacred".
15 *Brill's new Pauly* (2002–10), 9, 758–60, article "Nile" by S.J. Seidlmayer. See also (for example), with quoted texts in translation, Wilson (1949) 89–90; Bell (1971) 20–1.
16 Cary and Warmington (1963) 88.
17 Lydus, *De mensibus*, IV.68, summarising the lost part of Seneca, *NQ*, book IVa, and included in editions of *NQ*.
18 Alexander seems not to have travelled up the Nile further than Memphis; he entered Egypt in October or November 332 B.C. and left again the following spring (Arrian, *Anabasis*, III.1–5; of modern authors, e.g., Hammond [1981] 117–30; Bosworth [1988] 68–74). Callisthenes may possibly have travelled up the Nile while Alexander was in Egypt, as Hammond (1981) 128 suggests, but he cannot have seen summer rains in Ethiopia, since Alexander was never in Egypt in summer.
19 See Jacoby (1958) 199–201, no. 647, text 1.
20 Translation from Lee (1952).
21 Alexander, *In Mete.* p. 53.11–16 Hayduck (citing Aristotle, Περὶ τῆς τοῦ Νείλου ἀναβάσεως); cf. Philoponus, *In GC* p. 83.8–17 Vitelli.
22 Strabo also mentions him in connexion with wind-names (Strabo I.2.21 = Posidonius F137a EK).
23 Lydus, *De mensibus*, IV.68 (see n. 17 of this chapter).
24 Suggestions have been made: see Kidd (1988) 798.
25 See Aëtius IV.1.4, Diodorus I.39.1–6, and the "Anonymus Florentinus". The Scholiast on Apollonius Rhodius IV 269f attributes to Democritus a different theory, that the Nile flows from the southern ocean – a theory elsewhere attributed to Euthymenes (Aëtius IV.1.2; Seneca *NQ* IVa.2.22). (See DK 68A99 for Aëtius IV.1.4, Diodorus and the Scholiast.)
26 *Odyssey*, IV.477 and 581.
27 See above, Chapter 11 (pp. 103, 106–7) and Chapter 13 (p. 138).
28 See Proclus, *Comm. in Plato. Tim.*, ed. Diehl I p. 120.21–121.13 (= Aristotle fr. 687 in Gigon [1987]).
29 Page 750, col. 2,6ff in Gigon (1987). See also Photius, *Bibliotheca*, ed. Bekker cod. 249 p. 441a34–b14 = Aristotle fr. 686 Gigon.
30 Strabo XVII.1.5 refers to αὐτόπται, "eye–witnesses", of the summer floods.
31 Diodorus I.41.8.
32 Kidd (1988) 796.
33 See, for instance, Griffiths (1972) 374 (in chapter 11, "Ethiopian highlands"): "For most of the zone the heaviest rainfall is during the period June–September, for

instance Gondar 85%, Asmara 80%, Addis Ababa and Gambela 70%, Dire Dawa 60% and Jimma 55%".

34 XVII.1.5 = Posidonius F222.5-7 EK.
35 *Mete.* 349a8, and see 348b2ff (cited above, in the discussion of hail, Chapter 14 [pp. 146, and 150 n. 38]) for an explanation of ἀντιπερίστασις in this sense.
36 Page 750, col. 2,11ff in Gigon (1987).
37 Cf., in the "Anonymus Florentinus", συνωθουμένων τῶν νεφῶν, "the clouds being pushed together"; in Diodorus I.39.2–3, νέφη ... θραυόμενα, "clouds ... being shattered" (both referring to Democritus).
38 Posidonius F49.121–3 EK, with translation from Kidd (1999) 114.
39 Diodorus I.41.4–9.
40 Diodorus I.39.4–6.
41 *Aegyptius* 34.
42 See, e.g., Griffiths (1972) 370.
43 E.g., Pliny, *Nat.* 5.55; Lucan X.219–67; Aristides, *Aegyptius*.
44 Proclus (as cited in n. 28 of this chapter) accepts that rains on Ethiopian mountains cause the Nile to flood, and Alexander, *In Mete.* p. 53 9–16 Hayduck appears to do so.

16 Rainbows, haloes and mock-suns

Rainbows

Stoic theories of rainbows are summarised at Diogenes Laertius VII.152 (Posidonius F15 EK):

Ἶριν δ᾽ εἶναι αὐγὰς ἀφ᾽ ὑγρῶν νεφῶν ἀνακεκλασμένας ἤ, ὡς Ποσειδώνιός φησιν ἐν τῇ Μετεωρολογικῇ, ἔμφασιν ἡλίου τμήματος ἢ σελήνης ἐν νέφει δεδροσισμένῳ, κοίλῳ καὶ συνεχεῖ πρὸς φαντασίαν, ὡς ἐν κατόπτρῳ φανταζομένην κατὰ κύκλου περιφέρειαν

The rainbow is constituted by rays of light reflected from moist clouds, or, as Posidonius says in his *Meteorology*, a reflection of a section of the sun or moon in a dewy cloud that is hollow and continuous in appearance, the image showing itself as if in a mirror in the form of a circle's circumference.
(Trans. P. Mensch and J. Miller (2018, Oxford University Press)[1]

The first eight words are presumably an early Stoic theory. Further information on Posidonius' view is given by Seneca. At *NQ* I.5.10,[2] after describing a theory that the cloud in which a rainbow appears is actually coloured, not just acting as a mirror, he goes on:

Posidonius et hi qui speculari ratione talem effici iudicant visum hoc respondent: 'Si ullus esset in arcu color, permaneret et viseretur eo manifestius quo propius; nunc imago arcus, ex longinquo clara, interit, cum ex vicino ventum est'

Posidonius and those who consider that such a sight [i.e. a rainbow] is produced in the way of a mirror, make this reply: 'If there were any colour in the rainbow, it would remain and would be seen more clearly the nearer it was; but as it is, the image of the rainbow, clear at a distance, vanishes when one comes near'.

DOI: 10.4324/9780429399930-16

At *NQ* I.5.13 Seneca says:

> In eadem sententia sum qua Posidonius ut arcum iudicem fieri nube formata in modum concavi speculi et rotundi, cui forma sit partis e pila secta. Hoc probari, nisi geometrae adiuverint, non potest, qui argumentis nihil dubii relinquentibus docent solis illam esse effigiem non similem

> I hold the same opinion as Posidonius, so that I consider that a rainbow is produced when a cloud has been shaped into the form of a concave, round mirror, whose shape is that of part of a cut ball. This cannot be proved without the help of geometers, who show by arguments which leave no room for doubt that it is a reflection of the sun, but not like it.

At I.8.4 Seneca returns to this theory:

> Nostri, qui sic in nube quomodo in speculo lumen volunt reddi, nubem cavam faciunt et sectae pilae partem, quae non potest totum orbem reddere, quia ipsa pars orbis est

> Our school, who think that the light is reflected in a cloud in the way it is in a mirror, posit a hollow cloud and part of a cut ball, which cannot reflect a complete sphere because it is itself part of a sphere.

Before Posidonius, Aristotle's account of rainbows, in *Meteorologica* Book 3,[3] is by far the most detailed among those about which we have information. Seneca implies that there were two sorts of theory: those in which the rainbow is produced by reflection, with cloud just acting as a mirror, and those in which the cloud is actually coloured. Aristotle is clear that the cause is ἀνάκλασις, "reflection".[4] His account of rainbows is, in summary, as follows. At *Mete.* 373b14-33 he says that a rainbow is formed ἀπὸ ὕδατος ... ἀρχομένου γίγνεσθαι ... ἕκαστον γὰρ τῶν μορίων ἐξ ὧν γίγνεται συνισταμένων ἡ ψακὰς ἔνοπτρον ἀναγκαῖον εἶναι, "from water beginning to form: for each of the parts from which the raindrop is formed as they condense must be a mirror";[5] or, as he puts it just afterwards, ὅταν ἄρχηται ὕειν καὶ ἤδη μὲν συνίστηται εἰς ψακάδας ὁ ἐν τοῖς νέφεσιν ἀήρ, μήπω δὲ ὕῃ, "when it is beginning to rain, and the air in the clouds is already condensing into raindrops, but it is not yet raining".[6] Aristotle argues that under such conditions colour is reflected but not shape. At 374a3 to 375b11 he attempts, with digressions, to explain the colours of the rainbow; and at 375b16 to 377a28 he offers elaborate geometrical arguments, with the object of explaining why a rainbow is never more than a semicircle.[7] Nowhere does he try to explain why it is circular.

Posidonius' account appears to be different.[8] As Diogenes Laertius (quoted above) recounts his theory, a rainbow is seen "in a dewy cloud", and both he probably and Seneca explicitly talk about the *shape* of the cloud. Thus Posidonius spoke of reflection from a cloud, whose shape he described, while Aristotle spoke of reflection from water-droplets,[9] and said nothing about a cloud's shape. But the difference, if any, between Posidonius' νέφει δεδροσισμένῳ and Aristotle's water-droplets forming in a cloud, cannot be great. Aristotle says that the result of the air in a cloud condensing into rain-drops is that γίγνεσθαι ἔνοπτρον τὸ νέφος, "the cloud becomes a mirror" if the sun is opposite it,[10] and later he says that a rainbow appears ὅταν τοῦτον ἔχῃ τὸν τρόπον ὅ τε ἥλιος καὶ τὸ νέφος καὶ ἡμεῖς ὦμεν μεταξὺ αὐτῶν, "when the sun and the cloud are in this state and we are between them".[11] Despite what he says of water-droplets, Aristotle can still speak of cloud as a cause of rainbows. Posidonius is reported as saying that a rainbow is seen νέφει δεδροσισμένῳ, "in a dewy cloud". We cannot be sure that there is a serious difference between him and Aristotle on this point. Posidonius was saying something which Aristotle did not when he spoke of the hollow, circular shape of a cloud as cause of the circle of a rainbow, but he may have felt, not that he was diverging from Aristotle's theory, but that he was filling a gap in it.

That Posidonius, like Aristotle, supported his theory of rainbows with geometrical arguments is not definitely attested, but the latter part of my quotation (above) from Seneca *NQ* I.5.13 makes it highly likely:[12] we know that Posidonius was interested in mathematics.[13] But, if he did defend his rainbow theory with geometry, we have no evidence of what arguments he used.[14]

Aristotle consistently speaks of rainbows as a reflection not of sunlight, but of our sight, saying (for example) ἀνάκλασις ἡ ἶρις τῆς ὄψεως πρὸς τὸν ἥλιόν ἐστι, "the rainbow is a reflection of our sight to the sun".[15] (He speaks thus despite his denial, in the *De sensu* and *De anima*, that sight and reflection occur through anything going out from us.[16]) Direct evidence is lacking, but it seems probable that Posidonius, too, spoke of rainbows as a reflection of our sight. The Stoics did not reject the idea that rays from the eye are involved in sight: Chrysippus held that προχέονται ... ἐκ τῆς ὄψεως ἀκτῖνες πύριναι, οὐχὶ μέλαιναι καὶ ὁμιχλώδεις· διόπερ ὁρατὸν εἶναι τὸ σκότος, "fiery rays, not black and misty ones, pour forth from our sight; for which reason darkness is visible".[17] Aëtius says that Posidonius calls sight αὐγῶν ... σύμφυσιν, "a natural fusion of light rays",[18] apparently meaning, rather like Plato at *Timaeus* 45B-D, that sight requires rays both from our sight and from the object.[19] Diogenes' citation of Posidonius' rainbow theory, and my first two quotations from Seneca, do not say whether sunlight, or our sight, is reflected; in the last quotation, "lumen reddi" indicates that light is reflected, but Seneca is not always reliable about the details of other men's theories. Cleomedes, when he speaks of how we perceive things, normally uses expressions like "the rays poured out from the eyes", and hardly ever speaks of the sun's light causing vision;[20] the passages concerned do not mention Posidonius, but it is likely, in the absence of contrary evidence, that they reflect his usage. (Our understanding of the "mirror", the cloud in which we see the rainbow, is not affected by whether one speaks of sunlight, or our sight, being reflected.)

Others besides Aristotle, in his time and before him, regarded rainbows as phenomena similar to reflections in a mirror. His contemporary, Philip of

Opus, evidently believed this, since Alexander recounts his argument in proof of it, comparing the apparent movement of a rainbow as the observer moves with the movement of a reflected image when what is reflected moves.[21] Some pre-Socratics evidently held this view: reports of the views on rainbows of Anaxagoras and "some Pythagoreans" are in terms of mirror reflection,[22] and for Anaxagoras we have his own words in fragment 19: a rainbow is τὸ ἐν τῇσιν νεφέλῃσιν ἀντιλάμπον τῷ ἡλίῳ, "that in the clouds which shines back at the sun"[23] – a different way of referring to reflection. We have also reports, or a fragment, concerning the rainbow theories of three other pre-Socratics, Anaximenes,[24] Xenophanes[25] and Metrodorus of Chios,[26] but for them, the evidence does not explicitly mention reflection (or any comparable optical phenomenon) and this is not the place to discuss what their theories were; the same applies to the several explanations of rainbows suggested by Epicurus in his *Letter to Pythocles*, 109-10. The view that a rainbow is a reflection clearly goes back at least to Anaxagoras.

It is clear, then, that many ancient thinkers, including Posidonius, realised that a rainbow is merely an "appearance" (in Greek ἔμφασις[27]), which makes no substantial change in the cloud against which we see it, just as an image seen in a mirror makes no substantial change in the mirror. As so often, their explanation depended on an analogy, with mirrors in this instance, though neither the shape nor the colours of a rainbow resembles an ordinary mirror image. Aristotle's was not the first attempt to explain the colours of the rainbow: earlier attempts are attributed to Anaximenes and Metrodorus of Chios.[28] I do not discuss the ancient explanations here, because we lack evidence for Posidonius' view – if texts which we know influenced Posidonius and texts which we know he influenced were in agreement then we might say that he probably shared their view, but with the colours of the rainbow the relevant accounts, in Aristotle *Mete.* 374a3 to 375b11 and Seneca *NQ* I.3.12, have little resemblance.

Posidonius did try to explain the circular shape of the rainbow, a matter not discussed by Aristotle in his *Meteorologica*, nor, so far as we know, by any pre-Aristotelian. Epicurus does offer explanations:

τὸ δὲ τῆς περιφερείας τοῦτο φάντασμα γίνεται διὰ τὸ τὸ διάστημα πάντοθεν ἴσον ὑπὸ τῆς ὄψεως θεωρεῖσθαι, ἢ σύνωσιν τοιαύτην λαμβανουσῶν τῶν ἐν τῷ ἀέρι <ἀ>τόμων ἢ ἐν τοῖς νέφεσιν ἀπὸ τοῦ αὐτοῦ ἀέρος ἀποφερομένων ἀτόμων περιφέρειάν τινα καθίεσθαι τὴν σύγκρισιν ταύτην

its curved appearance is due either to the fact that the distance from every point is perceived to be equal, or to the fact that the atoms in the air are forced together in such a way, or because the atoms in the clouds, which are derived from the same air, have been united in such a way that their aggregate displays a sort of roundness.

(*Letter to Pythocles* 110, trans. P. Mensch and J. Miller [2018, OUP, p. 529], slightly modified.)

None of this resembles Posidonius' idea of "a cloud ... shaped into the form of a concave, round mirror, whose shape is that of part of a cut ball".[29] This concept was apparently original to Posidonius, and is another example of a surprising property which the desire to explain a meteorological phenomenon compelled an ancient author to ascribe to a cloud.

Haloes

The earliest Greek author recorded as trying to explain the occurrence of a halo round a heavenly body is Aristotle, whose explanation, Alexander tells us,[30] was followed by Posidonius. Aristotle's explanation[31] of a halo seen round the sun, moon, or a star is (very briefly) that a suitably even cloud or mist is formed between the observer and the heavenly body, and that then "our sight is reflected [sc. to the body] by the mist in the same way at every point" (ἀνακλᾶται ἀπὸ τῆς ... ἀχλύος ... ἡ ὄψις ... πάντοθεν ὁμοίως), so producing the appearance of a circle. Alexander,[32] commenting on this, adds that the cloud "perpendicularly below the [sc. heavenly] body above it becomes rarer" (τὸ κατὰ τὴν κάθετον τοῦ ὑπερφερομένου σώματος ... μανότερον ... γίνεσθαι), while what is not perpendicularly below "is not affected in this way" (οὐδὲν τοιοῦτον πάσχει) and remains uniformly even cloud; and this cloud forms "a continuous series of small mirrors" (συνεχῆ καὶ μικρὰ κάτοπτρα) in "a circle at the boundary of the well-rarefied vapour" (κύκλον ... τὸ πέρας τῆς ἐπιεικῶς διακεκριμένης ἀτμίδος) – this vapour must be the same as the rarefied cloud perpendicularly below the heavenly body. Reflection from these mirrors produces the circular halo. Alexander evidently envisages a circular cloud, of which the centre is rarefied and dissolved by the sun, while the periphery, less affected by the sun, remains cloud and forms a series of mirrors arranged in a circle. Alexander adds that, in reality, it is not our sight but "the light of the heavenly body striking the mirrors" (τὸ φῶς τὸ τοῦ ἄστρου προσπῖπτον τοῖς ... κατόπτροις) that is reflected. Such, says Alexander, was Aristotle's view, and

> ἐπηκολούθησε δὲ αὐτῷ καὶ Ποσειδώνιος, πάντων σχεδὸν τῶν ἄλλων οὐ κατὰ ἀνάκλασιν ἀλλὰ <κατὰ> κατακλάσεις ὄψεων αἰτιωμένων, ὡς ἐπὶ τῶν δι' ὕδατος ὁρωμένων γίνεται· ὑποτίθενται γὰρ σφαιροειδὲς καὶ κοῖλον τὸ νέφος, ἔπειτα τὸ ὑπερκείμενον ἄστρον αὐτοῦ κατὰ κύκλον φασὶ διεσπασμένον ἐν αὐτῷ ὁρᾶσθαι

Posidonius followed him; practically everyone else put the cause not as reflection, but as refraction of sight, as happens with objects seen through water. They assume the cloud to be global in shape and hollow, and then the star above it, they say, is seen spread in it as a circle.

(Alexander, *In Mete.*, p. 143.8-12 Hayduck,
trans. I.G. Kidd [1999, CUP, p. 189]) [33]

The mention of "objects seen through water" presumably refers to the familiar refraction by which a straight object half inserted in a bowl of water appears to be bent. The point, I take it, is that in κατάκλασις, "refraction", the ray of light, or sight, is deflected, but not actually bent backwards as it is in ἀνάκλασις, "reflection".

It is unclear whether Posidonius adopted the simpler form of the theory we find in Aristotle, or the more elaborate version described by Alexander; and it is also unclear who is meant in Alexander's "almost everyone else". Aristotle's was not the only explanation of haloes that was, or may have been, formulated before Posidonius' day. Epicurus, writing of lunar halces only, suggests three mechanisms by which air or cloud-like matter may come to be arranged in a circle round the moon.[34] The Syriac Meteorology explains a lunar halo as an effect of moonlight on thick air, its circular shape being compared to the circular ripples which form when a stone is thrown into water.[35] Seneca *NQ* I.2, in his account of haloes, uses this same analogy. None of these sources mentions reflection or refraction, or states clearly that a halo is an optical phenomenon. Alexander cannot have been thinking of them when he wrote that "practically everyone else" explained haloes as due to refraction.

Haloes are described as due to refraction (κατάκλασις) in an anonymous explanation given at Aëtius III.18.1 (= III.5a.1 in Mansfeld and Runia 2020):

μεταξὺ τῆς σελήνης ἤ τινος ἄλλου ἄστρου καὶ τῆς ὄψεως ἀὴρ παχὺς καὶ ὀμιχλώδης ἵσταται· εἶτα ἐν τούτῳ τῆς ὄψεως κατακλωμένης καὶ εὐρυνομένης ... κύκλος δοκεῖ περὶ τὸ ἄστρον φαίνεσθαι

thick and misty air is set between the moon or some other heavenly body; then in this situation, our sight being refracted and widened ... a circle seems to appear around the heavenly body.

Kidd quotes this in his comment on the above passage of Alexander,[36] and also a scholion on Aratus 811, which, in an explanation of haloes, appears to mention "refraction of sight" (ὄψις ... ἐγκατακλωμένη) as well as "reflection" (ἀνάκλασιν).[37] But this connection of haloes with κατάκλασις, "refraction", is not common in the surviving literature.[38]

There is evidence suggesting that Posidonius was interested in refraction in other contexts. Cleomedes writes of the sun being possibly visible when below the horizon:

δύναιτο δ' ἂν καὶ ἡ ἀπὸ τῶν ὀμμάτων ἀποχεομένη ἀκτὶς ἐνίκμῳ καὶ νοτερῷ τῷ ἀέρι ἐντυγχάνουσα κατακλᾶσθαι καὶ ἐντυγχάνειν τῷ ἡλίῳ ἤδη ὑπὸ τοῦ ὁρίζοντος κεκρυμμένῳ,

it would be possible for the ray poured out from the eye to encounter humid and moist air, be refracted, and encounter the sun already hidden below the horizon.[39]

As Posidonius was an important source for Cleomedes, this may be taken from Posidonius. However, the refraction which causes the sun to be seen when below the horizon produces a phenomenon which appears very different from the refraction which produces a halo round the sun or moon. Both phenomena involve the bending of a ray of light (or, in the common ancient usage, of sight), but, when the sun is seen when it is below the horizon, what is remarkable is what is beyond the bend; when we see a halo round a heavenly body, what is remarkable is light coming from the medium (in fact, ice crystals) which is causing the bend in the ray of light from (or, in ancient terms, of sight to) the body. It would not be surprising if some ancient thinkers, possibly including Posidonius, realised that a similar κατάκλασις, "refraction" must be occurring in both phenomena, but nevertheless have said that the light of the halo is reflected light, and it is in fact correct to say so. To produce a halo requires both refraction and reflection. The UK Meteorological Office website says "most optical phenomena ... arise through a combination of reflection, refraction and diffraction", and says of haloes "typically, sunlight or moonlight is reflected by ice crystals, producing a white halo ... The majority of ice crystals are hexagonal ... and the most common angle of refraction through such a crystal is about 22 degrees. This is the most frequent type of halo".[40]

Mock suns

A mock sun (also called a parhelion or, as in the following quotation, a sun dog) is

> an atmospheric optical phenomenon that consists of a bright spot to one or both sides of the sun ... The sun dog is a member of the family of halos caused by the refraction of sunlight by ice crystals in the atmosphere. Sun dogs typically appear as a pair of ... patches of light around 22° to the left and right of the sun and at the same altitude above the horizon ... Sun dogs are best seen ... when the sun is near the horizon.[41]

Posidonius' theory of mock-suns is described in the Scholia to Aratus 881:

> Ποσειδώνιος παρήλιόν φησι νέφος στρογγύλον περὶ τὴν τοῦ ἡλίου ἔκλαμψιν ἐκ τοῦ ἡλίου λάμπον· οὐ γὰρ ἰδίῳ φωτὶ κέχρηται, ἀλλὰ τῷ τοῦ ἡλίου, ὥσπερ καὶ ἡ σελήνη. ἡλιοειδὲς δὲ εἶπε τῷ τε στρογγύλον εἶναι καὶ τῷ ἐλλάμπεσθαι ὑπὸ τοῦ ἡλίου

> Posidonius says that a mock sun is a round cloud around [meaning "near" or "beside"?] the brightness of the sun, deriving light from the sun. For it does not have its own light, but that of the sun, as the moon does. And he said that it is like the sun by being round and by being illuminated by the sun. [42]

The text says nothing about the character of the cloud, except for its shape. A reference to ἀνάκλασις in what follows my quotation. if it is still reporting Posidonius, indicates that to Posidonius the mock-sun's appearance is produced by reflection.

We know of only two men before Posidonius who tried to explain this phenomenon. Earliest is Anaxagoras, who is simply recorded as saying that mock suns are caused in a similar way to rainbows:[43] presumably he thought that mock suns, like rainbows, are caused by reflection. The other theory is Aristotle's: that a mock-sun occurs ἀνακλωμένης τῆς ὄψεως πρὸς τὸν ἥλιον ... ὅταν ὅτι μάλιστα ὁμαλὸς ᾖ ὁ ἀὴρ καὶ πυκνὸς ὁμοίως· διὸ οαίνεται λευκός, "our sight being reflected to the sun ... when the air is as even as possible and uniformly dense; for which reason it [the mock sun] appears white".[44] He does not explain the mock sun's shape, which Posidonius evidently thought he must do, as he did with rainbows: to Posidonius the mock sun, as well as the rainbow, is due to the sun acting on (apparently, being reflected from) a cloud of a particular shape. However, while the rainbow is described in our sources as a reflection, the mock sun *is* a cloud and compared to the moon, and apparently has some substance. Seneca seems to confirm this, quoting a definition of "parhelion" given by "some people" ("quidam"): "nubes rotunda et splendida similisque soli", "a cloud, round and shining and like the sun".[45] The scholiast on Aratus probably expresses the view of Posidonius, and others, when he says that some meteorological phenomena, e.g. rainbows, exist only in appearance, some, e.g. comets, are substantial. "but mock suns are mixed" (μικτὰ δὲ παρήλιοι)[46] – are substantial as being clouds, but appearances as being reflections of the sun. I am reluctant to speculate about why this distinction was made between mock suns and rainbows. There is no suggestion in Alexander's commentary on Aristotle's account of mock suns in *Mete.* III.6 that anyone regarded mock suns as produced by refraction rather than reflection.

In what follows my quotation from the scholia to Aratus 881 at the beginning of this section, the scholiast gives details about mock suns which agree with those in Aristotle: that a mock sun is seen at the side of the sun (ἐκ πλαγίων τοῦ ἡλίου);[47] that it does not occur below the sun because clouds there are quickly dissolved,[48] nor far from it because reflection then fails;[49] that mock suns occur at sunset and sunrise.[50] This may be Posidonius following Aristotle, but it is also possible that the scholiast is no longer following Posidonius; also, as it is true that mock suns occur at the same height as the sun above the horizon and not far from it, and are most conspicuous when the sun is low,[51] some of it could be the result of independent observation. On mock suns it is clear that much of Posidonius' theory differed from Aristotle's. though he may have taken details from him.

Associated with παρήλιοι are phenomena called by Aristotle ῥάβδοι[52] and by Seneca "virgae",[53] i.e. "rods", presumably some variation of the mock sun phenomenon. It is likely that Posidonius had something to say about them, but I do not discuss them here as there is no evidence about what he may have said.

Conclusion

With rainbows, haloes and mock suns, some ancient thinkers were nearer to the truth than any of them were with some other meteorological phenomena, such as thunder and lightning. They realised that rainbows and the like were optical phenomena, that what we see has no substantial existence; and they saw that geometry should be able to aid their understanding. Posidonius' personal contribution, so far as we can tell, was not great. He probably supported his rainbow theory with geometrical arguments, but we have no idea what they were. I have suggested that he is likely to have been interested in the phenomenon of refraction, but definite evidence is lacking. His one clearly attested contribution is his hypothesis of specially shaped clouds as causing rainbows and mock suns. The shape of these phenomena, especially that of the rainbow, needed explanation, and his explanation of the rainbow's shape shows ingenuity; but it is hard to see it as bringing him nearer to the truth.

Notes

1 For discussion see Kidd (1988) 124-5.
2 *NQ* I.5.10 and 13 are part of F134 EK.
3 *Mete.* 371b18-372b11, 373a32-377a28.
4 E.g. *Mete.* 372a18, 373a32.
5 *Mete.* 373b14-16.
6 *Mete.* 373b19-20.
7 Some scholars regard a large part of this passage (376a7 to 376b7, and possibly more) as an interpolation: see (e.g.) Wilson (2013) 251ff.
8 For discussions see Gilbert (1907) 614-16; Kidd (1988) 499-502; Theiler (1982) II p. 206-7; Vimercati (2004) 573-5.
9 Kidd (1988) 501 (cf. Gilbert (1907) 615) points to this as a difference between Aristotle's and Posidonius' theories.
10 *Mete.* 373b22.
11 *Mete.* 373b30.
12 Seneca also refers at *NQ* I.4.1 to irrefutable geometrical arguments about rainbows, without saying what the arguments were.
13 For instance, he defended geometry against Epicurean attacks (Proclus, *In Euclidis Elementa* p. 199.15-200.2 and p. 216.20ff Friedlein = Posidonius F46.11-14 and 47.44ff EK), and tried to calculate the diameter of the sun (Cleomedes II.1.79-80 (II.1 lines 269-286 Todd) = Posidonius F115 EK) and the circumference of the earth (Cleomedes I.10.50-2 (I.7 lines 1-50 Todd) = Posidonius F202 EK).
14 On Seneca's statement at *NQ* I.5.13 that geometrical arguments leave no doubt about how rainbows are produced, Kidd (1978a) 14 comments that the rainbow theory "cannot be proved said Posidonius ... without the help of mathematics". I doubt we can be sure of this, even presuming that the arguments are those of Posidonius: the comment about "leaving no doubt" could well be Seneca's.
15 *Mete.* 373b33, with translation from Lee (1952).
16 *De sensu* 438a25-27; *De anima* 435a5-7.
17 Aëtius IV.15.3 (*SVF* II 866).
18 Aëtius IV.13.3 (Posidonius F194 EK), with translation from Kidd (1999).

19 See Kidd (1988) 699.
20 Cleomedes II.1 lines 257-8 Todd, αἱ ἀκτῖνες αἱ ἀποχεόμεναι ἀπὸ τῶν ὀμμάτων, "the rays poured out from the eyes"; cf. similar expressions at lines 31-2 and 59-60. At II.6 lines 171-3 Todd we find νέφους λαμπρυνομένου ὑπὸ τῶν ἡλιακῶν ἀκτίνων καὶ ἡλίου ἡμῖν φαντασίαν ἀποπέμποντος, "a cloud made bright by the sun's rays and sending out an appearance of the sun to us" (i.e. the eyes *receive* light), but immediately afterwards, lines 174-5, ἡ ἀπὸ τῶν ὀμμάτων ἀποχεομένη ἀκτίς, "the ray poured out from the eyes". Cf. lines 188-9; also lines 181 and 184-5, τοῦ ὁρατικοῦ πνεύματος, "the breath of sight", which proceeds from the observer to the perceived object.
21 Alexander, *In Mete*.151.32ff Hayduck. (I owe this reference to Merker (2003).)
22 For Anaxagoras see Aëtius III.5.11 (DK 59A86) which has ἀνάκλασιν, "reflexion" and κατοπτρίζοντος, "showing as in a mirror"; for Pythagoreans, Aëtius III.1.2 (DK 58B37c) which has ἀνακλῶντος, "reflecting" and κατοπτρικὴν φαντασίαν, "mirror image".
23 DK 59B19.
24 Aëtius III.5.10, Scholia to Aratus p. 515.27 Maass (both DK 13A18); Hippolytus, *Haer*. I.7.8 (DK 13A7).
25 DK 21B32 (partly quoted above, Chapter 4 (p. 23)).
26 Aëtius III.5.12, Scholia to Aratus p. 516 Maass (both DK 70A17).
27 For Posidonius see p. 160 above; for Aristotle, e.g. *Mete*. 373b24 and 31.
28 See DK 13A18 and 70A17.
29 See above, p. 161.
30 Alexander, *In Mete*., p. 142-3 Hayduck (F133 EK).
31 *Mete*. 372b12ff, with quote from 372b34-373a3.
32 Alexander, *In Mete*., p. 142.28-143.6 Hayduck.
33 I give the text with Kidd's insertion of κατὰ before κατακλάσεις. For discussion see Kidd (1988) 497-8.
34 *Letter to Pythocles* 110-11.
35 Daiber (1992) 269-70.
36 Kidd (1988) 498.
37 Maass (1898) 488.21ff. The explanation of haloes is complex and apparently unconnected to Posidonius, so is not discussed here.
38 I have checked this by searches (made 16/11/2020) on κατάκλασις and κατακλάω in the *Thesaurus linguae Graecae*.
39 Cleomedes II.6 lines 174-7 Todd (part of Posidonius F293 in Theiler (1982)).
40 See www.metoffice.gov.uk/weather/learn-about/weather/optical-effects, article "Haloes and coronas" (read 15/11/2020).
41 *Wikipedia*, article "Sun dog" (read 17/11/2020). Authorities differ about the colour. *Wikipedia* says "subtly colored patches", but the UK Meteorological Office website (see preceding note, article "Parhelion", read 20/11/2020) says "Most often they will appear ... with no discernible colour". Both articles include one or more images.
42 Maass (1898) 502.11 ff = Posidonius F121 EK. For a discussion of Posidonius' theory see Kidd (1988) 467-70; Vimercati (2004) 538-40.
43 Aëtius III.5.11 (DK59A86).
44 *Mete*. 377a31 and b 16-17.
45 *NQ* I.11.3.
46 Maass (1898) 488.14-18, on Aratus 811-18.
47 Compare *Mete*. 377b29.
48 Compare *Mete*. 378a7.
49 Compare *Mete*. 377b32.

50 Compare *Mete.* 377b28.
51 See my quotation at the beginning of this section from *Wikipedia* article "Sun dog" (p. 166 and n. 41).
52 *Mete.* III.6, 377a29ff.
53 *NQ* I.9-11.

17 Weather prediction and divination

Posidonius was a defender of divination in its various forms – that is, of ways of predicting future events other than (or additional to) weather – and he wrote a work in five books on the subject (Περὶ μαντικῆς)[1]; his theory of cosmic sympathy is a way of explaining relationships in the movements or changes of things distant from each other ("distantium rerum cognatio naturalis", "a natural relationship between distant things"),[2] such as the phases of the moon and the Atlantic tides. The truth of divination is proved not only from the existence of Providence but also by results.[3] Given Posidonius' interest in meteorology, one might have expected him to be interested in weather prediction, both for its own sake and because of its obvious similarity to divination; but there is not very much direct evidence. In this chapter I shall first look at ancient ways of predicting weather and at what evidence there is of Posidonius' concern with them, and then at the relation between Posidonius' interest in meteorological phenomena and his ideas about divination.

Weather prediction

There were two ways in which an ancient author could attempt to systematise weather prediction.[4] One way was to construct a *parapēgma*, a calendar (applicable to any year), in which, for each day mentioned, there was noted any astronomical event that would occur, for example, the rising of a star, or the expected weather, or both.[5] The other way was to collect and describe weather-signs, that is, phenomena supposed to be predictive of coming weather.[6] Pseudo-Theophrastus, *De signis tempestatum*, and the latter part of Aratus' *Phaenomena* are collections of such signs. There is no evidence that Posidonius compiled a work on weather prediction of either type, but there is some evidence of an interest in weather-signs. A knowledge of signs predictive of a meteorological phenomenon might illuminate discussion of the causes of the phenomenon, and this would have encouraged Posidonius' interest.

DOI: 10.4324/9780429399930-17

Of weather-signs in a strict sense I have noticed only one instance in fragments attributed by name to Posidonius, a part of his theory of comets:

κατὰ δὲ τὰς φαύσεις αὐτῶν καὶ πάλιν διαλύσεις τροπὰς γίνεσθαι συμβαίνει τοῦ ἀέρος· αὐχμούς τε γὰρ κἀκ τῶν ἐναντίων ῥαγδαίους ὄμβρους κατὰ τὴν διάλυσιν αὐτῶν γίνεσθαι, ἅτε δὴ ἐν ἀέρι τῆς συστάσεως αὐτῶν γινομένης

When they appear and again when they dissolve changes happen in the air; for there are droughts, and, from the opposite conditions, heavy rain showers as they dissolve, since the formation of them happens in air.[7]

Aratus says πολλοὶ γὰρ κομόωσιν ἐπ᾿ αὐχμηρῷ ἐνιαυτῷ, "many comets herald a season of drought".[8] Aristotle holds the same view and has an explanation of the cause (*Mete.* 344b19ff):

περὶ δὲ τοῦ πυρώδη τὴν σύστασιν αὐτῶν εἶναι τεκμήριον χρὴ νομίζειν ὅτι σημαίνουσι γιγνόμενοι πλείους πνεύματα καὶ αὐχμούς· δῆλον γὰρ ὅτι γίγνονται διὰ τὸ πολλὴν εἶναι τὴν τοιαύτην ἔκκρισιν, ὥστε ξηρότερον ἀναγκαῖον εἶναι τὸν ἀέρα

We may regard as a proof that their constitution is fiery that fact that their appearance in any number is a sign of coming wind and drought. For it is evident that they owe their origin to this kind of exhalation [i.e., the dry, fiery exhalation] being plentiful, which necessarily makes the air drier.

(Trans. H.D.P. Lee [1952, Harvard University Press, Loeb ed.])

Posidonius accepts that comets indicate drought, but adds that there are showers when they dissolve. This addition is presumably due to his theory of comets being rather different from Aristotle's[9]: his explanation of the weather-sign was, I suggest, that, while comets are shining, the available dense air is used up in feeding them, but when it ceases to do so, then the comets dissolve and the dense air condenses to rain.

At *Mete.* 344b19–345a5 Aristotle discusses the connection of comets with wind at some length, quoting specific instances, and he introduces the discussion, as my quotation shows, as a "proof" of his comet theory. Elsewhere. too, he deals with weather-signs, or the connection of certain weather-phenomena to other phenomena, but always as a necessary consequence, or an illustration, of some more general theory. At *Mete.* 360b27–361a4 he says that wind usually blows after rain, and ceases when rain begins, and that "these are necessary consequences of the principles we have stated".[10] At 361b23–362a7 he speaks of winds and calms associated with astronomical events such as the rising of Orion, as illustrating a principle that the sun both stops and arouses winds. In *Mete.* II.8 (366a5–368b12) and III.3 (372b17–34) he has much to say about weather phenomena associated with earthquakes, and with haloes, arguing that the phenomena are natural accompaniments of earthquakes and haloes, if earthquake and haloes occur in the way that he believes they do. Such discussions of weather-signs, produced as proofs or as corollaries of general theories, are details unlikely to be mentioned when a theory is briefly summarised, and very often brief summaries are all that we have of Posidonius' theories. Many more weather-signs may have been discussed in Posidonius' complete meteorological works.

One observation resembling a weather-sign probably comes from Posidonius' work on divination. At *De divinatione* I.130 Cicero implies that Posidonius approved a belief of the people of Ceos, reported by Heraclides Ponticus, that the dullness or brightness of the Dog Star at its rising was a sign of whether the coming year would be sickly or healthy – I discuss this further in the next section.

Meteorology and divination

Cicero[11] in his *De divinatione* defines "divination" as "the foreseeing and knowledge of future events" ("praesensionem et scientiam rerum futurarum"), or, more precisely, "the prediction and foreseeing of those events which are thought to happen by chance" ("earum rerum, quae fortuitae putantur, praedictio atque praesensio").[12] This latter definition Cicero puts in the mouth of his brother Quintus, who in this work defends divination; and Quintus immediately before it refers favourably to Stoic views on religion, and immediately afterwards offers an argument about divination and the existence of the gods which Cicero pronounces to be Stoic.[13] It is therefore a safe assumption that Quintus' definition of divination is one acceptable to the Stoics. (Here, and throughout this chapter, "Quintus" means "Quintus as a character in *De divinatione*".)

The Stoics were determinists.[14] Καθ' εἱμαρμένην ... τὰ πάντα γίνεσθαι, "all things happen in accordance with fate", is a view attributed by Diogenes Laertius to Posidonius and other leading Stoics.[15] If that is so, then everything that is going to happen is already fixed, and things happen "by chance" only in the sense that they are thought fortuitous ("fortuitae putantur"), because human beings have been unable to predict them. Hence the Stoics regarded chance, Simplicius says, as a cause (αἰτία) ἄδηλον ... ἀνθρωπίνη διανοίᾳ, "obscure to human understanding"[16] (or, in other sources, "to human reasoning", ἀνθρωπίνῳ λογισμῷ,[17] or "to human reason", ἀνθρωπίνῳ λόγῳ[18]).

The Stoics accepted all kinds of divination (μαντικὴν πᾶσαν) according to Diogenes Laertius,[19] nearly all ("omnia fere illa") according to Cicero, *Div.* I.6 (and the Diogenes passage names Posidonius). At *Div.* I.118 Cicero gives it as the Stoic view that "certis rebus certa signa praecurrerent, alia in extis, alia in avibus, alia in fulgoribus, alia in ostentis, alia in stellis, alia in somniantium visis, alia in furentium vocibus", "certain signs precede certain events, some in entrails, some in birds, some in lightning flashes, some in portents, some in stars, some in the visions of people dreaming, some in the utterances of people in frenzy".

In antiquity, most meteorological events could not be accurately predicted, and many weather-signs depend on the behaviour of birds or on the stars.[20] One might expect, therefore, that weather-signs would be regarded as a form of divination, but Cicero makes Quintus say that this is not strictly so: weather-signs "ex alio genere sunt, tamen divinationi sunt similiora", "are of another kind, but nevertheless are very similar to divination".[21] This is not explicitly said to be a Stoic view, but the Stoics are the main source in this book.[22] Weather-signs must be closely enough related to divination for it to be worth considering further this relationship, and other possible connections between divination and meteorological study.

To take the second point first: in two places in *Naturales quaestiones* Seneca's language suggests that Posidonius saw connections between meteorological events and the future which go beyond the observation of weather-signs. In Chapter 9[23] I discussed *NQ* VII.20.2, where Seneca writes of "marvels that Posidonius wrote about, burning columns and shields and other flames of remarkable novelty" which are seen in the sky, and Seneca suggests, without explicitly saying, that Posidonius was one of those who regarded such events as portents. I argued there that Posidonius probably did so regard them. Evidence just quoted reinforces that conclusion: Posidonius accepted all or nearly all the types of divination; of the types mentioned at *Div.* I.118 the "marvels" of *NQ* VII.20.2 would have counted as portents, *ostenta*; Seneca at *NQ* VII.21.1 writes of "tubas trabesque et alia ostenta caeli", "trumpets and beams and other portents of the sky". No doubt these phenomena would have been taken as portending happenings more exciting than changes in the weather.

In Chapter 14[24] I wrote of Seneca's scorn for certain Stoics who claimed that it was possible to recognise clouds from which hail would fall, and for the "hail-guards" of Cleonae who forecast imminent hail and called on the people to perform propitiatory sacrifices to ward it off, drawing blood from their own fingers if no better sacrifice was available; and I suggested that the irony with which Seneca presents Posidonius' hail theory might be due to Posidonius' endorsement of these claims and practices. The recognition of hail clouds might be merely a claim to have discovered a weather-sign, but the approval of propitiatory sacrifices does relate to the Stoic attitude to divination. Cicero presents Chrysippus as saying that the function of divination is to inform people in advance "dei erga homines mente qua sint quidque significent, quem ad modumque ea procurentur atque expientur", "the disposition of the gods towards men and what they are indicating, and in what manner those [sc. harmful] things can be averted and expiation made".[25] So Stoics believed that evils foretold by divination could be averted by the proper rites. Seneca also says that some people "suspicari ipsos aiunt esse in ipso sanguine vim quandam potentem avertendae nubis ac repellendae", "say they suspect that there is in blood itself a certain strong power for turning away cloud and driving it off".[26] This seems to tally with the Stoic belief that the gods are not immediately concerned with every divinatory sign,[27] so could have been the view of Posidonius or a follower of his.

This connection of meteorology with divination makes it the more worthwhile to consider the relation of Stoic divination to weather-signs. According to Quintus in *De divinatione*, "there are two types of divination, of which one belongs to art [i.e., technical skill], the other to nature" ("duo sunt ... divinandi genera, quorum alterum artis est, alterum naturae").[28] Divination from nature comes from "the power of the mind separated from bodily senses" ("animi vis seiuncta a corporis sensibus"), as in dreams and the utterances of people in frenzy.[29] Diviners by art "seek out new [i.e., forthcoming] events by inference; they have learned old ones by observation" ("novas res coniectura persequuntur, veteres observatione didicerunt")[30]; they know "by memory and hard work and the records of their predecessors" ("memoria et diligentia et monumentis superiorum")[31] that certain signs precede certain types of event, and so can predict the events. It is this latter type of divination which resembles predictions from weather-signs.

The value of divination is thus shown by its results[32] – this in addition to an argument that a beneficent Providence exists, and must, in its beneficence, allow human beings an insight into the future.[33] Diogenes Laertius confirms this, saying that the Stoics (including Posidonius) "say that divination exists ... if it is true that Providence exists; and they prove it to be a science as well through its results" (μαντικὴν ὑφεστάναι φασὶν ... εἰ καὶ πρόνοιαν εἶναι· καὶ αὐτὴν καὶ τέχνην ἀποφαίνουσι διά τινας ἐκβάσεις).[34] Cicero's Quintus is clear that he does not know the causes by which certain signs are predictive of certain events; he just knows that experience shows that they are, just as he knows that certain medicines benefit sufferers from certain diseases, and that certain weather-signs are predictive of certain kinds of weather, without knowing the causes why the medicines produce their benefits or the weather-signs lead to correct predictions.[35]

The Stoics believed that everything is determined by fate.[36] According to Aëtius I.28.4, a Stoic definition of the "substance of fate" (οὐσία εἱμαρμένης) was εἱρμὸν αἰτιῶν, τουτέστι τάξιν καὶ ἐπισύνδεσιν ἀπαράβατον, "a chain of causes, that is an unalterable arrangement and concatenation of them"; more briefly in Diogenes Laertius VII.149 fate is αἰτία τῶν ὅλον εἰρομένη, "a linked cause of everything in the universe". Similarly at *Div.* I.125 fate is "ordinem seriemque causarum", "an arrangement and series of causes". Divination is possible because "ut in seminibus vis inest earum rerum, quae ex eis progignuntur, sic in causis conditae sunt res futurae", "as in seeds there is the germ of the things that are brought into being from them, so the things that are to happen are hidden in their causes", and so can be divined.[37] However, as we have seen with regard to meteorology, the early Stoics were averse to speculation about causes[38]; they were presumably content to declare that all events have causes, without discussing what they were in particular instances.

At *Div.* I.13-15 Quintus quotes from Cicero's Latin translation of the weather-signs of Aratus, and asks

Quis igitur elicere causas praesensionum potest? Etsi video Boëthum Stoicum esse conatum, qui hactenus aliquid egit, ut earum rationem rerum explicaret, quae in mari caelove fierent. Illa vero cur eveniant, quis probabiliter dixerit?

Who can extract the causes of the foreknowledge [sc. of coming weather, from the weather-signs]? And yet I see that the Stoic Boëthus tried, and achieved something, to the extent that he explained the reason of those signs which happen in the sea or the sky. But who can explain with any probability why the following results come about?

And he goes on to quote signs which come from the behaviour of birds, frogs and oxen.Thus, when an ancient observer saw a particular phenomenon in the sea or the sky – for example, the appearance or disappearance of a comet – he might postulate, reasonably by ancient standards, that this marked a change in sky or sea, from which a particular change in the weather should follow; I have quoted an example in the first part of this chapter. But this applies only to some weather-signs; could this procedure be applied to any instances of divination proper? Cicero makes Quintus say that Posidonius thought it could.

At *Div.* I.125 Quintus says; "mihi videtur, ut Posidonius facit, a deo, de quo satis dictum est, deinde a fato, deinde a natura, vis omnis divinandi ratioque repetenda", "I think we should, as Posidonius does, trace the whole influence and rationale of divination first from god, about whom enough has been said, then from fate, and then from nature".[39] In I.125–8 the part played by fate is discussed, at I.129 he goes on to speak of nature:

A natura autem alia quaedam ratio est, quae docet quanta sit animi vis seiuncta a corporis sensibus, quod maxime contingit aut dormientibus aut mente permotis

From nature comes another particular rational explanation which teaches how great the power of the mind is when separated from the physical senses, which especially happens to men who are sleeping or inspired [or "frenzied"].[40]

As this sort of divination has previously been said to belong to "nature" (see above), there is no difficulty in associating it with nature here. After giving more details, Quintus goes on (I.130):

Hanc quidem rationem naturae difficile est fortasse traducere ad id genus divinationis, quod ex arte profectum dicimus, sed tamen id quoque rimatur quantum potest Posidonius. Esse censet in natura signa quaedam rerum futurarum

This rational explanation from nature is perhaps difficult to transfer to that kind of divination which we say proceeds from art; however, Posidonius examines that, too, as far as he can. He thinks that there are in nature some particular signs of future events.

Quintus in I.130–1 then gives two instances: first, according to Heraclides Ponticus, the Ceans carefully watch each year the rising of the Dog Star,

> coniecturamque capere … salubrisne an pestilens annus futurus sit: nam
> si obscurior et quasi caliginosa stella extiterit, pingue et concretum esse
> caelum ut eius aspiratio gravis et pestilens futura sit

> and get an indication … whether the year will be healthy or pestilent: for
> if the star rises rather dim and more or less dark, then the atmosphere is
> thick and dense, so that the breath from it will be heavy and pestilent;

but if the star appears "bright and radiant" ("inlustris et perlucida"), it is a sign that the atmosphere is "thin and pure and therefore healthy" ("tenue purumque et propterea salubre"). Second, Democritus thought it sensible that the entrails of sacrificed animals should be examined: "from their condition and colour" ("ex habitu atque ex colore") signs can be seen of health or pestilence, sometimes also of "the future barrenness or fertility of the land" ("sterilitas agrorum vel fertilitas futura") – the idea presumably is that the state of health of sacrificed animals is a sign of the healthiness of their environment. "Quintus" concludes this argument:

> Quae si a natura profecta observatio et usus agnovit, multa afferre potuit
> dies quae animadvertendo notarentur

> If observation and custom have recognised that these proceed from
> nature, the passage of time has been able to bring many things to be
> noted and recorded.[41]

It is unclear what "these" ("Quae") are. Wardle (2006) has "these techniques", but it could equally easily be "these predictions". It is surely implied that there have been other examples similar to those quoted from Heraclides and Democritus.

There is doubt whether both these examples are from Posidonius. EK print only the first as part of their F110,[42] but Theiler (1982) includes both in his F378,[43] and Wardle (2006) 417 supports this, saying that the last sentence which I quote "formally ends the Posidonian argument, so this [the example from Democritus] should be part of it". I incline to accept this argument. However, Struck (2016) 212–13 calls the two examples "unhelpful", implying that neither is from Posidonius, and says that they "seem not to need any particular divinatory explanation. They do not depart from scientifically observable phenomena".[44] Just afterwards, he writes that, to Posidonius, "the information we gather in divination is different from that which we draw out from the other signs we might observe, and from which we might draw conclusions, precisely and by definition because its workings lie beyond our regular inferential reasons", and, referring to the definition of divination as the prediction of things thought to occur by chance, he writes "to say that an event is thought to occur by chance is to say that it has a cause that ... lies outside the reach of our calculating, reasoning powers".

Up to this point I have generally followed Struck, but here I think he goes beyond the evidence and is probably mistaken. It is true that Simplicius, after quoting the Stoic view that chance is a cause "obscure to human understanding" (see p. 173 above), adds ὡς θεῖόν τι οὖσαν καὶ ... τὴν ἀνθρωπίνην γνῶσιν ὑπερβαῖνον, ὥσπερ οἱ Στωϊκοὶ δοκοῦσι λέγειν, "as being something divine and ... beyond human investigation, as the Stoics seem to say". A cause "beyond human investigation" is perhaps as near as one can get to "chance" in a strictly deterministic system; but there is surely no reason to insist that this understanding of "chance" must apply to *all* divination because divination is defined as "the prediction ... of those events which are thought to happen by chance". People using divination may *think* that a coming event is a matter of chance when others, better informed, can predict it by reasoning.

The only evidence we have which should show what Posidonius meant when he connected divination "by art" with nature is in the two examples in *Div.* I.130–1; if those examples are in fact irrelevant, we had best admit that we do not know what Posidonius was getting at. But this conclusion seems to me unnecessary, because I think we can derive from those two examples a reasonable interpretation of Posidonius' meaning. The "ratio naturae", "the rational explanation from nature" that is illustrated by the two examples is, I suggest, that in these instances we can understand the natural processes which produce both the signs and the future events predicted from them; we can see the underlying causes. If we had better understanding, we could no doubt do the same for other instances of divination "by art".

The objection may be made that, if this is so, the predictions are not really divination. But the question whether a prediction is divinatory or not must surely depend on the point of view of the predictor. The Ceans' observation of the Dog Star must have been a traditional ritual, as evidence quoted by Wardle helps to confirm.[45] The original observers probably had no philosophical theories about the state of the atmosphere through which they observed the star; that has surely been provided by Heraclides or Posidonius. The examination of entrails was a typical divinatory act. To the ordinary person both Cicero's examples were pieces of divination. It is the trained philosophical mind which can, in these instances, see the underlying cause which produces both the diviner's observation and the fulfilment of the prediction which he derives from it.

At *Div.* I.127 Quintus says that "if there could be a mortal who saw in his mind the linking together of all causes, then certainly nothing would deceive him" ("si quis mortalis potest esse, qui colligationem causarum omnium perspiciat animo, nihil eum profecto fallat") – he would foresee everything. Quintus says, reasonably enough, that such complete knowledge is possible only for a god. But he does not suggest that it is impossible for a mortal to see the linking together of *some* causes, to see in *some* instances both the divinatory signs which give warning of a particular event and the underlying cause which is producing the signs and will produce the event.

I suggest, then, that Posidonius, keen as he was to describe the causes of meteorological events, was probably also keen to find the links which connected weather-signs to the weather that men foretold from them,[46] and to find, if only in a few instances, the underlying causes which, he believed, connected divinatory signs to the events of which they were signs. He may have thought that meteorology provided a promising line of approach to such enquiries. The prediction derived from the Ceans' observations of the Dog Star is in a broad sense a meteorological prediction. Democritus' prediction of poor or abundant crops from the examination of entrails surely involved assumptions about the climate and weather of the place concerned. Both involve meteorology; and Posidonius, I suggest, believed that something had been discovered about the underlying cause in these two cases.

Notes

1 Diogenes Laertius VII 149 (Posidonius F7 and F27 EK), also Cicero, *Div.* I.6 (F26 EK).
2 Cicero, *Div.* II.33–5 = Posidonius F106 EK (F106.13 quoted). Kidd (1988) 423 stresses that Cicero is here speaking of Stoics generally, not just of Posidonius.
3 See below, p. 175.
4 On Posidonius and weather prediction see Kany-Turpin (2003), especially pp. 373–7.
5 An example of a *parapēgma* is included at the end of Geminus, *Isagoge*.
6 For discussions of these two types of weather prediction (*parapēgmata* and weather-signs), with extensive reference to earlier literature, see Taub (2003) 15–69.
7 *Scholia in Aratum* 1091 (Posidonius F131a.33–7 EK).
8 Aratus 1093 (with translation by G.R. Mair in Mair and Mair [1955] 293).
9 See above, Chapter 9 (p. 78–80).
10 *Mete.* 360b30, ταῦτα γὰρ ἀνάγκη συμβαίνειν διὰ τὰς εἰρημένας ἀρχάς, translation from Lee (1952).
11 This section owes much to Struck (2016) 171ff.
12 *Div.* I.1 and I.9.
13 *Div.* I.8–10.
14 This paragraph largely follows Struck (2016) 199–200.
15 Diogenes Laertius VII.149 (Posidonius F25 EK). So too Cicero *Div.* I.125, immediately after citing Posidonius, "Fieri ... omnia fato ratio cogit fateri", "reason compels us to admit that all things happen by fate" (tr. Falconer [1923]).
16 Simplicius, *In Phys.* p. 333.1 Diels (*SVF* II.965).
17 Aëtius I.29.7 (*SVF* II.966), cf. Alexander, *De fato* 8, p. 174.2 Bruns (*SVF* II.970).
18 Theodoretus, *Graec. affect. cur.* p. 87.41 (*SVF* II.971).
19 Diogenes Laertius VII.149 (Posidonius F7 EK).
20 E.g., Aratus 913ff (birds), 740ff and 892ff (stars).
21 Cicero, *Div.* I.13.
22 See, for instance *Div.* II.8, where Cicero, as a speaker in the dialogue, says to "Quintus", obviously referring to Book I, "Stoice Stoicorum sententiam defendisti", "you have defended the opinion of the Stoics in a Stoic way". On the (mainly Stoic) sources of *Div.* Book I see Struck (2016) 174–7.
23 See above, Chapter 9 (p. 76).
24 See above, Chapter 14 (p. 146–7).

25 *Div.* II.130.
26 *NQ* IVb.7.1.
27 *Div.* I.118.
28 *Div.* I.11–12, also I.34, I.129–30, II.26.
29 *Div.* I.129.
30 *Div.* I.34.
31 *Div.* I.127. Compare *Div.* I.12 "Observata sunt haec tempore immenso et eventis(?) animadversa et notata", "These [divinatory signs] have been observed over an immense period of time and noticed and recorded because of their results(?)". There is doubt about the exact text, but the meaning is clear. See also *Div.* I.109.
32 On this aspect of the Stoics' treatment of divination – arguing from "the collection of examples" – see Struck (2016) 175–6.
33 So in most detail at *Div.* I.82–3; see also I.9–10, I.117, II.41 (cited by Struck [2016] 186).
34 Diogenes Laertius VII.149. I give the text as printed at F7 EK, with the translation of Kidd (1999).
35 *Div.* I.12–16; see also I.35, I.84–6, I.109. (Passages listed by Struck [2016] 176n.)
36 On the relation of divination to the Stoics' doctrines of fate and of causation see Struck (2016) 195–208. It seems unnecessary, for present purposes, to discuss the complications, dealt with by Struck, which arise from various other aspects of Stoic physics.
37 *Div.* I.128.
38 See above, Chapter 4 (p. 27–9).
39 Posidonius F107 EK, with translation from Kidd (1999).
40 Translation from Kidd (1999). *Div.* I.129–30 = Posidonius F110 EK.
41 Translation from Wardle (2006), modified.
42 Kidd (1988) 435 says that the Heraclides example is "probably" from Posidonius, but does not mention the example from Democritus. Vimercati (2004) also omits the Democritus example (see his A109). His analysis (pp. 565–7) of the passage seems to me mistaken. In Cicero's text it is clear that divination "a natura", as distinct from divination "ex arte", comes to the mind without use of the senses, e.g., in dreams; but Vimercati includes divination from observed (e.g., meteorological) phenomena in the "natural" class.
43 Theiler (1982) II.302 expresses no doubts about both instances being from Posidonius.
44 Struck is not the only scholar to doubt whether the examples from Heraclides and Democritus are true examples of divination: see Wardle (2006) 416.
45 Wardle (2006) 417 quotes a report from the scholia to Apollonius Rhodius II.498 that the Ceans' custom was to await the rising of the Dog Star in full armour and to sacrifice to it.
46 *Div.* II.47 (Posidonius F109 EK) suggests this.

18 Meteorology and Providence

In Chapter 4 I commented on the scarcity or absence of mentions of Providence or of final causes in the meteorology of Aristotle and in what few fragments we have of early Stoic meteorology, and I suggested that the complete absence of final causes from Aristotle's *Meteorologica* I–III is due to pre-Socratic tradition: a full account of the physical world had to include an explanation of the material and efficient causes of meteorological phenomena (to use Aristotelian terms), but Aristotle did not need to say anything about the final causes because most pre-Socratics had said nothing about such things.[1] I suggest that a similar feeling may have actuated Posidonius: that he felt that his account of the cosmos had at least to match those of his predecessors and his Epicurean rivals by explaining the material and efficient causes of meteorological events, but that he was not required to discuss the aspect of the subject which even Aristotle had ignored: the final causes of those events and their part in the designs of Providence. In the texts, discussed in previous chapters, which name Posidonius as the author of meteorological ideas or descriptions, we have found "cosmic sympathy" or the Stoic Providence invoked only in Strabo's quoting him as saying that tides occur "in sympathy with the moon" (συμπαθῶς τῇ σελήνῃ),[2] and in the idea, attributed to Posidonius by Macrobius, that the purpose of the Milky Way is to warm parts of the universe not warmed by the sun (and perhaps he regarded the Milky Way as an astronomical, not a meteorological, phenomenon).[3]

Posidonius, as a Stoic, believed that the cosmos is governed by Providence.[4] In Chapter 4 I quoted texts which show that the early Stoics regarded as a proof of this the sequence of the seasons, which guarantee the fertility of the earth.[5] Cicero confirms that Posidonius shared this view: in *De natura deorum* he makes Balbus, as the defender of Stoic theology, say that, if the world were not governed by "reason and wisdom" ("ratione et sapientia"),

> possetne uno tempore florere, deinde vicissim horrere terra, aut tot rebus ipsis se inmutantibus solis accessus discessusque solstitiis brumisque cognosci, aut aestus maritimi fretorumque angustiae ortu aut obitu lunae commoveri?

DOI: 10.4324/9780429399930-18

would it be possible for the earth at one definite time to be gay with flowers and then in turn all bare and stark, or for the spontaneous transformation of so many things about us to signal the approach and the retirement of the sun at the summer and winter solstices, or for the tides to flow and ebb in the seas and straits with the rising and setting of the moon?

(*De natura deorum* II.19, trans. H. Rackham [1933,
Harvard University Press, Loeb ed.])[6]

Thus the proofs of Providence include events which are, in a broad sense, meteorological: changes of season, and the tides. The mention of tides strongly suggests that this is derived from Posidonius, or at least is based on his ideas.

Did Posidonius say more than I have quoted about meteorology and Providence? Seneca occasionally touches on the subject, and Posidonius could be his source. At *NQ* II.46 Seneca asks why Jupiter passes by the guilty or strikes the innocent, and answers that this must be discussed elsewhere, but he will say in the meantime that Jupiter does not himself send thunderbolts, but "nevertheless they do not occur without reason, which belongs to him" ("tamen sine ratione non fiant, quae illius est"). This seems orthodox Stoic doctrine, since it is consistent with Cicero, *Div*. I.118, which says it is the Stoic view that the gods are not themselves concerned with individual divinatory signs, but that the world is so ordered that certain signs precede certain events. At *NQ* V.18.1–4 Seneca speaks of the providential benefits produced by wind, preventing the air from being stagnant, distributing rain around the earth, facilitating travel, and so on. At *NQ* VI.3.1 he says that sky and earth are not shaken due to the wrath of the gods, "sed quibusdam vitiis, ut corpora nostra turbantur", "but, like our bodies, they are upset by certain defects".[7] The implication of this, that the power of the gods and of Providence is not absolute, that the influence of another factor or factors is also at work, recalls Plato's doctrine at *Timaeus* 47E–48A that our world has its origin in the combined work of Reason and of Necessity; and there is evidence that Posidonius was influenced by the *Timaeus*.[8] The theme of Seneca's *De providentia* is that people become morally and physically stronger through suffering, and this includes the experience of a harsh climate.[9]

It seems quite possible that Seneca found some, if not all, of these ideas in Posidonius; but, if he did, it seems to me most likely that he found them as incidental comments by Posidonius, not as part of a treatise on the subject of meteorology and Providence. Such a treatise would have been an unusual one on a controversial subject. The surviving literature would surely have contained some reference to it.

I think we can infer, at least in outline, the position which Posidonius held. Despite his belief in Providence, he seems always to have proposed naturalistic explanations of meteorological phenomena of just the same character as those put forward by other philosophers, including the Epicureans. His view must surely have been that meteorological events have naturalistic causes, in a chain of causes[10] so arranged by Providence that the events occur on the appropriate occasions. For example: at *Div*. I.109 Cicero makes his brother Quintus include lightnings ("fulgoribus") in a list of phenomena from which divinatory predictions may be made; and at II.44 Cicero replies that this cannot be so, because thunder and lightning occur "vi naturae … nullo rato tempore", "by force of nature [so not by divine action] … at no fixed time". But in Posidonius' system everything occurs in accordance with fate,[11] and if this is strictly true then every thunderstorm, and every human act and intention, is fixed by fate from the beginning of the cosmos; lightning therefore does occur at fixed times, and it is not impossible that fate (Providence) has so fixed the times that the lightning conveys a divinatory warning appropriate to the current human situation, which is also fixed by fate. Cicero's second objection remains: that, to all appearances, many lightning flashes do not give warnings[12]; but their doing so is not impossible.

There is also the possibility that, in Posidonius' cosmos, some things occur which are not part of the providential plan; that Posidonius agreed with Seneca's statement at *NQ* VI.3.1 that earthquakes and the like are caused not by the wrath of the gods but by faults in bodies. If so, Posidonius must surely have believed that such events are nevertheless predetermined – to suppose otherwise would nullify the claim that everything occurs in accordance with fate.

Notes

1　See above, Chapter 4 (pp. 24, 29–30).
2　See above, Chapter 13 (p. 133, cf. p. 136).
3　See above, Chapter 9 (p. 84).
4　See Diogenes Laertius VII.138 and 149 (Posidonius F21 and F7 EK).
5　See above, Chapter 4 (p. 29).
6　Theiler (1982) prints this as a fragment of Posidonous (his F356), and in vol. II p. 257 draws attention to language which he thinks typical of Posidonius. Vimercati (2004) 392 prints this as B5 among "Frammenti attribuibil:", i.e., fragments 'attributable' to Posidonius though he is not named.
7　Tr. Hine (2010).
8　See below, Chapter 20 (p. 193); Chapter 21 (p. 201).
9　*De providentia* 4.14–15.
10　See above, Chapter 17 (p. 175).
11　Diogenes Laertius VII.149 (Posidonius F25 EK).
12　*Div*. II.45.

19 Epicurean meteorology compared with that of the Stoics

We have accounts of Epicurean meteorology in Epicurus' *Letter to Pythocles* and in Book VI of Lucretius, *De rerum natura* (probably written just later than Posidonius' work on meteorology). A comparison of Epicurean meteorology with that of the Stoics and of Posidonius should throw light on both systems.

The Epicurean denial of Providence in relation to meteorology

The Epicureans denied the existence of Providence, and rejected divination. Epicurus' explanations of meteorological phenomena never involve divine action, and he is clear that signs of changes of weather (ἐπισημασίαι) have naturalistic causes. At *Letter to Pythocles* 98 he says that ἐπισημασίαι can happen κατὰ συγκυρήσεις καιρῶν, καθάπερ ἐν τοῖς ἐμφανέσι παρ' ἡμῖν ζῴοις, "by coincidences of times, as with animals obvious among us" (was he thinking of such things as the appearance of swallows in spring?) and παρ' ἑτεροιώσεις ἀέρος καὶ μεταβολάς, "by alterations and changes of air". At §115 he states that there is no necessity and no "divine nature" (θεία φύσις) involved in signs of weather-changes (ἐπισημασίαι) derived from the behaviour of animals, but that this occurs "by a coincidence of time" (κατὰ συγκύρημα τοῦ καιροῦ).

In principle this is very different from the Stoic view; but, as we have seen, Stoic weather-signs, too, have naturalistic causes, in Providence's chain of causes.

The similarity of Epicurean and non-Epicurean meteorological theories

We have seen that Posidonius, and other ancient believers in Providence or final causes, usually had little to say about such things when they offered explanations of meteorological phenomena. So, unsurprisingly, most of the explanations of such phenomena in the *Letter to Pythocles* and in Lucretius VI (usually the same in both works, but the relationship is not a simple one) were the same as, or resemble, explanations propounded by non-Epicurean thinkers. I give some examples from Lucretius' explanations of thunder. These include the sound of clouds broken by wind (VI.137–41), the sound of clouds colliding (VI.96ff), and the sound of clouds rubbing together (VI.116–20); all three are

DOI: 10.4324/9780429399930-19

among the explanations of thunder and lightning in the *Letter to Pythocles* 100–1, and are attributed to at least one pre-Socratic (the breaking of cloud to Anaximander, collision and rubbing to Democritus); also, all three are attributed to the early Stoics, and the breaking and the rubbing of clouds are found in Seneca's account of Posidonius' theory.[1] Lucretius also has the old explanation of thunder as the sound of a cloud burst by wind, like the sound of a burst bladder (VI.121–31), combining this with the idea of a spinning whirlwind (line 126, "turbine versanti") inside a cloud, which is first recorded in Epicurus and is attributed to Posidonius[2]; this is one idea which Posidonius may have taken from Epicurus.[3] Lucretius also explains thunder as due to a process like the plunging of red-hot iron into cold water (VI.145–9), which recalls the theory of Anaxagoras, seriously discussed by Aristotle.[4]

There are also explanations not found in or attributed to earlier authors: in the *Letter to Pythocles* 100, thunder may be somehow due to "clouds that have become solid like ice" (νεφῶν … πῆξιν εἰληφότων κρυσταλλοειδῆ[5]); Lucretius VI 156–9 seems to be a modification of this idea. First recorded in Lucretius is the theory that thunder may be the sound of wind blowing through "branched clouds" ("ramosa nubila"), like the sound of wind blowing through a wood (VI.132–6), or the sound of clouds breaking in the way that waves break in rivers and the sea (VI.142–4). Epicurus and Lucretius seem to have been readier with new ideas of this kind than Posidonius was,[6] doubtless because they sought every possible explanation of a meteorological phenomenon (see below), but Posidonius only the one, or perhaps sometimes the few, that seemed most probable.

Of course, many Epicurean explanations are distinct from those of most ancient thinkers in that they mention atoms. Lucretius' clouds contain "seeds of fire" ("semina ignis", e.g., VI.206-7), and "seeds of water" ("semina aquai", VI.497), by which they produce lightning and rain, and are held together because formed from "bodies [i.e. atoms] entangled by small projections" ("corpora moris indupedita exiguis", VI.451-4) – compare Epicurus' clouds formed by περιπλοκὰς ἀλληλούχων ἀτόμων, "entanglements of atoms holding each other".[7] However, Lucretius, and very likely Epicurus in a less summary work than the *Letter to Pythocles*, generally has an analogy with some familiar phenomenon as a basis for each explanation – a normal feature, as we have seen, of ancient meteorological theories – which means that his explanations had to be translatable into a form which non-atomists could accept, since every physical system had to allow for the occurrence of familiar phenomena. Many meteorological explanations, in the *Letter to Pythocles* and in Lucretius, make no mention of atoms.

Multiple explanations of meteorological phenomena

The distinctive feature of Epicurean meteorology is that the Epicureans, as a matter of principle, give several explanations for each phenomenon – in an extreme case, Lucretius VI.96–159 gives nine explanations of thunder.

Bakker has shown that this is an original feature in the meteorology and astronomy of Epicurus: Aristotle, Theophrastus and possibly Democritus occasionally give two or three explanations of a phenomenon, but they do not do so regularly (here I agree with Bakker's arguments that the Syriac Meteorology is not a reliable source for Theophrastus).[8] Our evidence indicates that Posidonius, like Aristotle, may sometimes have described two or three causes of one phenomenon (thunder[9] and earthquakes[10] are possible examples), but did not do so regularly.

The Epicureans had a theoretical justification for their giving of multiple explanations which has been analysed by Bakker.[11] In summary: for the basic principles of physics only the atomic theory is in harmony with the observed phenomena, but for celestial and meteorological phenomena (τῶν μετεώρων) there are multiple causes that are consistent with what is observed.[12] This is at least in part because of "the appearance of things seen at a distance" (τὴν ἐκ τῶν ἀποστημάτων φαντασίαν).[13] However, "signs" (σημεῖα) of how things occur "in the regions above us" (ἐν τοῖς μετεώροις) are provided by "some of the phenomena occurring among us, which are observed as they happen" (τῶν παρ' ἡμῖν τινα φαινομένων, ἃ θεωρεῖται ἦ ὑπάρχει)[14] – presumably referring to the use of familiar analogies to provide explanations of meteorological phenomena. But this process does not provide us with a single explanation of any such phenomenon: "the appearance of each must be observed and distinguished with regard to the things connected with it, the occurrence of which in various ways is not contradicted by the evidence of things which happen among us" (τὸ … φάντασμα ἑκάστου τηρητέον καὶ ἐπὶ τὰ συναπτόμενα τούτῳ διαιρετέον, ἃ οὐκ ἀντιμαρτυρεῖται τοῖς παρ' ἡμῖν γινομένοις πλεοναχῶς συντελεῖσθαι)[15]; as he puts it elsewhere, "we must seek the cause of phenomena in the region above us and of everything obscure taking into account in how many ways the like effect occurs among us" (παραθεωροῦντας ποσαχῶς παρ' ἡμῖν τὸ ὅμοιον γίνεται, αἰτιολογητέον ὑπέρ τε τῶν μετεώρων καὶ παντὸς τοῦ ἀδήλου).[16]

Error arises when we form opinions about what our sense-organs have perceived; if an opinion "is not confirmed or is contradicted by evidence, it is false; but if it is confirmed or not contradicted by evidence, it is true" (ἐὰν μὲν μὴ ἐπιμαρτυρηθῇ ἢ ἀντιμαρτυρηθῇ, τὸ ψεῦδος γίνεται, ἐὰν δὲ ἐπιμαρτυρηθῇ ἢ μὴ ἀντιμαρτυρηθῇ, τὸ ἀληθές).[17] That an opinion not contradicted by evidence is possibly true is uncontroversial, but Epicurus apparently says, in the latter part of this quotation from the *Letter to Herodotus*, that such an opinion is actually true. Bakker accepts this, with reservations: in an infinite Epicurean universe, with an infinite number of worlds comparable to ours, every possibility must be true somewhere, even if not in our world.[18]

Bakker shows that (apart from one reference in a late author, Diogenes of Oenoanda) there is no evidence that the Epicureans had a rule by which they might decide that any one explanation of a meteorological phenomenon is more probable than any other; and, although there are a number of passages which have been taken as evidence that Epicurus or Lucretius thought that one explanation of a particular phenomenon was more probable than other possibilities, Bakker argues that they have been misinterpreted and do not actually show this.[19] If this is right, then it was the Epicurean view that all the possible explanations of meteorological phenomena that they had formulated were as probable as each other.

So, to the Epicureans, every meteorological and astronomical phenomenon may be produced by a variety of causes; or, even if only one of the possible causes applies in our world, we do not know which it is, and other causes apply in other worlds. Lucretius V.526–33 speaks of the many possible causes of the movements of heavenly bodies which may apply "through the universe in various worlds" ("per omne in variis mundis"), and then goes on (531–3)

e quibus una tamen siet hic (?) quoque causa necessest
quae vegeat motum signis; sed quae sit earum
praecipere haudquaquamst pedetemptim progredientis

of these, however, there must indeed be one cause here [presumably meaning "in our world"[20]] which imparts motion to the heavenly bodies; but which of them it is, is not for [sc. men] to pronounce in their lumbering progress.

(Lucretius does sometimes suggest that a particular cause produces a particular sort of effect; for instance, the thunder produced by a cloud breaking in the way that a wave breaks is "quasi murmur", suggesting a continuous rumble,[21] but he does not suggest that one can confidently identify the cause on a particular occasion by such a feature.)

For the Epicureans, therefore, all their meteorological theories are certainly true somewhere in the universe, but we cannot tell which theory, or theories, apply in our world. This cannot satisfy someone who is curious to know what causes, in our world, a particular meteorological phenomenon. If such an enquirer was considering phenomena like thunder and lightning, of which the ancients had no hope of finding the true cause, it might be no bad thing to be reminded how many possible causes had been proposed, and so realise how uncertain all the available explanations were; but with a matter like the Nile floods, where increased geographical knowledge had made clear an important part of the cause, the Epicurean insistence on multiple causes gave no encouragement to seek the relevant information. Lucretius VI.712–37 mentions four causes of the Nile floods, of which the true cause – rain on equatorial mountains – is the third and is given no special prominence.[22]

The aim of Epicurean meteorology

This, however, is to mistake the aim of Epicurean meteorology. The Epicureans are clear that the purpose of their astronomical and meteorological studies is to obtain peace of mind. Epicurus says: Μὴ ἄλλο τι τέλος ἐκ τῆς περὶ μετεώρων γνώσεως ... νομίζειν εἶναι ἤπερ ἀταραξίαν καὶ πίστιν βέβαιον, "do not think that there is any end in view from knowledge of things in the region above us, other than peace of mind and firm belief".[23] "The greatest disturbance to human minds" (τάραχος ὁ κυριώτατος ταῖς ἀνθρωπίναις ψυχαῖς) comes from fear of the gods.[24] At the beginning of his account of meteorological phenomena Lucretius writes:

> quae fieri in terris caeloque tuentur
> mortales, pavidis cum pendent mentibu' saepe
> et faciunt animos humilis formidine divum ...
> ignorantia causarum conferre deorum
> cogit ad imperium res

> things which mortals see happening on land and in the sky, when they are often perplexed with fearful minds and make their spirits humble in fear of the gods ... Ignorance of causes compels them to attribute what is happening to the power of the gods.[25]

So the object of meteorological study is to free us from the fear that harmful meteorological events – lightning-strikes, storms, earthquakes – are the instruments of divine punishment or divine ill-will. The Epicureans sought to establish this by arguing that such intervention in what happens in our world is inconsistent with the blessed immortality which is an essential attribute of divinity,[26] and that there are numerous naturalistic explanations of meteorological phenomena, making any belief in divine causation supererogatory.

Epicurean meteorology compared with that of the Stoics

As I explain in Chapter 21, the precise cause of any particular meteorological phenomenon was not crucial to the fundamental teachings of any philosophical school,[27] and it was obvious, around 300 B.C., that more than two centuries of discussion had failed to reach a consensus about the causes of many of them. Hence the early Stoics, like the Epicureans when they spoke of our world, thought it impossible to be certain about causes in meteorology. They could not ignore meteorology altogether: their system included an explanation of the physical world, and in Greek philosophical tradition this included an account of meteorological phenomena. But they preferred to deal with them briefly, and to say as little as possible about their causes. When they did speak about their causes, then, like other philosophers, they spoke of naturalistic causes, and when Cleanthes wrote, in traditional terms, of Zeus wielding the thunderbolt, he was speaking allegorically. However, in the Stoics' pantheistic system, this does not rule out divine involvement altogether.[28] They surely believed that meteorological phenomena were, in some way, part of the providential plan for the good of the world.

It seems unlikely that any philosophically educated Greek in the 3rd century B.C. believed that meteorological phenomena were due to divine action in the way described (say) by Homer – at any rate, no philosopher is recorded as defending such a view. But, if anyone interested in Stoicism did hold that traditional belief, then Stoic teachers had no compelling reason to contradict it: they did believe that divinity was involved in some sense. For the Epicureans the denial of divine action was essential, and that was why they described every possible naturalistic cause of meteorological phenomena, while the Stoics preferred to say nothing about causes. The letters to Herodotus and to Pythocles are, Epicurus says, summaries, for people unable to study his physical doctrines in detail[29] – people, presumably, who might have had little or no prior philosophical training, and might accept traditional beliefs about the gods and their control of weather.

Posidonius was unwilling to accept this refusal to seek definite answers. There is something unsatisfactory in systems which, like those of the early Stoics and the Epicureans, claim to provide an explanation of the physical world, and yet can give no definite account of conspicuous and familiar phenomena such as those of the weather. Other thinkers (including Aristotle, a man Posidonius admired) had given single explanations, or at most two or three, for each phenomenon, presumably believing they had found the true explanation, or at least the most probable. Posidonius resolved to follow them. What his motives were, what methods he used, what degree of certainty he thought he had achieved, are matters I shall discuss further in the following chapters.

Notes

1 See above, Chapter 8: p. 68–70 for Epicurus, the early Stoics and Posidonius, p. 65–7 for the pre-Socratics.
2 See above, Chapter 8 (pp. 68, 69, 71).
3 This is only a possibility. Thunder as due to the rotation of wind within a cloud is mentioned in two other sources which *may* antedate Posidonius: the Syriac Meteorology 1.6 (see Daiber [1992] 261) and *De mundo* 395a12.
4 See above, Chapter 8 (p. 66), and below, Chapter 21 (p. 205–6).
5 Probably, thunder is the sound of κατάξεις, "breakings" of the clouds, but the word is doubtful (see Dorandi [2013]). The mention of "ice-like" clouds is in any case unprecedented in this context,
6 Bakker (2016) 5, 8, and 58–62 is rather dismissive of this aspect of Epicurean meteorology, saying that the individual explanations "very likely" derive from a doxographical source (p. 62). This is not the place to discuss Epicurus' source or sources, but one should not ignore his, and Lucretius', originality.
7 *Letter to Pythocles* 99. See above, Chapter 4 (p. 25).
8 See Bakker (2016) 63–74.
9 See above, Chapter 8 (p. 69–70).
10 See above, Chapter 12 (p. 87–8).
11 Bakker (2016) 8–74.
12 *Letter to Pythocles* 86.
13 *Letter to Herodotus* 80.
14 *Letter to Pythocles* 87.
15 *Letter to Pythocles* 88.
16 *Letter to Herodotus* 80. Cf. *Letter to Pythocles* 94; Bakker (2016) 35, citing these two passages and *Letter to Pythocles* 87.

17 Epicurus, *Letter to Herodotus* 51. (The same was possibly also said in §50, but the text has to be amended. Dorandi [2013] excises the lines involved.) See also Diogenes Laertius X.34. Sextus Empiricus *Adv. Math.* VII.211 says that to Epicurus opinions are true if they are "confirmed *and* not contradicted by evidence" (my italics; in Greek ἐπιμαρτυρούμεναι καὶ οὐκ ἀντιμαρτυρούμεναι), but presumably Epicurus' own words are superior testimony.
18 See Bakker (2016) 20, 28–31.
19 Bakker (2016) 37–58.
20 "Hic" ("here") is an emendation, but the general sense is clear.
21 Lucretius VI.142.
22 See above, Chapter 15 (p. 157).
23 *Letter to Pythocles* 85.
24 *Letter to Herodotus* 81.
25 Lucretius VI.50–2, 54–5.
26 See, e.g., *Letter to Herodotus* 76–7, 81; *Kyriai doxai* 1 (Diogenes Laertius X.139).
27 Chapter 21 (p. 206).
28 On early Stoic meteorology see above, Chapter 4 (p. 26–30).
29 *Letter to Herodotus* 35, and *Letter to Pythocles* 84–5.

20 The place of meteorology among the different branches of knowledge

Posidonius agreed with the earlier Stoics that there are three parts of philosophy: physics (or "natural philosophy"), ethics and logic.[1] ("Natural philosophy" expresses the Stoics' meaning better than "physics", but for convenience I use both terms.) Posidonius and his teacher Panaetius differed from earlier Stoics about the order in which these subjects should be taught: Zeno and Chrysippus put logic first, Panaetius and Posidonius put "physics" first.[2] More interestingly, Posidonius used a different image from his predecessors to illustrate the relationship of physics, ethics and logic. Sextus Empiricus says:

Οἱ ἀπὸ τῆς Στοᾶς ... ὁμοιοῦσι τὴν φιλοσοφίαν παγκάρπῳ ἁλωῇ, ἵνα τῇ μὲν ὑψηλότητι τῶν φυτῶν εἰκάζηται τὸ φυσικόν, τῷ δὲ νοστίμῳ τῶν καρπῶν τὸ ἠθικόν, τῇ δὲ ὀχυρότητι τῶν τειχῶν τὸ λογικόν ... ὁ δὲ Ποσειδώνιος, ἐπεὶ τὰ μὲν μέρη τῆς φιλοσοφίας ἀχώριστά ἐστιν ἀλλήλων, τὰ δὲ φυτὰ τῶν καρπῶν ἕτερα θεωρεῖται καὶ τὰ τείχη τῶν φυτῶν κεχώρισται, ζῴῳ μᾶλλον εἰκάζειν ἠξίου τὴν φιλοσοφίαν, αἵματι μὲν καὶ σαρξὶ τὸ φυσικόν, ὀστέοις δὲ καὶ νεύροις τὸ λογικόν, ψυχῇ δὲ τὸ ἠθικόν

The Stoics ... liken philosophy to a garden rich in its variety of fruit, comparing natural philosophy to the height of the plants, ethics to the abundance of the crop, and logic to the strength of the walls. ... Posidonius differed: since the parts of philosophy are inseparable from each other, yet plants are thought of as distinct from fruit and walls are separate from plants, he claimed that the simile for philosophy should rather be with a living creature, where natural philosophy is the blood and flesh, logic the bones and sinews, and ethics the soul.[3]

Sextus mentions, as the reason for Posidonius' comparison, the fact that it demonstrates the inseparability of the three parts of philosophy, but one might add that it is also a more dynamic image than that of the garden. The rules of logic are unchanging, as an animal's bones appear to be, but natural philosophy as pursued by Posidonius requires the study of bodies constantly in motion, as the heavenly bodies revolve, as the tides rise and fall, and thunderstorms and other meteorological phenomena occur: an appropriate

DOI: 10.4324/9780429399930-20

comparison to the movements of an animal's flesh, as it breathes, as food passes through its digestive system, and so on.

Aëtius says that, to the Stoics, we are concerned with "physics" (φυσικόν) "when we enquire about the cosmos and the things in the cosmos" (ὅταν περὶ κόσμου ζητῶμεν καὶ τῶν ἐν κόσμῳ).[4] Diogenes Laertius VII.132 gives a more elaborate account, which certainly shows the influence of Posidonius, although he is not named:

> Τὸν δὲ φυσικὸν λόγον διαιροῦσιν εἴς τε τὸν περὶ σωμάτων τόπον καὶ περὶ ἀρχῶν καὶ στοιχείων καὶ θεῶν καὶ περάτων καὶ τόπου καὶ κενοῦ. Καὶ οὕτω μὲν εἰδικῶς, γενικῶς δὲ εἰς τρεῖς τόπους, τόν τε περὶ κόσμου καὶ τὸν περὶ τῶν στοιχείων καὶ τρίτον τὸν αἰτιολογικόν.

> Their physical doctrine they divide into the section about bodies; and one about principles; and about elements, and gods, and limits, and place, and void. This is a division into species; but generically they divide it into three topics, the one about the cosmos, and the one about the elements, and third the study of causes.[5]

I have previously quoted Strabo's statement that Posidonius' study of causes (τὸ αἰτιολογικόν) is Aristotelian, and something avoided by the Stoics.[6] The inclusion of τὸν αἰτιολογικόν in Diogenes' list is surely taken from Posidonius, or at least is due to his influence.

Diogenes next gives more information about the part of "physics" which is about the cosmos, and about the study of causes.[7] In the Stoic view, he says, the study of the cosmos is in two parts, one of which is shared by mathematicians, and concerns such questions as whether the sun and moon are the size they appear to be, and their revolutions; the other part belongs only to the philosophers who study nature (τοῖς φυσικοῖς), and concerns such things as the substance of the cosmos, whether or not it had a beginning, whether it is or is not alive, whether it is governed by Providence. This is closely similar to the view described as that of Posidonius by Simplicius, *In Aristotelis Physica* II.2.[8]

Diogenes then describes the study of causes:

> Τόν τε αἰτιολογικὸν εἶναι καὶ αὐτὸν διμερῆ. Μιᾷ δ᾽ αὐτοῦ σκέψει ἐπικοινωνεῖν τὴν τῶν ἰατρῶν ζήτησιν, καθ᾽ ἣν ζητοῦσι περί τε τοῦ ἡγεμονικοῦ τῆς ψυχῆς καὶ τῶν ἐν ψυχῇ γινομένων καὶ περὶ σπερμάτων καὶ τῶν τούτοις ὁμοίων· τοῦ δ᾽ ἑτέρου καὶ τοὺς ἀπὸ τῶν μαθημάτων ἀντιποιεῖσθαι, οἷον πῶς ὁρῶμεν, τίς ἡ αἰτία τῆς κατοπτρικῆς φαντασίας, ὅπως νέφη συνίστανται, βρονταὶ καὶ ἴριδες καὶ ἅλως καὶ κομῆται καὶ τὰ παραπλήσια.

> The topic concerned with causation also has two branches. In one of its branches medical enquiries have a share, the one in which they investigate the ruling principle of the soul, psychological processes, seeds and the like. The second part is also claimed by the mathematicians, the one in

which they investigate how we see, what causes the image in the mirror, and the origin of clouds, thunder, rainbows, haloes, comets, and the like.[9]

The sequence Ruling part – Sight – Mirrors is reminiscent of Plato's discussion of the causes of things at *Timaeus* 44D–46D,[10] and suggests that the *Timaeus* was a source for the ideas recounted in this passage of Diogenes. This is further evidence that Posidonius was here a source for Diogenes, since, though the early Stoics shared Posidonius' interest in the "ruling part" of men and in sight, they did not share his interest in Plato.[11] Neither Plato nor any earlier Stoic shared Posidonius' interest in meteorology. We have already seen reasons to connect with Posidonius the statements of Diogenes which immediately precede these sentences on causes; these sentences with their stress on meteorology must surely be derived from Posidonius or due to his influence.

That philosophers and mathematicians should be said to share in the study of optics and catoptrics is not surprising: surviving ancient studies of these subjects treat them mathematically.[12] Also, Aristotle's accounts of haloes and rainbows include geometrical arguments,[13] and Seneca mentions geometrical arguments about rainbows which are probably those of Posidonius,[14] so mathematics were thought relevant to those subjects. The interrelation of physics, mathematics, optics and astronomy had been pointed out by Aristotle.[15] However, ancient writers are not recorded as treating clouds, thunder or comets mathematically.

With this passage of Diogenes we may contrast Galen's *Institutio logica*, Chapter XIII, another passage thought to have been influenced by Posidonius.[16] This chapter discusses syllogisms related to the different Aristotelian categories. Galen says that searches for causes (αἱ τῶν αἰτιῶν ζητήσεις) are in the categories of ποιεῖν καὶ πάσχειν, "doing and experiencing", and instances ἐν ἰατρικῇ μὲν οὖν ἐκ τίνος αἰτίας γίγνονται φωνὴ καὶ ἀναπνοή ... ἐν φιλοσοφίᾳ δὲ σεισμὸς κεραυνὸς ἀστραπή τε καὶ βροντή, "in medicine from what cause there is voice and breath [and other phenomena], and in philosophy earthquake, thunderbolt, lightning and thunder".[17] No examples are given of syllogisms on these subjects, but study of the causes of these meteorological phenomena is assigned to philosophy, with no mention of mathematics. This reflects the actual practice of ancient authors.

It would not be surprising if Posidonius, or a follower of his, sought to extend the use of mathematics in meteorology: Greek investigators successfully used mathematics in several fields besides astronomy and optics; for instance in mechanics and mathematical geography.[18] There is a way in which ancient thinkers could have applied mathematics to some meteorological happenings other than optical ones. Many meteorological events are caused by, or correlated with, the movements of the sun, moon and stars, or in antiquity were thought to be so. Changes of season are one obvious example; the cycle of celestial events are important in ancient weather-signs[19], sea tides depend precisely on the movements of moon and sun. Insofar as celestial movements could be calculated mathematically, and meteorological events are correlated with them, then those meteorological events could be calculated mathematically also. We have seen that Posidonius did occasionally record figures, estimated or roughly measured, that are relevant to meteorology: we have his figures for the height to which clouds and wind occur,[20] for the deepest measured depth of the sea,[21] and – what is relevant here – we are told that, in his study of tides, he recorded the height to which a tide rose at Cadiz (which he said was measured), and the extent of areas in Spain affected by an exceptional tide (these last figures presumably supplied by people he spoke to).[22] He knew that the tides depend on the movements of moon and sun, which could be calculated mathematically. Did he perhaps record these figures because he thought that it might be possible to use those mathematical calculations in the study of the tides?

However, if Posidonius did think this, there is no evidence that he tried to do it, and we know nothing of the mathematical arguments that he probably used in studying rainbows. All that we have of Posidonius' meteorology was to him, presumably, philosophy.

Notes

1 Diogenes Laertius VII.39; Sextus Empiricus, *Adv. Mathematicos* VII.16–19 (Posidonius F87 and F88 EK).
2 Diogenes Laertius VII.40–1 (Posidonius F91 EK).
3 Sextus Empiricus, *Adv. Mathematicos* VII.16–19 (Posidonius F88 EK), with translation from Kidd (1999). Diogenes Laertius VII.40 in his account of Stoicism, without naming Posidonius, also reports this comparison of philosophy to an animal, but with the "more fleshy parts" (τοῖς σαρκωδεστέροις) corresponding to ethics and natural philosophy corresponding to the soul. Sextus is surely right: it must be the soul, not the flesh, which is connected to right conduct.
4 Aëtius I Proem. 2 = *SVF* II.35.
5 Diogenes Laertius VII.132, with a fairly literal translation based on that of Hicks (1925). Diogenes Laertius VII.132–3 is Posidonius F254 in Theiler (1982).
6 Strabo II.3.8, quoted above, Chapter 4 (p. 27).
7 Diogenes Laertius VII.132–3.
8 Page 291.22–292.31 Diels = Posidonius F18 EK.
9 Diogenes Laertius VII.133, with the translation of Mensch and Miller (2018); but I put "mathematicians" for their "scientists".
10 I must thank David Sedley for pointing this out to me.
11 For Posidonius' views on the soul and its workings see Posidonius F139 ff. EK. On the renewed interest in Plato taken by Posidonius and his teacher Panaetius see Frede (1999) 782–4.
12 See, for instance, the *Optica* and *Catoptrica* of Euclid.
13 *Mete.* III.3, 373a1ff, on haloes, and III.5, 375b16ff, on rainbows. At *APo.* 79a11–13 Aristotle says that the study of the rainbow is the concern partly of the natural philosopher (φυσικός), and partly of the student of optics or mathematics.
14 *NQ* I.5.13 (see above, Chapter 16 [pp. 161, 162]).
15 *Physics* II.2, 193b22–194a12. It is Simplicius' comment on this passage which preserves the fullest surviving account of Posidonius' ideas on the subject (F18 EK, cited in note 8 above).
16 So Kidd (1978b) 277.
17 *Institutio logica* XIII.9.

18 Lloyd (1979) 120–1.
19 See (e.g.) [Theophrastus] *De signis* 1–9; Aratus 740 ff.
20 See above, Chapter 6 (pp. 45, 48–9).
21 See above, Chapter 13 (p. 129).
22 See above, Chapter 13 (p. 132–3).

21 Sources and methods in Posidonius' meteorology

The sources of Posidonius' meteorological theories

Posidonius, when considering the cause of a meteorological phenomenon, would probably have begun by considering what his predecessors had said about it. I have discussed this aspect of his meteorology in Chapter 5. Subsequent chapters have slightly modified the provisional conclusions there reached.

Though earlier Stoics had been uninterested in meteorology, their works may have suggested to Posidonius alternative causes of thunder and lightning,[1] and he accepted his teacher Panaetius' view that the equatorial regions are habitable.[2]

Though some of his theories have pre-Socratic origins, there are few indications of his studying pre-Socratic meteorology: he criticised Parmenides on climatic zones,[3] he cited the obscure Thrasyalces on wind-names and the Nile floods,[4] and his works may be the source of what is said in the Scholia to Aratus on the comet theories of the Pythagoreans, Anaxagoras and Democritus.[5] There is ample confirmation of his use of Aristotle's *Meteorologica*. Besides the wind-rose, climatic zones, thunder and lightning, and haloes (all mentioned in Chapter 5), there is evidence of Aristotle's influence on Posidonius' theories of comets[6] and their significance as weather-signs[7]; on the Milky Way[8]; and on varieties of earthquakes[9]; and he need not have looked outside the *Meteorologica* to find a source for his theories of rain, snow and hail,[10] and perhaps rainbows.[11] Only on tides does the evidence suggest that Posidonius misunderstood or misrepresented Aristotle.[12] But he did not follow Aristotle uncritically, rejecting Aristotle's views where they conflicted with the principles of Stoic physics[13] or with discoveries made by the expansion of Greek geographical knowledge.[14]

Among philosophers younger than Aristotle, only Theophrastus and Epicurus wrote at length on meteorology before Posidonius. For Theophrastus certainty is impossible; this study has found no clear evidence that Posidonius used his meteorological work.[15] One scrap of evidence suggests that he may have taken individual ideas from Epicurus' meteorology.[16]

Posidonius also used Hellenistic authors who were not philosophers. Three not mentioned in Chapter 5 are Timosthenes (naval commander under Prolemy II and author of a book *On harbours*) and Bion, an astronomer, both

DOI: 10.4324/9780429399930-21

cited by Posidonius on the wind-rose,[17] and also the historian Agatharchides, possibly consulted on the Nile floods.

Posidonius also liked to quote his earliest possible source, Homer, for he had a belief in ancient wisdom.[18] He maintained that there had been a golden age in which men had been ruled by sages[19]; other ancient philosophers who envisaged a past golden age seem generally to have held that in that age men had avoided vicious lives without the need for philosophy. But, if the whole passage of Galen which is printed as Posidonius F156 EK represents Posidonius' thought,[20] he advocated caution; one should, the passage says, call Homer or other more ancient thinkers (τοὺς πρεσβυτέρους) as witnesses, not at the beginning of a discussion (ἐν ἀρχῇ τῶν λόγων), but when one has adequately settled the question before one (ἐπειδὰν ἱκανῶς ἀποδείξῃ τις τὸ προκείμενον), and one should cite them not about obscure matters, but about things that are clearly evident (οὔτε περὶ πραγμάτων ἀδήλων ... ἀλλὰ ... περὶ φαινομένων ἐναργῶς). His citations of Homer on meteorological matters illustrate this sort of policy: he claims that his meteorological knowledge was already known to Homer, but to do this he has to interpret Homer in ways that would never have occurred to anyone who lacked Posidonius' knowledge of meteorology and his determination to find that knowledge in Homer: this is especially clear when he finds in Homer a knowledge of the sea's tides, and of the rain in Ethiopian mountains which causes the Nile to flood[21] – facts evident to anyone who is in a position to observe them. There was a Stoic tradition of interpreting poets' stories and myths about the gods as allegories of Stoic doctrines,[22] and Posidonius was in his way following this.

Besides written sources, he drew information from his own observations and experience (including, obviously, the everyday experience of familiar phenomena that we all have, besides specially made observations), and he had oral testimony from people he spoke to. To the examples mentioned in Chapter 5 one might add the enquiries he evidently made about the voyages of Eudoxus,[23] which gave him the opportunity to enquire also about the geography of land at the equator.[24] He may also have enquired about earthquakes from people who remembered them.[25]

How ancient thinkers devised and defended their meteorological theories

Thus equipped, Posidonius devised some new theories. His classification of earthquakes seems to have been partly original – we know so little about it that we cannot tell how interesting his innovation may have been.[26] Also likely to have been original to him is the theory that there is exhalation from earth as a material which provides nourishment for the stars[27]; the explanation of the shape of the rainbow[28]; the explanation of mock-suns[29]; and, among phenomena unknown to the Greeks until after Aristotle's time, the theory of the causes of the tides,[30] and some of the reasons suggested to explain why equatorial regions are habitable.[31] More often he accepted a theory previously devised by someone else, but presumably he first considered what reasons there were to accept it.

We rarely have records of the reasons which led ancient thinkers to adopt the meteorological theories which they held or of any arguments by which they defended them. Even with Aristotle – a systematic thinker whose *Meteorologica* survives intact – we have, as we have seen, to guess the reasons which led him to postulate that there is a dry exhalation from earth as well as the wet one, the formation of water-vapour, from water.[32] Having decided that there is a dry exhalation, Aristotle then sometimes offers very little in the way of argument to show that it causes a particular phenomenon. For instance, in *Mete.* II.4, on wind, he restates his two-exhalation theory, indicating that the two exhalations occur together.[33] He then continues:

ἡ μὲν ὑγροῦ πλέον ἔχουσα πλῆθος ἀναθυμίασις ἀρχὴ τοῦ ὑομένου ὕδατός ἐστιν ... ἡ δὲ ξηρὰ τῶν πνευμάτων ἀρχὴ καὶ φύσις πάντων. Ταῦτα δὲ ὅτι τοῦτον τὸν τρόπον ἀναγκαῖον συμβαίνειν, καὶ ἐξ αὐτῶν τῶν ἔργων δῆλον· καὶ γὰρ τὴν ἀναθυμίασιν διαφέρειν ἀναγκαῖον, καὶ τὸν ἥλιον καὶ τὴν ἐν τῇ γῇ θερμότητα ταῦτα ποιεῖν οὐ μόνον δυνατὸν ἀλλ' ἀναγκαῖόν ἐστιν

The exhalation containing the greater amount of moisture is ... the origin of rain water: the dry exhalation is the origin and natural substance of winds. That this must be the case is evident from the facts. For the exhalation must differ [i.e., sometimes it is wetter and sometimes drier], and that the sun and the warmth in the earth should produce these effects is not only possible but necessary.[34]

Let us allow that, on Aristotle's principles, there must be a dry as well as a wet exhalation; he does not give us the slightest evidence that the dry one causes winds. (Of course he is not always so arbitrary: I have quoted above his argument for his theory of earthquakes.[35])

With most other writers on ancient meteorology, we have the disadvantage of not having their own words, but only reports of their views (possibly inaccurate) by other authors, often in a very summary form; or, as with the *De mundo*, what the author has written is itself a summary. Normally, we have a report of their conclusions – their preferred explanations of the different phenomena – but little or nothing of the evidence or the mental processes on which their explanations were based. We are rather better off with Posidonius than with most ancient writers on meteorology, since Seneca does in a few places record an argument for, or against, a meteorological theory and attributes the argument to Posidonius; or else, from the context, the argument seems likely to be his. These I shall examine later. For the most part we have to rely on common sense and on the explanations themselves, to deduce what we can about the evidence and the previous ideas which were the basis for ancient thinkers' explanations of meteorological phenomena.

One basis was direct observation, but that could rarely provide information about the cause of a meteorological phenomenon. Sometimes, a suggestion might be derived from a thinker's physical system. For instance, as I have argued, Aristotle's theory of dry and wet exhalations was a natural deduction from his element theory, and, the idea, probably original to Posidonius, that the stars are nourished by exhalation from earth as a material, seems likely to have been devised to meet a difficulty in the Stoic physical system.[36]

However, the ideas derivable from such a source were limited. A commoner way to find a cause was by analogy. Ancient enquirers could not enter the interior of a cloud and see how raindrops form there; but it would be easy to see an analogy between this process and the familiar domestic one of clouds of "steam" rising from heated water and condensing back to water when it meets a cold surface.

Numerous examples of analogies used or probably used by Posidonius are mentioned in, or can be inferred from, sources quoted in previous chapters. Aristotle's dry exhalation, and the exhalation from earth probably hypothesised by Posidonius, are not only required by their general physical systems but must also be based on an analogy with the observed evaporation of water, as "steam" rises from heated water in domestic contexts and ponds and streams dry up in hot weather. Posidonius' theory of thunder and lightning, and that of several other ancient thinkers, involves analogies with the lighting of fires by friction and the bursting of heated bladders.[37] The idea that heavenly bodies are "nourished" (Greek τρέφεσθαι, Latin "nutriri") by exhalations from the earth is an analogy with the food required by people and animals on the earth.[38] Seneca's theory that the power of wind causes earthquakes involves an analogy with the power of wind arousing fires and waves, and he may well have taken this from Posidonius.[39] Posidonius (according to Priscianus) drew an analogy between the moon causing tides to rise and the rising of water in a moderately heated kettle.[40] The belief that the rainbow is a reflection involves an analogy with mirrors and with reflections from still water, and Seneca says that Posidonius explained the rainbow's circular shape by supposing that the reflecting cloud has the shape of part of a cut (presumably hollow) ball.[41] Even an idea which one might call simply common sense – for instance, the idea that the effect of the moon and sun in raising the tides is strongest when they are acting together in the same direction – may, when analysed, be found to depend on a familiar analogy, for instance with two strong men acting together.

This frequent use of analogies, in a broad sense, is not surprising. "Philosophers", says G.E.R. Lloyd (with supporting quotations from Hume and J.S. Mill) "have ... drawn attention to the element of analogy in all reasoning"[42]; and he details the numerous uses of analogy in the works of early Greek authors up to and including Aristotle, who uses many analogies in his meteorology, biology and psychology, and derives many interesting ideas from them, especially in biology.[43] Even in modern scientific research, the use of analogies is arguably of value to suggest hypotheses, which must then be tested and verified[44]; but, as Lloyd writes, "analogies ... in early Greek science ... were generally treated as not so much a source of preliminary hypotheses, as the basis and justification of definitive accounts". Aristotle is "more cautious" than his predecessors; nevertheless, "the differences between his own use of analogy and that of earlier writers ... are less than he appears to claim".[45]

So Aristotle, in the *Meteorologica* and other works, bases explanations of many things on analogies; but when he discusses *methods* of demonstration, he treats such arguments as less than probatory. In his *Organon* and *Rhetoric* he discusses arguments by "example" (παράδειγμα), in which a single similar example is cited as evidence in support of the conclusion – a procedure equivalent to an argument from an analogy – and he is clear that such an argument may be persuasive, but does not amount to proof, which depends on syllogisms and "perfect" inductions; that is, inductions in which all the particular instances are detailed in support of a generalisation (something which Aristotle rarely, if ever, does in practice in his surviving works).[46] (In the *Organon* he shows no interest in the heuristic use of analogy to suggest new hypotheses.[47])

There is, therefore, an apparent clash between Aristotle's theory of how scientific truth should be established and his practice. I shall next argue that this also applies to Posidonius' meteorology, and point out the one way by which they could reconcile this inconsistency: by admitting the uncertainty of their meteorological theories.

Meteorology and the theory of knowledge in Posidonius

In the Stoic theory of knowledge the criterion of truth is the φαντασία καταληπτική, "cognitive presentation". This is not the place for a discussion of this much-debated concept, but it does seem clear that φαντασία, "presentation", was thought of primarily – not exclusively – as a matter of sense-perception. Diogenes Laertius VII.49 (quoting Diocles the Magnesian) says: ἀρέσκει τοῖς Στωϊκοῖς τὸν περὶ φαντασίας καὶ αἰσθήσεως προτάττειν λόγον, "the Stoics agree to put in the forefront the doctrine of presentation and sensation".[48] Cicero *Acad.* I.40 makes Varro, as spokesman for the Old Academy, give it as the view of Zeno that the things of which we have presentations are "haec quae visa sunt et quasi accepta sensibus", "these things which have been seen and, as it were, accepted by the senses". Sextus Empiricus discusses and criticises the concept of φαντασία καταληπτική at considerable length: nearly all his examples are in terms of what can be apprehended by the senses.[49]

Posidonius' view seems rather different. From his book Περὶ κριτηρίου, "On criterion [sc. of truth]" just one fragment survives, in Diogenes Laertius VII. 54[50]:

Ἄλλοι δέ τινες τῶν ἀρχαιοτέρων Στωϊκῶν τὸν ὀρθὸν λόγον κριτήριον ἀπολείπουσιν, ὡς ὁ Ποσειδώνιος ἐν τῷ Περὶ κριτηρίου φησί

Some others of the older Stoics leave right reason as criterion, as Posidonius says in On criterion.

> ("Others" must at least mean other than Chrysippus, mentioned in the previous sentence)

This statement contradicts the general view of our sources that the Stoics, from the beginning, regarded φαντασία καταληπτικὴ as the criterion of truth, and it seems fairly certain that Posidonius has misleadingly attributed to early Stoics what was really his own view.[51]

This does not imply that Posidonius abandoned the concept of φαντασία καταληπτικὴ. The Stoics never held that it was always the product of sense-perception. Diogenes Laertius VII. 52 says:

Ἡ δὲ κατάληψις γίνεται κατ᾽ αὐτοὺς αἰσθήσει μὲν λευκῶν καὶ μελάνων ... λόγῳ δὲ τῶν δι᾽ ἀποδείξεως συναγομένων, ὥσπερ τὸ θεοὺς εἶναι

According to them [the Stoics] cognition happens by perception of black and white ... but by reason when things are inferred through demonstration, as that gods exist.

But Posidonius' emphasis seems to have been different. Diogenes Laertius is not alone in recording him as stressing the importance of λόγος, "reason". Sextus Empiricus *Adv. Math.* VII.93 says:

Καὶ ὡς τὸ μὲν φῶς, φησὶν ὁ Ποσειδώνιος τὸν Πλάτωνος Τίμαιον ἐξηγούμενος, ὑπὸ τῆς φωτοειδοῦς ὄψεως καταλαμβάνεται, ἡ δὲ φωνὴ ὑπὸ τῆς ἀεροειδοῦς ἀκοῆς, οὕτω καὶ ἡ τῶν ὅλων φύσις ὑπὸ συγγενοῦς ὀφείλει καταλαμβάνεσθαι τοῦ λόγου

And as light, says Posidonius in expounding Plato's *Timaeus*, is grasped by sight that is luminous and sound by hearing that is airy, so too the nature of all that there is should be grasped by the *logos* that is kin to it.[52]

Here Posidonius seems to have devised a special interpretation of *Timaeus* 46D–E in order to find support for his own view, for what he says about λόγος goes beyond what Plato says.[53]

Posidonius' emphasis on right reason as a criterion of truth,[54] where earlier Stoics had emphasised sense-perception, must be connected to his special concern with finding causes. Strabo speaks of his concern with τὸ αἰτιολογικόν in the context of geography and natural philosophy generally.[55] Simplicius mentions his concern with causes in astronomy,[56] Seneca does this for ethics[57]; in psychology, Galen tells us, Posidonius explained the causes of things where Chrysippus admitted ignorance.[58] The ancients could rarely, if ever, directly observe the causes of happenings in fields like meteorology, astronomy and psychology; one could only hope to infer them by reasoning.

Strabo calls Posidonius ἀποδεικτικός, "master of logical demonstration";[59] Galen says he was τεθραμμένος ἐν γεωμετρίᾳ καὶ μᾶλλον τῶν ἄλλων Στωϊκῶν ἀποδείξεσιν ἕπεσθαι συνειθισμένος, "trained in geometry and ... accustomed to follow demonstrative proof more than any other Stoic".[60] Aristotle defines ἀπόδειξις as συλλογισμὸν ἐπιστημονικόν, "scientific deduction".[61] To the Stoics, ἀπόδειξις, "demonstration", was λόγον διὰ τῶν μᾶλλον καταλαμβανομένων τὸ ἧττον καταλαμβανόμενον παριστάντα, "an argument that infers [or "proves" ?] things less well apprehended from things better apprehended".[62] Sextus Empiricus says that the Stoics distinguished a syllogism which is ἀποδεικτικός ("demonstrative" or "probative") from other syllogisms, by the fact that it establishes a conclusion which is both true and "not evident" (ἄδηλος) – as opposed to syllogisms like "If it is day, then it is light; but it is day, therefore it is light", which are correctly formed but reach a conclusion which normally is already obvious.[63]

Ἀπόδειξις, therefore, is a way of inferring, by logical deduction, things which could not otherwise be known; and this is what Posidonius was trying to do in his search for causes.[64]

In astronomy we can see how his system worked. He distinguished, Seneca tells us,[65] philosophy (including the philosophy of nature) from subordinate studies such as mathematics, which are its necessary tools. In his view, says Simplicius, the philosopher studies the substance (οὐσία), power (δύναμις) and so on of the heavenly bodies, and the true causes of astronomical events, while the astronomer studies mathematically their order, sizes, distances and the like and proposes hypotheses to explain apparent anomalies in their movements.[66] And so (it is implied) the astronomer's hypothesis of a moving earth and stationary sun, although it "saves the phenomena", must be rejected, because the philosopher knows that the nature of earth is to be at rest. The astronomer must take his first principles (ἀρχάς) from the philosopher.

Posidonius, however, did not disdain work on mathematical astronomy. Cleomedes describes his attempts to calculate the circumference of the earth[67] and the diameter of the sun.[68] Kidd's analysis of these calculations[69] shows that the figures which Posidonius used are mostly very approximate measurements, and one of them completely arbitrary, but that the mathematical calculations based on them are correct. Posidonius evidently realised that his conclusions might well be inaccurate: Cleomedes reports him as saying that the earth's circumference is 240,000 stades if (as is assumed in the calculation) Rhodes is 5,000 stades from Alexandria, εἰ δὲ μή, πρὸς λόγον τοῦ διαστήματος, "but if not, in proportion to the [sc. actual] distance".[70] After giving the figure for the sun's diameter he says: ἐνδέχεται δὲ καὶ μείζονα αὐτὸν ὄντα ἢ πάλιν μείονα ἡμᾶς ἀγνοεῖν, "but it is possible that it is greater or again less: we don't know".[71]

This suggests that Posidonius, in his attempts to establish the nature and the causes of phenomena, regarded the correctness of his reasoning as more important than precise and accurate observations.[72] (And if so, there is some justification: inaccurate though his figures are, his calculations give a much better idea of the size of earth and sun than could be attained without mathematical reasoning.[73]) What reasonings did he use in his meteorology?

To argue from a simple analogy (A resembles B, B is caused by C, therefore A is caused by C) is a weak argument. A stronger argument is one based on a general rule (all Bs are caused by C, A is a B, therefore A is caused by C; or, negatively, everything caused by C is B, A is not B, therefore A is not caused by C).

Seneca provides some evidence of arguments of this stronger kind used by Posidonius in his meteorology – he is not a reliable guide to the details of other people's theories, but what he says should serve as an indication of what is likely. At *NQ* I.5.10[74] he says that Posidonius and others argued that, if a rainbow were actual colour in the cloud where we see it, then the colour would become clearer as we approached it, which it does not – a generalisation drawn from ordinary experience: we see things more clearly the closer we are to them.

At *NQ* II.54.1-3[75] Seneca describes Posidonius' theory of thunder: dry, smoky exhalation does not tolerate being shut up in clouds, and also "whatever is rarefied in the air itself is at the same time dried and heated; this too if it has been shut up equally seeks escape and comes out with noise. ... Therefore this wind forces out thunder". Posidonius seems to have set out a general rule about how heated air behaves when confined (in a cloud, or a sealed bladder or utensil, or whatever), and to have drawn from this a conclusion about a cause of thunder.

There is some indication that two similar inferences drawn from general rules, in passages of *NQ*, may be from Posidonius. At *NQ* II.22.1–2 Seneca says of lightning and thunderbolts:

> Quoniam constat utramque rem ignem esse, videamus quemadmodum ignis fieri soleat apud nos; eadem enim ratione et supra fiet. <Fit> duobus modis, uno si excitatur sicut e lapide; altero si attritu invenitur, sicut cum duo ligna inter se diutius fricta sunt ... Potest ergo fieri ut nubes quoque ignem eodem modo vel percussae reddant vel attritae

> Since it is agreed that both are fire, let us see how fire is usually produced with us; for it will be produced in the same way also above us. It is produced in two ways: in one way if it is ignited as from a stone [sc. by striking]; in the other, if it is found by rubbing, as when two sticks are rubbed against each other for a long time. ... It therefore can happen that clouds too should give fire, by being struck or rubbed.

Here we have a general rule about the causes of fire and an inference drawn from it about the cause of lightning; and it is part of a passage which is probably what Seneca is referring to when he says, at II.54.1, "let me now return to the opinion of Posidonius".[76]

At *NQ* VI.21.1 Seneca says, as evidence that "spiritus" ("air", "wind") causes earthquakes, "nihil est in rerum natura potentius, nihil acrius, sine quo ne illa quidem quae vehementissima sunt valent", "nothing in nature is more powerful, nothing is keener, without it not even the most violent things have force" – he cites, as examples, wind arousing fire, and setting water in motion (presumably by making waves). Here we have a general rule about the exceptional power of wind, and an inference from it about the cause of earthquakes; and immediately afterwards, in VI.21.2, Seneca cites Posidonius for a different point about earthquakes. (The argument in *NQ* VI.21.1 seems an improved version of one used by Aristotle.[77])

In both these passages there is a good chance that Seneca is using an argument derived from Posidonius. However, we cannot conclude that Posidonius regularly used arguments of this kind in his meteorology. Frequently we have no evidence of arguments which Posidonius may have used in support of his theories, and sometimes he seems to have used none: at *NQ* IVb.3.2 Seneca implies that Posidonius propounded his theory of hail arbitrarily, with no supporting evidence or arguments.[78] And no ancient source names Posidonius as someone who took special care to defend his meteorological theories with logical arguments.

However, it seems to me highly probable that Posidonius in his meteorology did at least on occasion use arguments of the kind I have quoted from Seneca. As M.R. Johnson has done for Aristotle on haloes,[79] one could convert the arguments I suggest Posidonius used into syllogisms; perhaps, for instance (assuming he would have used the Stoics' preferred, hypothetical form of syllogism):

> If a fire has been started in a particular place, there has been in that place an impact, or rubbing, to cause it.
> Lightning is a fire started in that cloud.
> Therefore there has been an impact, or rubbing, in that cloud to cause it.

This is a correctly formed syllogism, but are the premises sound? Presumably, in Posidonius' meteorology, as in his astronomy, one should work from first principles, ἀρχαί. Kidd writes that "the philosopher, according to Posidonius, works by ἀπόδειξις – deductive proof from first principles, ... The important science for him was ... pure mathematics".[80] This is hardly consistent with Posidonius' view that mathematics is subordinate to philosophy[81]; but is surely true in the sense that Posidonius' ideal in natural philosophy was a system like that of Euclidean geometry, a series of strict logical deductions from axioms that can hardly be challenged. (This was evidently Aristotle's ideal: for him, demonstration requires first principles, ἀρχαί, and "if equals are removed from equals, the remainders are equals" is an example.[82])

When Posidonius argued that the astronomer's hypothesis of a moving earth must be rejected because the philosopher knows that the nature of earth is to be at rest,[83] he may well have thought that he was arguing from a scarcely challengeable axiom. It was Stoic doctrine that the earth is at the centre of the cosmos,[84] therefore does not change its position. Aristotle says that the earth is at the centre of the cosmos and immobile[85]; Plato in the *Timaeus* implies this.[86] Immobility of the earth appears to accord with everyday experience. If anyone disagreed, that could be regarded as an eccentricity comparable to the Epicureans' rejection of geometry.[87]

In meteorology Posidonius must have been aware that he was rarely, if ever, proving his theories by syllogisms based on unchallengeable first principles. If he supported a theory by no argument, or a simple analogy, he was not arguing syllogistically. The syllogisms which I have suggested he may have used do not have unchallengeable first principles as their premises, but inductions based on everyday experience; and he was assuming that the everyday experiences were replicated in locations he could not examine closely or directly: clouds and their interiors, and the interior of the earth. Like the figures he used in his astronomical calculations, these premises might be termed "rough and ready".

There is evidence suggesting that the Stoics generally did not put a high value on inductive arguments. In Philodemus *De signis* the Epicureans are defenders of arguments from "similarity" (ὁμοιότης), meaning, roughly, analogy and induction,[88] while their opponents, usually identified as Stoics,[89] reject such arguments, citing exceptional cases which invalidate apparently obvious generalisations,[90] and even questioning a proposition as familiar as "all men are mortal".[91]

The Stoics cannot have rejected induction altogether. They regarded the study of syllogisms as "highly useful" (εὐχρηστοτάτην),[92] and must surely have realised that, if their syllogisms were to be more than an exercise with symbols ("If A, then B", and so forth), and were to reveal anything about the world we live in, they must use premises which are, or imply, generalisations based on our experience.[93] Aëtius IV.11 (*SVF* II.83) preserves a Stoic account of how we form such general mental concepts (ἔννοιαι) when we have "many memories of the same kind" (ὅταν ὁμοειδεῖς πολλαὶ μνῆμαι γένωνται), and of how with time we learn to reason from them.

There is, therefore, doubt about the Stoics' attitude to induction; they seem to have preferred to argue syllogistically. There is no evidence of any theoretical view expressed by Posidonius about induction, but he must have realised that inductive arguments – such as I suggest he may have used in his meteorology – were not the arguments from undeniable first principles which would surely have been his ideal; and that he was still further from his ideal if he used weaker forms of argument, such as simple analogies, or proposed meteorological theories not supported by any argument at all.

The admitted uncertainty of ancient meteorological theories

Both Aristotle and Posidonius must surely have realised that the arguments and analogies which they used in their meteorology fell short of the strict reasoning from unchallengeable premises by which they would ideally have wished to prove their theories. For Posidonius we lack the evidence to know whether he ever discussed the relation of his meteorological theories to his view of the criterion of truth, or whether he had any theoretical concept of methods of inference which yield probable but not certain conclusions.[94] Aristotle's *Analytica posteriora*, his account of how knowledge should be acquired (or, perhaps better, be presented[95]), does refer to meteorology, but not in connection with this point.

He twice refers to the rainbow: at *APo.* 79a10–13 it is a subject studied by two sciences,[96] and at *APo.* 98a28f it is an example of reflection (ἀνάκλασις). More interesting here is *APo.* II.8–10 (93a1–94a19), where Aristotle describes a type of definition which is λόγος ὁ δηλῶν διὰ τί ἔστιν, "an account which shows why something exists", and gives as an example a definition of thunder as ψόφος ἀποσβεννυμένου πυρὸς ἐν νέφεσιν, "a noise of fire being extinguished in the clouds".[97] This is a quite different explanation of thunder from that in *Meteorologica*,[98] and even if he changed his mind about the cause of thunder between writing the two works, he must have known, when he wrote *Analytica posteriora*, that the extinguishing-of-fire theory was not the only one to have been proposed.[99] At *APo.* 93b14 he seems to admit this: after speaking of the extinguishing of fire as a middle term between cloud and thunder in explaining the latter phenomenon, he adds ἂν δὲ πάλιν τούτου ἄλλο μέσον ᾖ, ἐκ τῶν παραλοίπων ἔσται λόγων, "if there is another middle term for this, it will come from the remaining accounts", apparently admitting that other explanations of thunder are possible.[100] It looks as though Aristotle meant, in this passage, that "thunder is a noise of fire being extinguished in the clouds" is an example of a correctly formed definition, although its actual correctness depends on the uncertain assumption that this is indeed how thunder is produced.

Although Aristotle here cites a meteorological example and apparently admits uncertainty about it, he does not attempt to describe a method for dealing with problems, such as meteorological ones, where certainty is impossible, nor, so far as we know, did Posidonius or other ancient meteorologists. But they must surely have been aware that very few of their meteorological theories commanded general assent, or met any strict standards by which the truth of an argument or a theory might be judged; and they had no strong reason for refusing to admit such an awareness. The precise explanation of any particular meteorological phenomenon – unlike, say, the existence or non-existence of Providence, or the nature of the human soul – was not a matter crucial to any philosophical system. Probably all the philosophers who discussed the causes of meteorological phenomena would have rejected any mythological concept that such phenomena were simply due to the action of a divinity, but they usually had no reason to insist on any one naturalistic explanation in preference to another.

Unsurprisingly, therefore, at least some ancient thinkers were ready on occasion to admit the uncertainty of their meteorological theories. We cannot tell how many, because it is not something likely to be recorded when we depend, as for most authors we do, on fragments and summaries. However, Aristotle does admit to uncertainty: at the beginning of *Mete.*, in the course of setting out the subjects with which he is going to deal, he says ἐν οἷς τὰ μὲν ἀποροῦμεν, τῶν δὲ ἐφαπτόμεθά τινα τρόπον, "of … these phenomena, some we find inexplicable, others we can to some extent understand".[101] At *Mete.* 344a5–8, beginning his own account of comets, he says:

Ἐπεὶ δὲ περὶ τῶν ἀφανῶν τῇ αἰσθήσει νομίζομεν ἱκανῶς ἀποδεδεῖχθαι κατὰ τὸν λόγον, ἐὰν εἰς τὸ δυνατὸν ἀναγάγωμεν, ἔκ τε τῶν νῦν φαινομένων ὑπολάβοι τις ἂν ὧδε περὶ τούτων μάλιστα συμβαίνειν

We consider that we have given a sufficiently rational explanation of things inaccessible to observation by our senses if we have produced a theory that is possible, and on the evidence available one may assume that what happens with these phenomena is surely as follows.[102]

Elsewhere in the *Meteorologica* Aristotle is much less tentative (I have quoted on page 198 a specimen of his more dogmatic manner), but these two passages admit uncertainty.

Seneca, too, expresses doubts. At *NQ* VI.16.1 he says of the earthquake theory which he favours: "Etiamnunc dicendum est quod plerisque auctoribus placet et in quod fortasse fiet discessio", "and now I must say what is preferred by most [or "very many"] authors and in favour of which perhaps the vote will go". At the end of his account of comets (VII.29.3) he says:

> Haec sunt quae aut alios movere ad cometas pertinentia aut me. Quae an vera sint, dii sciunt, quibus est scientia veri. Nobis rimari illa et coniectura ire in occulta tantum licet, nec cum fiducia inveniendi nec sine spe

> These are the things relating to comets which have made an impression either on others or on myself. Whether they are true the gods know: they have knowledge of the truth. To us it is only allowed to investigate them and to proceed by conjecture into things that are hidden, neither confident of making discoveries nor without hope of it.

When Aristotle, who influenced Posidonius, and Seneca, who was influenced by him, say the same thing, then the probability is that Posidonius said it also, especially as he surely accepted the general Stoic principle that only the ideal wise man (which he would not have claimed to be) truly knows anything.[103] That Posidonius was willing on occasion to admit that a meteorological view of his was merely supposition is confirmed by Strabo III.5.8, which reports Posidonius' observations of tides in summer at Cadiz, and adds that Posidonius "conjectures" (εἰκάζει) that tides are exceptionally high at the winter solstice and lower at the equinoxes – something observable, but which Posidonius had had no opportunity to observe. It seems clear, from the way Strabo reports the matter, that Posidonius himself spoke of his view as a supposition.[104]

There are other passages where Strabo uses εἰκάζειν, "conjecture" of views put forward by Posidonius: three which relate to matters discussed in this book are two passages on the wanderings of the Cimbri,[105] and one about passages in Homer which Posidonius suggested were references to tides.[106] It is possible that the suggestion of conjecture in these passages is due to Strabo, but Kidd says of these and some similar passages "it looks as if he is taking care to echo guarded statements on the part of Posidonius".[107] It does seem likely that Posidonius was willing to admit to uncertainty on a number of subjects, and meteorology was surely one of them.

Notes

1　See above, Chapter 8 (pp. 68, 69, 71).
2　See above, Chapter 7 (p. 58).
3　See above, Chapter 7 (p. 55).
4　See above, Chapter 11 (p. 102, on wind names), Chapter 15 (pp. 153, 155–6, on Nile floods).
5　See above, Chapter 9 (p. 80–1).
6　See above, Chapter 9 (p. 80).
7　See above, Chapter 17 (p. 172).
8　See above, Chapter 9 (p. 84).
9　See above, Chapter 12 (p. 118–9).
10　See above, Chapter 14 (pp. 142–6).
11　See above, Chapter 16 (p. 160–2).
12　See above, Chapter 13 (pp. 131, 137).
13　See above, (e.g.) Chapter 5 (p. 37) and Chapter 10 (p. 91–2).
14　See above, Chapter 7 (p. 55–6).
15　On Theophrastus *De ventis* see above, Chapter 11 (p. 107).
16　See above, Chapter 19 (p. 185).
17　See above, Chapter 11 (p. 102). On Timosthenes see *Brill's new Pauly* (2001–10) 14, 706, article "Timosthenes [2]" by G.A, Gärtner; this Bion seems to be otherwise unknown, but Strabo calls him an astronomer, and mentions him after Aristotle and Timosthenes, suggesting his date was later.
18　What follows in this paragraph owes much to Boys-Stones (2001) 3–44.
19　See passages of Seneca, *Epistulae* 90, printed as Posidonius F284 EK.
20　Galen, *De placitis Hippocratis et Platonis*, V.502, p. 486.14–487.10 M. Only the last sentence names Posidonius, but Kidd (1988) 566–7 argues that the whole passage "could all derive from Posidonius" and "is at least relevant to his practice".
21　See above, Chapter 13 (p. 138) and Chapter 15 (p. 156). (See also Chapter 11 (106–7) for Posidonius on Homeric wind-names.)
22　Cicero, *De natura deorum*, I.40–1 and II.63 ff. (References from Brill's new Pauly (2002–10) 1, 512, in article "Allegoresis" by C. Cancik-Lindemaier and D. Mohr-Sigel.)
23　See Kidd (1988) 247.
24　See above, Chapter 7 (p. 59).
25　See above, Chapter 12 (p. 120).
26　See above, Chapter 12 (p. 118–9).
27　See above, Chapter 10 (p. 93–4).
28　See above, Chapter 16 (pp. 160–1, 168).
29　See above, Chapter 16 (p. 167).
30　See above, Chapter 13 (p. 137–8).
31　See above, Chapter 7 (p. 59–61).
32　See above, Chapter 10 (p. 91–2).
33　*Mete.* II.4, 359b27–360a10.
34　*Mete.* II.4, 360a11–17. The translation down to "from the facts" is from Lee (1952).
35　Chapter 12 (p. 115).
36　See Chapter 10 (pp. 93–4).
37　See Chapter 8 (pp. 65–70).
38　See Chapter 10 (p. 93).
39　See Chapter 12 (p. 118).
40　See Chapter 13 (p. 134).
41　See Chapter 16 (p. 160–3). Aristotle *Mete.* 372a30 cites reflection "from water" (ἀφ' ὕδατος) as an analogy in his account of rainbows.
42　Lloyd (1966) 172.
43　See Lloyd (1966) 360–80.
44　See Lloyd (1966) 175.

45 Lloyd (1966) 382–3.
46 See *APr* 68b38–69a19; *Rh* 1393a23–1394a18; *APo* 71a1–11; Lloyd (1966) 405–14. (On perfect inductions *APr* 68b28–9; Lloyd (1966) 410–1).
47 Lloyd (1966) 413.
48 Tr. Hicks (1925).
49 Especially *Adv. Math.* VII.226–260, 402–35.
50 Posidonius F42 EK; *SVF* I.631.
51 This is the conclusion of Kidd (1989). He finds a parallel in Posidonius' claims – dubious claims, in Kidd's view – in his work on psychology, to be going back, beyond Chrysippus, to Zeno or Cleanthes or both (see the passages of Galen's *De placitis Hippocratis et Platonis* printed as Posidonius F32, F165 lines 2–5 and F166 EK). Sedley (1992) 27–33 reaches a similar conclusion on different grounds. He argues that Posidonius Περὶ κριτηρίου is the source of Sextus Empiricus *Adv. Math.* VII.89–140, which discusses the views of various thinkers said to have regarded λόγος as the criterion of truth, and "re-interprets" the view of λόγος held by several of them in the way that Posidonius evidently did in the passage cited from Diogenes Laertius VII.54.
52 Posidonius F85 EK, with translation from Kidd (1999).
53 At *Timaeus* 46D–E Plato says only that physical forces like cooling and heating cannot have reason or thought (λόγον ... οὐδὲ νοῦν), and goes on to say τὸν δὲ νοῦ καὶ ἐπιστήμης ἐραστὴν ἀνάγκη τὰς τῆς ἔμφρονος φύσεως αἰτίας πρώτας μεταδιώκειν, "and the lover of thought and knowledge must needs pursue first the causes which belong to the intelligent nature" (tr. Bury [1929]).
54 The importance, in some Stoic thought, of right reason in assessing "presentations" is also shown by Diogenes Laertius VII.47, τὴν δὲ ἀματαιότητα ἕξιν ἀναφέρουσαν τὰς φαντασίας ἐπὶ τὸν ὀρθὸν λόγον, "earnestness is a disposition to refer presentations to right reason" (tr. Mensch and Miller [2018], substituting "presentations" for their "impressions"). But there is no evidence of which Stoic said this.
55 Strabo II.3.8: ὅσα γεωγραφικά· ὅσα δὲ φυσικώτερα, "things geographical; and things more concerned with natural philosophy".
56 Simplicius *In Ph.* II.2 (p. 292.21–293.31 Diels) = Posidonius F18 EK.
57 Seneca *Epistulae* 95.65 = Posidonius F176 EK.
58 Galen, *De placitis Hippocratis et Platonis*, IV, p. 395–6 M. = Posidonius F165.70ff EK.
59 Strabo II.3.5 = Posidonius T46 EK, with tr. from Kidd (1999).
60 Galen, *De placitis Hippocratis et Platonis*, IV.390, p. 362.5–9 M. = Posidonius T83 EK, with tr. from Kidd (1999).
61 Aristotle *APo.* 71b17, with translation from Barnes (1994).
62 Diogenes Laertius VII.45 = *SVF* II.235. The translation is that of Mensch and Miller (2018).
63 Sextus Empiricus *Adv. Math.* VIII.411–423 = *SVF* II.239. On ἀπόδειξις see also Cicero *Acad. Pr.* II.26 = *SVF* II.111.
64 On the importance of logic to Posidonius see Kidd (1978b)
65 Seneca, *Epistulae* 88.21ff = Posidonius F90 EK. (I accept the interpretation of this passage by Kidd [1988] 359–65).
66 Simplicius *In Ph.* II.2 (p. 291.21–292.31 Diels) = Posidonius F18 EK.
67 See Cleomedes I.10.50–2 (I.7 lines 1–50 Todd) = Posidonius F202 EK.
68 See Cleomedes II.1.79–80 (II.1 lines 269–286 Todd) = Posidonius F115 EK.
69 Kidd (1988) 719–29 and 443–54.
70 Posidonius F202 EK lines 47–9.
71 Posidonius F115 EK, lines 19–20, with translation from Kidd (1999) 172.
72 Kidd (1978a) reaches a similar conclusion, though I would not accept all that he says about Posidonius' dismissive attitude towards empirical observation.

73 Assuming, for a rough calculation, that one stade = 200 metres, then 240,000 stades for the earth's circumference equals 48,000 kilometres, (and the 180,000 stades attributed to Posidonius in Strabo's manuscripts (see Chapter 7 [p. 56]) equals 36,000 kilometres). Posidonius' figure for the sun's diameter, 3,000,000 stades, equals 600,000 kilometres. By modern computation, the earth's circumference at the equator is 40,075 kilometres and the sun's diameter is about 1,390,000 kilometres. (Figures from Wikipedia articles "Sun" and "Earth's circumference", read 12 January 2022.) I have discussed Posidonius' calculation of the sun's diameter, from a different point of view, in Chapter 6 (p. 44).

74 See above, Chapter 16 (p. 160).

75 See above, Chapter 8 (p. 69).

76 See above, Chapter 8 (p. 70–1).

77 See above, Chapter 12 (pp. 118, 120).

78 See above, Chapter 14 (p. 144–5).

79 See Johnson (2020) 168–9, where he formulates syllogisms Aristotle might have used to prove his theory of haloes, based on the account of haloes at *Mete.* 372b12ff and on *APo.* 79a10–13, where the study of rainbows involves both the student of nature (*physikos*) and the student of optics (*optikos*).

80 Kidd (1978a) 12.

81 See Seneca, *Epistulae* 88.25–28 (Posidonius F90.21ff EK).

82 *APo.* 76a31–41, with tr. from Barnes (1994).

83 See above, p. 202.

84 Diogenes Laertius VII.137, 155.

85 *De caelo* 296b22.

86 *Timaeus* 38D speaks of heavenly bodies revolving around the earth; at 55D–E earth is the most immobile (ἀκινητοτάτη) of the four elements.

87 Epicurean arguments against geometry were attacked by Posidonius (Proclus, *In Euclidis Elementa* pp. 199.3ff and 214.15ff Friedlein = Posidonius F46 and F47 EK).

88 See especially *De signis* Chapter 18.

89 See Sedley (1982), especially pp. 240–1; also works cited by Allen (2001) 207 n.8 (though Allen himself is doubtful).

90 For instance, the magnet refutes any generalisation that stones do not attract iron (*De signis* Chapter 3).

91 *De signis* Chapter 5.

92 Diogenes Laertius VII.45, with tr. from Mensch and Miller (2018).

93 Their favourite example is "If it is day, it is light" (e.g. Diogenes Laertius VII.76; Sextus Empiricus *Adv. Math.* VIII.413ff). Diogenes Laertius VII.77 quotes as an example "If Plato lives, Plato breathes". Sextus *Adv. Math.* VIII.423, in his account of the Stoic concept of ἀπόδειξις, gives the example "If a woman has milk in her breasts, she has conceived". One might say that the first two, though surely not the third, are a priori truths, if one defines a day as "a period of light" and a living man as "a man who is breathing". But it is surely evident that if syllogisms using these premises are to reveal anything about the real world, all three premises must be based on human experience, of day and night, of living, breathing people, and of pregnant women.

94 Allen (2001) 184–8 argues for a Stoic concept of "signs" (σημεῖα) which provide "evidence which satisfies only weaker … standards". I have checked the occurrences of σημεῖον in the remains of Posidonius listed in *Thesaurus linguae Graecae* (which for Posidonius uses Theiler [1982]). Apart from texts omitted by EK as not naming Posidonius, σημεῖον does not occur at all in the sense of "evidence" or "sign" (except in Strabo II.3.5 = F49.260 EK, which is Strabo criticising Posidonius). "Signum", the Latin equivalent of σημεῖον, does occur in Cicero

Div. I.130 = F110 EK, "[Posidonius] esse censet in natura signa quaedam rerum futurarum", "[Posidonius] thinks that there are in nature some signs of things that are to be". We cannot, on the evidence, attribute Allen's theory about Stoic "signs" to Posidonius.

95 See Barnes (1994) xii.
96 Cited above, p. 210 n. 79.
97 *Apo.* 93b38–94a7, with translations from Barnes (1994).
98 See above, Chapter 8 (p. 67).
99 See above, Chapter 8 (p. 65–7).
100 Translation from Barnes (1994); I follow the interpretation suggested by Barnes (1994) 220–1, though he finds the sentence "very obscure'.
101 *Mete.* I.1, 339a2, with translation from Lee (1952).
102 The translation down to "that is possible" is from Lee (1952).
103 Cicero, *Academica*, II.145; Sextus Empiricus, *Adv. Math.* VII.432.
104 See above, Chapter 13 (p. 130).
105 Strabo II.3.6 and VII.2.2 (Posidonius F49.303 and F272.32 EK). See above, Chapter 12 (p. 125 n. 46).
106 Strabo I.1.7 (Posidonius F216.11 EK). See above, Chapter 13 (p. 138).
107 Kidd (1988) 77. One might also cite Cleomedes I.6.32 (I.4 lines 107–9 Todd; Posidonius F210.20–23 EK), quoted above, Chapter 15 (p. 153), which says that it is conjectured (ὑπονοεῖται) that summer rains in Ethiopia cause the Nile to flood. But here it may well be that the suggestion of conjecture is supplied by Cleomedes.

22 Assessment of the meteorology of Posidonius and his successors

According to Clement of Alexandria, Posidonius held that the purpose (τέλος) of human life is, in part, "to live contemplating the truth and order of all things together" (τὸ ζῆν θεωροῦντα τὴν τῶν ὅλων ἀλήθειαν καὶ τάξιν).[1] For him, to know the truth was evidently an end in itself. Or one might prefer to say that he was intellectually curious: he liked to know about things. Πάντες ἄνθρωποι τοῦ εἰδέναι ὀρέγονται φύσει, says Aristotle, "all men by nature desire knowledge".[2] This is doubtfully true as a generalisation about the human race, but it is certainly true of Aristotle, and surely of Posidonius also. Clear indications of this are, I suggest, the details which we are told that he recorded on botany and zoology, subjects remote from his usual interests: on curious trees in Spain, on beavers and horses there, on the blackness of Spanish crows, on monkeys observed on the coast of Africa[3] – things he evidently saw, or was told about, on his visit to Spain and the return voyage to Italy. Finding a subject, meteorology, to which so many earlier philosophers had paid so much attention, he was not content to abandon it, like the early Stoics, as too obscure for a search for the causes of the phenomena to be worthwhile; nor was he likely to be content with the apparent Epicurean position, that the phenomena have many possible causes, and that we cannot tell which one operates in any particular instance.[4]

Accordingly, I suggest, Posidonius set out to provide a Stoic's account of the causes of meteorological phenomena. His aim was to describe, if not the demonstrably true cause of each phenomenon, at any rate the most probable cause (or perhaps, in some instances, causes).[5] The causes he described were naturalistic. They are fitted into a chain of causes devised by Providence from the beginning of the world; but when he was writing about meteorology he usually ignored this, did as his predecessors, most notably Aristotle, had done, and described only the material and efficient causes of the phenomena. One might regard Posidonius' meteorology as an updating of Aristotle's, modified so as to accord with the principles of Stoic physics – I have already summarised his debts to Aristotle and the early Stoics.[6] But he also had regard to developments and discoveries in astronomy and geography, unrelated to the teachings of any philosophical school, which had been made since Aristotle's day. I would stress the importance of these latter developments, which are not based, as so

DOI: 10.4324/9780429399930-22

much ancient meteorology is, simply on unverifiable analogies with familiar phenomena, or unverifiable deductions from assumed first principles.

I summarise the instances that I have noticed:

(i) Posidonius (to judge from Pliny) related the celestial realm to the meteorological in a new way, by combining in one account estimates of the distance of sun and moon from the earth with an estimate of the height to which weather phenomena occur – estimates which depend at least in part on approximate measurements and mathematical calculations from them.[7] His estimate of the height up to which weather phenomena occur – probably "less than 40 stades" (but presumably not much less) – seems to be based partly on measurements of Greek mountains and partly on a guess about the greater height of the Alps, but is not far from the truth.[8]

(ii) Posidonius' work evidently included correct information about the equatorial regions: (*a*) the habitability of land at the equator; (*b*) the low altitude of (some) land there; (*c*) the beneficial effects on India of the south-west monsoon; and (*d*) the rain in Ethiopian mountains which causes the Nile floods. These facts must have been ultimately derived from the reports of people who had seen those regions.[9]

(iii) Posidonius' description of tides (though not his explanation of their cause) was mostly correct. One source was the work of the astronomer Seleucus, but what Posidonius recorded about tides in Spain was nearly all derived from his own observation or what he was told by those with local knowledge.[10]

(iv) He recorded, presumably from personal observation or someone with local knowledge, rather more information about the Stony Plain in southern France than Aristotle had done.[11]

(v) He recorded, partly from his own experience and partly (we can assume) from what he was told, information about an annual easterly summer wind in the western Mediterranean, apparently the wind now called "Levanter".[12]

(vi) His wind-rose was an improvement on Aristotle's, in that it consisted of twelve winds blowing from equally spaced directions, avoiding Aristotle's awkward assertion that there are directions from which no wind blows. His sources, according to Strabo, were Timosthenes, a naval commander, and Bion, an astronomer.[13]

(vii) He collected details about particular earthquakes and volcanic eruptions – mostly, I assume, from historical records, some of them, perhaps, from talking to people who remembered them.[14]

These instances are all at least partly correct (or, if now unverifiable, as with the seismic and volcanic details, they at least appear plausible) and they must be based, at least in part, on experience and observation, though the observations were rarely made with much care. (Of course, there are also ideas which

Posidonius evidently claimed were supported by observation but which are clearly wrong, for instance that comets are signs of coming weather, and that the observed fulfilment of past predictions proves the truth of "artificial" divination.[15])

The occasional recording of numerical data relevant to meteorology, roughly measured or estimated – of the heights to which tides rose, of the height to which weather phenomena occur, of the maximum depth of the sea[16] – seems to be an innovation by Posidonius, suggesting a belief that mathematical calculations might illuminate other aspects of the subject besides the optical ones. If so, this was a correct insight, but one that was never fulfilled in antiquity.

Posidonius, with his interest in causes, probably regarded the finding of the causes of meteorological phenomena as the most important part of the subject. That most of the causes he described were not original is unsurprising. All authors surveying so broad a subject must find, and be conscious, that most of the theories they record are not original to themselves. I have listed in Chapter 21 the theories and explanations that seem likely to be original to Posidonius.[17]

The meteorological causes described by Posidonius, even when original, were generally of the same character as those described by earlier authors. However, there are indications in Seneca's *Naturales quaestiones* that Posidonius sometimes – not always – set out formal arguments in defence of meteorological theories,[18] and may sometimes have stated a theory in terms more plausible than had been used previously, as with earthquakes,[19] or in terms closer to what we now know to be the truth, as with wind[20] (though the credit in both those instances may be due to Seneca). This does not amount to a major change. I discuss in my final section the difficulties which kept Posidonius and his successors from a deeper understanding of causes in meteorology.

Posidonius' successors

Posidonius gave a new impetus to the study of meteorology. We know of at least two pupils who wrote, or probably wrote, about aspects of the subject: Asclepiodotus, who wrote about earthquakes and volcanoes, and Athenodorus, probably the man Strabo mentions as an expert on tides[21] – possibly also the Diodorus who wrote on the Milky Way.[22] Works on meteorology survive from the two centuries after Posidonius' death, most of which he clearly influenced: the poem *Aetna*; a large part of Book II of Pliny's *Naturalis historia*[23]; Seneca's *Naturales quaestiones*; the meteorological fragments of Arrian; the meteorological section of the *De mundo* (assuming that is later than Posidonius); and the astronomical work of Cleomedes, which includes some meteorology as defined in this book. I have discussed the latter four as possible sources for Posidonius' own views, and have shown that none of them followed Posidonius blindly: they all agree with Posidonius on some points and disagree with him on others.[24] This is a problem when one seeks to use their work as a source for the views of Posidonius, but it shows that meteorology was a live subject.

This is particularly clear with Seneca, whose relevant surviving work is much the longest. Writing in the early 60s A.D.,[25] he several times refers to observations made and happenings experienced in the preceding decade: to a fight observed between crocodiles and dolphins, not before 55 A.D.[26]; to hearing a report by two centurions of what they had seen when sent by Nero to seek the source of the Nile[27]; to details of an earthquake which had just occurred in Campania (probably in 62 A.D.)[28]; to another that occurred in Achaea and Macedonia in the previous year[29]; to comets seen in 54 and in 60 A.D., and the differences between them.[30] He is ready to question commonly accepted beliefs: at *NQ* IVa.2.27 he denies that caves and wells are hot in winter and cold in summer, saying that they just do not admit cold and heat at those seasons.[31] At *NQ* VII.2.3 he says it is worth considering whether it is the earth which rotates while the heavens are at rest, rather than the other way round. Mainly in *NQ* VII, on comets, he devotes considerable attention to the opinions of men unknown, or scarcely known, from other sources: Epigenes,[32] Apollonius of Myndus,[33] Artemidorus of Parium.[34] His motive must be that their theories were subjects of serious discussion in his own day; it is surely likely that they were his contemporaries.[35]

So there were people thinking about meteorological subjects in the 1st and 2nd centuries A.D. However, where Posidonius had discovered for himself, or found in the work of Hellenistic predecessors, what we now know to be better knowledge or ideas than Aristotle and his contemporaries had had, these improvements were not always accepted in later antiquity. The facts which Posidonius had recorded about tides were known, and the one error attributed to him was corrected.[36] This is not surprising: when Roman rule had spread to the Atlantic coasts of Iberia and Gaul, and across the sea to Britain, then an understanding of tides became of practical importance to Roman generals and administrators. Julius Caesar attributes damage suffered by his ships on his first expedition to Britain to Roman ignorance of spring tides.[37] The belief (correct, but not obvious) that tides are especially high at equinoxes (and not, as Posidonius reportedly said, at solstices)[38] may well be due to the experience of the Roman army in 15 A.D., recorded by Tacitus.[39] At the conclusion of a campaign in Germany, two legions were marching by the shore of the North Sea. At first their journey was easy, "the tide flowing moderately" ("modice adlabente aestu"), but later the north wind blew "at the same time as the equinox, when the ocean swells most greatly" ("simul sidere aequinoctii, quo maxime tumescit oceanus"), the land was flooded, the legions were thrown into confusion and men, animals and baggage were lost. The wind was presumably more important than the equinoctial tide; but it may well have been this event which led later roman writers, like Seneca and Pliny, to state that equinoctial tides are the greatest.[40]

Other Hellenistic or Posidonian discoveries were doubted or rejected in later antiquity: some writers did not accept it as established that the Nile floods are due to summer rain in Ethiopia[41]; many continued to maintain that there was, or might well be, an uninhabitable zone at the equator.[42] Posidonius thought that clouds and wind occur, or might occur, up to, or nearly up to, a height of 40 stades[43] – a figure not far from the truth, but probably based on an exaggerated estimate of the height of the Alps. Throughout later antiquity the orthodox view seems to have been that of Eratosthenes, that 10 stades is the maximum height of mountains (and presumably of weather), which is far too low. Some, however, preferred a higher figure: Arrian thought that clouds occur up to a height of 20 stades, Cleomedes said that the highest mountain is not more than 15 stades high, and Philoponus in the 6th century said 12 stades. We may, I suggest, allow credit to Posidonius for persuading these later writers that 10 stades was too low a figure. But, if they knew and had reflected on the facts that 15 or 20 stades had been put forward as the height of the highest Greek mountains, and that the Alps are certainly much higher, they might have realised that the earth's highest mountain must be higher than 20 stades.[44] So difficult was it, with ancient means of communication, to make new ideas and discoveries generally known, if they were not of immediate practical importance.

The difficulty of progress in meteorology

Although there was discussion of meteorological causation in the 1st and 2nd centuries A.D., the methods of investigation were unchanged from those used in earlier centuries, which, as we have seen, could not yield certain results.[45] So long as meteorological theories were expressed in purely qualitative terms, it was generally impossible to confirm or to refute them. For instance, the theory might be proposed that a phenomenon is due to a moderate, not extreme, degree of heat: the moon's effect on the sea in causing tides and the habitability of equatorial regions are examples attributed to Posidonius.[46] Human sensations of heat are an unreliable way of detecting small variations of heat, as Aristotle realised[47]; there was therefore no way of confirming or refuting these theories without some means of measuring temperature. Equally, the true cause of wind could not be understood without an understanding of, and some means of measuring, atmospheric pressure.

"The invention of the barometer and thermometer marks the dawn of the real study of the physics of the atmosphere", says a famous 20th-century meteorologist.[48] The ancients lacked both, although at least a primitive device for measuring temperature could have been made in later antiquity. The earliest "thermoscope", an invention (commonly attributed to Galileo) of around 1600,[49] resembles quite closely a device described by Philo of Byzantium (probably 2nd century B.C.).[50] In both devices a ball containing air is connected to a tube, the other end of which is in a vase containing water. If the ball is cooled, the air in it contracts and water is sucked up the tube; if the ball is heated, the air expands, water in the tube is pushed back, and air bubbles out of the end of it if the heat is sufficient. In Philo's device the ball, and apparently the tube, are opaque, so the effect of heat can only be seen when air bubbles out, but Galileo's ball and tube are of glass, so that the level of liquid in the tube could be seen and measured. Glass-blowing was invented in the 1st century B.C., and after that a glass ball and tube could have been made[51] – probably too late for Posidonius, but not for his successors. Seneca, for one, would have welcomed such an invention, for he shows an interest in technical devices: in water-filled glass globes which magnify small written characters; in glass rods which must have been some sort of prism, since they produce the colours of the rainbow when the sun shines through them; in observing a solar eclipse by examining its reflection in a bowl filled with oil or pitch.[52]

The principle of the barometer might also have been thought of in antiquity. The ancients knew that, if a vessel is immersed in a liquid and filled with it, and then is raised from the liquid in such a way that its open end remains beneath the surface, it raises part of the liquid with it and so remains full, unless air can get in.[53] It might have occurred to someone in antiquity to test whether there is a limit to the height to which liquid can be raised in this way, as was done in a similar experiment by Torricelli in 1643, his "vessel" being a long glass tube closed at one end and his liquid, mercury.[54] He found that mercury can be raised in this way only to a height of about 76 cm, any space in the tube above that height being left empty, that is, there is a vacuum there. As he realised, this is, of course, because the atmosphere has weight and presses down on the mercury exposed to it, but this pressure is only enough to support a column 76 cm high of mercury not exposed to atmospheric pressure. Torricelli's experiment was the beginning of the barometer.

This experiment should have been possible in antiquity. Mercury would have been needed for its weight (a water barometer would need a tube more than 10 metres long), but it was known to Aristotle[55]; however, making a glass tube of the necessary length might have been a challenge.

Torricelli's experiment, if correctly performed and interpreted, should have led Aristotle and the Stoics to modify their view that there is no void, or none within the cosmos,[56] and the Stoics to modify their view that air is "upward moving" (ἀνωφερής), or else has neither weight nor lightness[57] (Aristotle says that air has weight in its own region,[58] which fits the experimental result.) A thermoscope should have settled the question of whether caves are genuinely warmer in winter than in summer, and might have led Posidonius and others to question some theories about the meteorological effects of heat, for example, that the moon's heat causes tides.

However, ancient thinkers lacked the habit of experimentation, and had no expectation, or even conception, of steady scientific progress. In any case, meteorology is a very complex science. The invention of simple instruments to measure atmospheric temperature and pressure would not on its own have led to major discoveries. Other knowledge was also needed. For example: the cause of wind is the natural tendency of air to flow from where its pressure is greater to where it is less; but it would not be easy to prove this by the use of barometers, because normally the wind does not blow directly from where the pressure is greater to where it is less; it is diverted by the effect of the earth's rotation.[59] Therefore, understanding of the cause of winds required a knowledge of the earth's rotation. The effect of this on wind seems to have been first established, in the 17th and 18th centuries, by a study of the trade winds of the tropics[60] – a part of the world hardly known to the ancient Greeks and Romans. And for a proper understanding they needed to know physical laws about the behaviour of gases which could only be found by experiment in a laboratory, such as the laws of Boyle and Charles about the relation of the volume of a mass of gas to its pressure and temperature.[61]

This need for prior knowledge applies to other branches of meteorology also. To explain thunder and lightning, the ancients needed to know about electricity; to explain the tides, they needed to know about gravity; to realise the modern explanation of earthquakes, they needed to know about plate tectonics – a theory not generally accepted until the second half of the 20th century.[62] It was not perverseness on the part of ancient thinkers that kept them from seeing what pieces of research – like the invention of thermometers and barometers – were possible for them, and would in time lead to serious progress in understanding meteorological causation. The way to progress was genuinely difficult to find. As Koestler wrote, "much depends on asking the right question at the right time".[63] Antiquity was not the time to ask, with any hope of finding the right answer, "What is the cause of wind?" or "What is the cause of thunder and lightning?" Great thinkers of the 17th and 18th centuries (men such as those of whom Koestler wrote "first and foremost they were giant question-masters") managed to find, where the ancients had failed, the problems which they could solve and which would lead to progress in meteorology. And, unlike the ancients, they had the advantage of the printing press with which to circulate their ideas to each other.[64]

Notes

1 Clement, *Stromateis* II.xxi.129.1–5 (Posidonius F186 EK), with translation from Kidd (1999).
2 *Metaphysics A*, 980a22.
3 Strabo III.5.10. III.4.15, XVII.3.4 (Posidonius F241, 243 and 245 EK).
4 Whether there is evidence of Posidonius attacking the Epicurean position seems to me doubtful. Verde (2016) suggests, somewhat tentatively, that in the Posidonian arguments reported by Simplicius, *In Aristotelis physica*, II.2, p, 291.21ff. Diels (Posidonius F18 EK) "Epicurus and his doctrine of multiple explanations [are] among Posidonius' potential polemical targets" (see Verde [2016] 437, and 447 where he calls this proposal "just a hypothesis"). This seems to me unlikely: the arguments of F18 are about the limited value of mathematical hypotheses for discovering the movements of celestial bodies; the Epicureans avoided using such hypotheses.
5 As just described, Chapter 21 (pp. 201–2, 205–7). On multiple causes see p. 186.
6 See above, Chapter 21 (p. 196).
7 See above, Chapter 6 (p. 43).
8 See above, Chapter 6 (p. 49).
9 See above, Chapters 7 and 15.
10 See above, Chapter 13.
11 See above, Chapter 12 (p. 123).

12 See above, Chapter 11 (p. 108).
13 See above, Chapter 11 (p. 103–5).
14 See above, Chapter 12 (p. 119–20).
15 See above, Chapter 17 (pp. 172, 176–7).
16 See above, Chapter 20 (p. 193–4).
17 See above, Chapter 21 (p. 197).
18 See above, Chapter 21 (p. 203–4).
19 See above, Chapter 12 (pp. 118, 121).
20 See above, Chapter 11 (pp. 102, 109).
21 See above, Chapter 2 (p. 5).
22 See above, Chapter 9 (p. 84).
23 Pliny *Nat.* I names Posidonius as a source for Book II.
24 See above, Chapter 3 (p. 11–14).
25 See Oltramare (1961) vi–vii; Hine (2010) 10.
26 *NQ* IVa.2.13. For the date see Oltramare (1961) 184n; Hine (2010) 199n29.
27 *NQ* VI.8.3–4. Nero became emperor in 54 A.D.
28 *NQ* VI.1.1–3, VI.27.1, VI.31. For the date see Oltramare (1961) vi–vii; Hine (2010) 193–4n13.
29 *NQ* VI.1.13.
30 *NQ* VII.17.2, VII.21.3–4, VII.28.3; VII.29.3. For the date see Oltramare (1961) 319n and 331n; Hine (2010) 206n27.207n42.
31 However, at VI.13.2–6 he does not repeat this when describing Strato's earthquake theory, in which cold in winter does enter caves and concentrates the heat there.
32 *NQ* VII.4–10.
33 *NQ* VII.4.1 and VII.17.1.
34 *NQ* I.4.3 and VII.13–15.
35 I thus incline to accept the view of Oltramare (1961) 298–9 that Epigenes and Apollonius were contemporaries of Seneca. Hine (2010) 206n9 says just that they are "of uncertain date". Neither hazards a guess at the date of Artemidorus: see Oltramare (1961) 25n5, Hine (2010) 206n21 and 210n41.
36 Tides are clearly referred to by (for instance) Julius Caesar (see next note); Varro, *De lingua Latina*, IX.26; Seneca, *NQ* III.28.6; Pliny, *Nat.* II.215; Silius Italicus III.60; Tacitus, *Annals*, I.70; in Greek, *De mundo* 396a25; Cleomedes II.1 (lines 388–92 Todd) and II.3 (lines 64–5 Todd).
37 *De bello Gallico* IV.29.
38 See above, Chapter 13 (p. 130).
39 *Annals* I.70.
40 Seneca, *NQ* III.28.6; Pliny, *Nat.* II.215.
41 See above, Chapter 15 (p. 157).
42 See above, Chapter 7 (p. 62).
43 For this and the following sentences see above, Chapter 6 (p. 48–9 and notes).
44 See above, Chapter 6 (p. 47): on the assumptions made there, 20 stades is between 3,000 and 4,000 metres; the highest of the Alps is 4,810 metres.
45 This section is partly based on p. 422–33 of my Ph.D. dissertation, Hall (1969).
46 See above, Chapter 13 (p. 134), and Chapter 7 (p. 60).
47 Aristotle *De partibus animalium* 648b15ff.
48 Shaw (1926) 115.
49 Described and illustrated by (e.g.) Sherry (2011). Wikipedia article "Thermoscope" (read 14 June 2021).
50 Philo, *Pneumatics* 7 (Philo's work survives only in Arabic and a partial Latin translation from the Arabic. I have used the edition of the Latin in Schmidt (1899) 458–89: see pp. 474–7).
51 See Israeli (1991).
52 Seneca, *NQ* I.6.5, I.7.1, I.12.1.

53 As happens when liquid is raised with a *clepsydra*, as described by Empedocles (DK 31B100). See also Philo, *Pneumatics* 6 (Schmidt (1899) 472–3), describing an experiment illustrating this raising of liquid.
54 See, e.g., Shaw (1933) 85; Wikipedia article 'Evangelista Torricelli' §2.1 (read 12 February 2022).
55 *De anima* 406b19.
56 See Aristotle, *Ph.* 213a12–217b28; for the Stoics, Diogenes Laertius VII.140.
57 Plutarch, *De Stoicorum repugnantiis* 1053E (*SVF* II.434, 435).
58 *De caelo* 311b8ff, 312b3ff.
59 Sutton (1960) 36–8.
60 See Halley (1686), Hadley (1735); discussed by Shaw (1926) 288ff.
61 On the early modern meteorologists' need for laboratory research see Shaw (1926) 115f.
62 See Wikipedia article "Plate tectonics" (read 19 June 2021).
63 My quotations from Koestler are from Koestler (1964) 260n and 265.
64 For my view on this see Hall (1983).

Bibliography

N.B. This bibliography does not include internet sites (e.g. *Thesaurus linguae Graecae; Wikipedia*), nor does it include ancient texts as such; when a particular edition or translation of an ancient text is included, it is listed under the editor or translator.

Algra, K., et al., eds. (1999). *The Cambridge history of Hellenistic philosophy*. Cambridge: Cambridge University Press.

Allen, J., (2001). *Inference from signs*. Oxford: Oxford University Press.

Arnaud, P. (2005). *Les routes de la navigation antique*. Paris: Errance.

von Arnim, H.F.A., ed. (1903–24). *Stoicorum veterum fragmenta*. 4 vols. Lipsiae [Leipzig]: Teubner. [Cited as *SVF*]

Aujac, G., ed. & tr. (1975). *Géminos, Introduction aux phénomènes*. (Coll. des univ. de France … Budé.) Paris: Belles Lettres.

Bailey, C., ed. (1926). *Epicurus, the extant remains*, Oxford: Oxford University Press.

Bakker, F.A. (2016). *Epicurean meteorology: sources, method, scope and organization*. Leiden: Brill.

Barnes, J., et al., eds. (1982). *Science and speculation: studies in Hellenistic theory and practice*. Cambridge: Cambridge University Press.

Barnes, J., tr. (1994). *Aristotle's Posterior Analytics, translated with a commentary*. 2nd ed. Oxford: Oxford University Press.

Bauslaugh, R.A. (1979). "The text of Thucydides IV 8.6 and the South Channel of Pylos", *Journal of Hellenic Studies*, 99, 1–6.

Beaujeu, J., ed. & tr. (1950). *Pline l'Ancien, Histoire naturelle, livre II*. (Coll. des univ. de France … Budé.) Paris: Belles Lettres.

Bell, B. (1971). "The dark ages in ancient history. I. The first dark age in Egypt", *American Journal of Archaeology*, 75, 1–26.

Bertrand, E. and R. Compatangelo-Soussignan, eds. (2015). *Cycles de la nature, cycles de l'histoire: de la découverte des météores à la fin de l'age d'or*. Pessac: Ausonius.

Betegh, G., and P. Gregoric (2014). "Multiple analogy in Ps.-Aristotle *De mundo* 6", *Classical Quarterly, new series*, 64, 574–591.

Beullens, P. (2014). "Facilius sit Nili caput invenire: towards an attribution and reconstruction of the Aristotelian treatise *de inundatione Nili*", in de Leemans, P., ed. (2014), 303–329.

Boechat, E. (2016). "Stoic physics and the Aristotelianism of Posidonius", *Ancient Philosophy*, 36(pt 2), 425–463.

Bosworth, A.P. (1988). *Conquest and empire: the reign of Alexander the Great*. Cambridge: Cambridge University Press.

Boys-Stones, G.R. (2001). *Post-Hellenistic philosophy*. Oxford: Oxford University Press.

Brill's New Pauly (2002–2010). *Encyclopaedia of the ancient world, ed. by H. Cancik and H. Schneider. English ed. Antiquity*. 15 vols. Leiden: Brill.

Burri, R. (2014). "The geography of De Mundo", in Thom, J.C. ed. (2014), 89–106.

Burstein, S.M. (1976). "Alexander, Callisthenes and the sources of the Nile", *Greek, Roman and Byzantine Studies*, 17, 135–146.

Bury, R.G., ed. and tr. (1929). *Plato with an English translation. VII. Timaeus [etc.]*. (Loeb Classical Library.) Cambridge, MA: Harvard University Press.

Capelle, W. (1905). "Der Physiker Arrian und Poseidonios", *Hermes* 40, 614–635.

Capelle, W. (1916). *Berges- und Wolkenhöhen bei griechischen Physikern*. (Stoicheia, 5), Leipzig: Teubner.

Cartwright, D.E. (1999). *Tides: a scientific history*. Cambridge: Cambridge University Press.

Cary, M., and E.H. Warmington (1963). *The ancient explorers*. Revised ed. Harmondsworth: Penguin.

Compatangelo-Soussignan, R. (2013). "Un tsunami antique à l'embouchure du Guadalquivir? Sources documentaires et archives paléoenvironnementales", in Daire, M.-Y., et al., eds. (2013), 595–603.

Compatangelo-Soussignan, R. (2015). "La théorie des marées de Poséidonios d'Apamée et les cycles de la nature dans les traditions philosophiques des IVe–Ier siècles a. C.", in Bertrand, E., and R. Compatangelo-Soussignan, eds. (2015), 83–96.

Compatangelo-Soussignan, R. (2016). "Poseidonios and the original cause of the migration of the Cimbri: tsunami, storm surge or tides?", *Revue des études anciennes*, 118(2), 451–468.

Concise Oxford English dictionary (1999). 10th ed. by J. Pearsall. Oxford: Oxford University Press.

Coutant, V., and V.L. Eichenlaub, ed. and tr. (1975). *Theophrastus De ventis*. Notre Dame: University of Notre Dame Press.

Cusset, C., ed. (2003). *La météorologie dans l'antiquité: entre science et croyance: actes du Colloque international interdisciplinaire de Toulouse 2-3-4 mai 2002*. (Centre Jean Palerne, mémoires, 25). Saint-Etienne: Publications de l'Université de Saint-Etienne.

Daiber, H. (1992). "The *Meteorology* of Theophrastus in Syriac and Arabic translation", in W.W. Fortenbaugh and D. Gutas, eds. (1992), 166–293.

Daire, M.-Y., et al., eds. (2013). *Anciens peuplements littoraux et relations homme/milieu sur les côtes de l'Europe atlantique = Ancient maritime communities and the relationship between people and environment along the European Atlantic coasts*. Oxford: Archaeopress.

Darwin, G. (1911). *The tides and kindred phenomena in the solar system*. 3rd ed. London: John Murray.

Diels, H., ed. (1879). *Doxographi Graeci*. Berolini [Berlin]: G. Reimer.

Diels, H., and W. Kranz, eds. (1951-2 and later editions). *Die Fragmente der Vorsokratiker. 6te Aufl.* [and later editions]. Berlin: Weidmann. [Cited as DK]

Diez Minguito, M., et al. (2012). "Tide transformation in the Guadalquivir estuary ...", *Journal of Geophysical Research: Oceans*, 117(C3).

Dorandi, T. (1999). "Chronology", in Algra, K., et al. eds. (1999), 31–54.

Dorandi, T., ed. (2013). *Diogenes Laertius, Lives of eminent philosophers*. Cambridge: Cambridge University Press.

Edelstein, L., and I.G. Kidd, eds. (1989). *Posidonius I: the fragments*. 2nd ed. Cambridge: Cambridge University Press. [Cited as EK]

Evelyn-White, H.G., ed and tr. (1914). *Hesiod, The Homeric hymns, and Homerica*. (Loeb Classical Library.) Cambridge, MA: Harvard University Press.

Falcon, A. (2012). *Aristotelianism in the first century BCE: Xenarchus of Seleucia*. Cambridge: Cambridge University Press.

Falconer, W.A., ed. and tr. (1923). *Cicero De senectute, De amicitia, De divinatione, with an English translation*. (Loeb Classical Library.) Cambridge, MA: Harvard University Press.

Forster, E.S., tr. (1927). *Problemata*. (The works of Aristotle translated into English, vol. 7.) Oxford: Oxford University Press.

Forster, E.S., and D.J. Furley, ed. and tr. (1955). *Aristotle On sophistical refutations, On coming-to-be and passing-away, On the cosmos*. (Loeb Classical Library.) Cambridge, MA: Harvard University Press.

Fortenbaugh, W.W., et al., eds. (1992). *Theophrastus of Eresus: sources... Part one*. Leiden: Brill.

Fortenbaugh, W.W., and D. Gutas, eds. (1992). *Theophrastus: his psychological, doxographical and scientific writings*. New Brunswick: Transaction Publishers.

Fowler, R.L. (2000). "P. Oxy. 4458: Poseidonios", *Zeitschrift für Papyrologie und Epigraphik*, 132, 133–142.

Frankfort, H., et al. (1949). *Before philosophy: the intellectual adventure of ancient man*. Harmondsworth: Penguin.

Frede, M. (1999). "Epilogue", in Algra, K., et al., eds. (1999), 771–797.

Geiger, J. (1992). "Julian of Ascalon", *Journal of Hellenic Studies*, 112, 31–43.

Gigante, M. (1979). *Catalogo dei papiri ercolanesi*. Napoli: Bibliopolis.

Gigon, O., ed. (1987). *Aristotelis Opera* (*ex recensione I. Bekkeri, ed. 2*), III: *Librorum Deperditorum Fragmenta*. Berlin: De Gruyter.

Gilbert, O. (1907). *Die meteorologischen Theorien des griechischen Altertums*. Leipzig: Teubner.

Graham, D.W., Z. Herzog and M. Williams (2021). "Earth, wind and fire: Aristotle on violent storm events, with reconsideration of the terms ἐκνεφίας, τυφῶν, κεραυνός and πρηστήρ", *Apeiron*, 55(3), 417–442.

Grewe, C.-V. (2008). *Untersuchung der naturwissenschaftlicher Fragmente des stoischen Philosophen Poseidonios*. Frankfurt-am-Main: Peter Lang.

Griffiths, J.G., ed. (1972). *Climates of Africa*. Amsterdam: Elsevier.

Gschnitzer, F. (1973). "Prytanis", *Realencyclopädie der classischen Altertumswissenschaft ... Supplementband* XIII [= S XIII], 730–815.

Guthrie, W.K.C., ed. and tr. (1939). *Aristotle On the heavens, with an English translation*. (Loeb Classical Library.) Cambridge, MA: Harvard University Press.

Guthrie, W.K.C. (1962). *A history of Greek philosophy. Volume 1. The earlier Presocratics and the Pythagoreans*. Cambridge: Cambridge University Press.

Hadley, G. (1735). "Concerning the cause of the general trade-winds", *Philosophical Transactions*, 39, 58–62.

Hall, J.J. (1969). *Ancient theories of wind, and the physical principles on which they are based, from the earliest times to Theophrastus*. Cambridge University Ph.D. dissertation. (*Available on Open Access from the Cambridge University repository, with the link* https://doi.org/10.17863/CAM.91015)

Hall, J.J. (1977). "Seneca as a source for earlier thought (especially meteorology)", *Classical Quarterly, New Series*, 27, 409–436.

Hall, J.J. (1983). "Was rapid scientific and technical progress possible in antiquity?", *Apeiron*, 17, 1–13.

Halley, E. (1686). "An historical account of the trade winds, and monsoons, observable in the seas between and near the tropicks, with an attempt to assign the physical cause of the said winds", *Philosophical Transactions*, 16, 153–168.

Hammond, N.G.L. (1981). *Alexander the Great*. London: Chatto & Windus.

Heath, T. (1913). *Aristarchus of Samos*. Oxford: Oxford University Press.

Heiberg, J.L., ed. (1912). *Heronis opera quae supersunt omnia. Vol. IV.* Leipzig: Teubner.

Hicks, R.D., ed. and tr. (1925). *Diogenes Laertius, Lives of eminent philosophers, with an English translation.* 2 vols. (Loeb Classical Library.) Cambridge, MA: Harvard University Press.

Hine, H.M. (1981). *An edition with commentary of Seneca Natural questions book two.* New York: Arno Press.

Hine, H.M. (2002). "Seismology and vulcanology in antiquity?", in Tuplin, C.J., and T.D. Rihll, eds., (2002), 56–75.

Hine, H.M., tr. (2010). *Seneca Natural questions, translated.* Chicago: University of Chicago Press.

Huby, P., and G. Neal, eds. (1989). *The criterion of truth: essays written in honour of George Kerferd.* Liverpool: Liverpool University Press.

Hughes, D., ed. (1998). "[Oxyrhynchus papyrus] 4458", *Oxyrhynchus papyri* 65, 66–71.

Hülsen, – (1894). "Aitne (1)", *Paulys Real-Encyclopädie der classischen Altertumswissenschaft.* Bd 1, 1111–1112.

Hultsch, F. (1882). *Griechische und Römische Metrologie.* Berlin: Weidmann.

Hultsch, F. (1903). "Diodoros (53) aus Alecandreia", *Paulys Real-Encyclopädie der classischen Altertumswissenschaft.* Halbbd 9 [=V.1], 710–712.

Ideler, J.L. (1832). *Meteorologia veterum Graecorum et Romanorum.* Berolini [Berlin]: G.C. Nauck.

Israeli, Y. (1991). "The invention of blowing", in Newby, M., and K. Painter, eds., (1991), 46–55.

Jacoby, F. (1907). "Euthymenes (4) von Massilia", *Paulys Real-Encyclopädie der classischen Altertumswissenschaft.* Halbbd 11 [=VI.1], 1509–1511.

Jacoby, F. (1929). *Fragmente der griechischen Historiker (F Gr Hist). 2. Teil, B.* Berlin: Weidmann.

Jacoby, F. (1958). *Fragmente der griechischen Historiker (F Gr Hist). 3. Teil, C, 1. Bd.* Leiden: Brill.

Jakobi, R., and W. Luppe (2000). "P. Oxy. 4458 col. I: Aristoteles redivivus", *Zeitschrift für Papyrologie und Epigraphik*, 131, 15–18.

Johnson, M.R. (2020). "Meteorology", in Taub, L., ed. (2020), 160–184.

Jones, A.H.M. (1971). *The cities of the eastern Roman provinces.* 2nd ed. Oxford: Oxford University Press.

Jones, H.L., ed. and tr. (1917–32). *Strabo, Geography.* 8 vols. (Loeb Classical Library.) Cambridge, MA: Harvard University Press.

Jones, W.H.S., ed. and tr. (1923). *Hippocrates, with an English translation. Vol. 1.* (Loeb Classical Library.) Cambridge, MA: Harvard University Press.

Kany-Turpin, J. (2003). "Météorologie et signes divinatoires dans le De divinatione de Cicéron", in Cusset, C., ed. (2003), 367–378.

Kidd, I.G. (1978a). "Philosophy and science in Posidonius", *Antike und Abendland*, 24, 7–15.

Kidd, I.G. (1978b). "Posidonius and logic", in *Stoïciens et leur logique* (1978), 273–282.

Kidd, I.G. (1988). *Posidonius II: the commentary*. 2 vols. Cambridge: Cambridge University Press.

Kidd, I.G. (1989). "Orthos logos as a criterion of truth in the Stoa", in Huby, P., and G. Neal, eds. (1989), 137–150.

Kidd, I.G. (1999). *Posidonius III: the translation of the fragments*. Cambridge: Cambridge University Press.

Kirk, G.S. (1954). *Heraclitus: the cosmic fragments*. Cambridge: Cambridge University Press.

Kirk, G.S., J.E. Raven and M. Schofield (1983). *The Presocratic philosophers*. 2nd ed. Cambridge: Cambridge University Press.

Koestler, A. (1964). *The sleepwalkers: a history of man's changing vision of the universe. Penguin Books ed*. Harmondsworth: Penguin.

König, R., & G. Winkler, ed. & tr. (1974). *C. Plinius Secundus d. A., Naturkunde lateinisch-deutsch. Buch 2*. München: Heimeran.

Kühn, C.G., ed. (1821–33). *Claudii Galeni opera omnia*. 20 vols. Lipsiae [Leipzig]: K. Knobloch.

Lane, F.W. (1968). *The elements rage*. Sphere books ed. 2 vols. London: Sphere Books.

Lasserre, F., ed. (1966). *Die Fragmente von Eudoxos von Knidos*. Berlin: de Gruyter.

Lee, H.D.P., ed. and tr. (1952). *Aristotle Meteorologica*. (Loeb Classical Library.) Cambridge, MA: Harvard University Press.

de Leemans, P., ed. (2014). *Translating at the court: Bartholomew of Messina and cultural life at the court of Manfred, King of Sicily*. (Mediaevalia Lovanensia, Ser. I, 45.) Leuven: Leuven University Press.

Lehmann-Haupt, F. (1929). "Stadion (Metrologie)", *Paulys Real-Encyclopädie der classischen Altertumswissenschaft. Zweite Reihe, Bd 3* [=III A.2], 1931–1963.

Liddell, H.G., and R. Scott, rev. H.S. Jones (1940). *A Greek-English lexicon*. New ed. Oxford: Oxford University Press. [Cited as LSJ]

Lloyd, G.E.R. (1966). *Polarity and analogy: two types of argumentation in early Greek thought*. Cambridge: Cambridge University Press.

Lloyd, G.E.R. (1973). *Greek science after Aristotle*. London: Chatto & Windus.

Lloyd, G.E.R. (1979). *Magic, reason and experience: studies in the origins and development of Greek science*. Cambridge: Cambridge University Press.

Long, A.A. (1974). *Hellenistic philosophy*. London: Duckworth.

Long, H.S., ed. (1964). *Diogenis Laertii Vitae philosophorum*. Oxonii [Oxford]: Oxford University Press.

Maass, E., ed. (1898). *Commentariorum in Aratum reliquiae*. Berolini [Berlin]: Weidmann,.

Mair, A.W., and Mair, G.R., ed. and tr. (1955). *Callimachus, Hymns and epigrams, [and] Lycophron, with an English translation by A.W. Mair; Aratus, with an English translation by G.R. Mair*. 2nd ed. (Loeb classical library), Cambridge, MA: Harvard University Press.

Mansfeld, J., and D.T. Runia (1997). *Aëtiana*, vol. 1. Leiden: Brill.

Mansfeld, J., and D.T. Runia, ed. (2020). *Aëtiana V: an edition of the reconstructed text with a commentary and a collection of related texts*. Leiden: Brill.

Mayhoff, K., ed. (1906). *C. Plini Secundi Naturalis historiae libri*. Vol. 1. Leipzig: Teubner.

Mensch, P., tr., ed. by J. Miller (2018). *Diogenes Laertius, Lives of the eminent philosophers*. New York: Oxford University Press.

Merker, A. (2003). "La théorie de l'arc-en-ciel dans les *Météorologiques* d'Aristote", in Cusset, C., ed. (2003), 317–330.

Mugler, C. (1963). *Les origines de la science grecque chez Homère.* (Etudes et commentaries, 46.) Paris: Klincksieck.

Müller, C., ed. (1855–61). *Geographi Graeci minores.* 2 vols. Parisiis [Paris]: Didot.

Neugebauer, O. (1975). *A history of ancient mathematical astronomy.* Berlin: Springer.

Newby, M., and K. Painter, eds. (1991). *Roman glass: two centuries of art and invention.* London: Society of Antiquaries.

Oltramare, P., ed. and tr. (1961). *Sénèque Questions naturelles. 2ᵉ éd.* 2 vols. (Coll. des univ. de France ... Budé.) Paris: Belles Lettres.

Oxford classical dictionary (2012). *4ᵗʰ ed. General editors, S. Hornblower and A. Spawforth.* Oxford: Oxford University Press.

Oxford English Dictionary (1989). 2nd ed. *Prepared by J. A. Simpson and E. S. C. Weiner.* Oxford: Oxford University Press.

Oxford Latin Dictionary (1982). *Edited by P. G. W. Glare.* Oxford: Oxford University Press.

Pajón Leyra, I. (2013). "The Aristotelian Corpus and the Rhodian tradition: new light from Posidonius on the transmission of Aristotle's works", *Classical Quarterly, New Series,* 63, 723–733.

van Raalte, M. (2003). "God and the nature of the world: the theological excursus in Theophrastus Meteorology", *Mnemosyne,* ser. 4, 56, 306–342

Rackham, H., ed. and tr. (1933). *Cicero De natura deorum, Academica.* (Loeb classical library.) Cambridge, MA: Harvard University Press.

Rackham, H., ed. and tr. (1938). *Pliny, Natural history. Vol. 1.* (Loeb classical library.) Cambridge, MA: Harvard University Press.

Rome, A., ed. (1936). *Commentaire de Pappus et de Théon d'Alexandrie sur l'Almageste. Tome II: Théon d'Alexandrie, Commentaire sur les livres 1 et 2 de l'Almageste.* Città del Vaticano: Biblioteca apostolica vaticana.

Roos, A.G., ed., rev. G. Wirth (1968). *Flavius Arrianus. Vol. II: Scripta minora et fragmenta.* Lipsiae [Leipzig]: Teubner.

Ross, W.D., ed. (1936). *Aristotle's Physics: a revised text with introduction and commentary.* Oxford: Oxford University Press.

Sambursky, S. (1959). *The physics of the Stoics.* London: Routledge & Kegan Paul.

Sandbach, F.H. (1975). *The Stoics.* London: Chatto and Windus.

Sandbach, F.H. (1985). *Aristotle and the Stoics.* Cambridge: Cambridge Philological Society.

Schmidt, W. (1899). ed. *Heronis Alexandrini opera quae supersunt omnia. Vol. I.* Leipzig: Teubner.

Schütrumpf, B. (2008). *Heraclides of Pontus: texts and translations.* Piscataway, NJ: Transaction Publishers.

Sedley, D. (1982). "On signs", in J. Barnes et al., eds. (1982), 239–272.

Sedley, D. (1992). "Sextus Empiricus and the atomist criteria of truth", *Elenchos,* 13, 21–56.

Sharples, R.W. (1998). *Theophrastus of Eresus: sources for his life, writings, thought and influence. Commentary volume 3.1: sources on physics (texts 137-223).* Leiden: Brill.

Shaw, W.N. (1926). *Manual of meteorology. Vol. 1: Meteorology in history.* Cambridge: Cambridge University Press.

Shaw, W.N. (1933). *The drama of weather.* Cambridge: Cambridge University Press.

Shcheglov, D.A. (2006). "Posidonius on the dry west and the wet east: fragment 223 reconsidered", *Classical Quarterly, New Series,* 56, 509–527.

Sherry, D. (2011). "Thermoscopes, thermometers and the foundations of measurement", *Studies in History and Philosophy of Science*, 42 part 4, 509–524.

Stadter, P.A. (1980). *Arrian of Nicomedia*. Chapel Hill: University of North Carolina Press.

Steinmetz, P. (1962). "Zur Erdbebentheorie des Poseidonios", *Rheinisches Museum für Philologie, N.F.* 105, 261–263.

Stoïciens et leur logique (1978). *Actes du colloque de Chantilly, 18-22 septembre 1976*. Paris: J. Vrin.

van Straaten, M. (1946). *Panétius: sa vie, ses écrits et sa doctrine avec une édition des fragments*. Amsterdam: H.J. Paris.

van Straaten, M., ed. (1962). *Panaetii Rhodii fragmenta. Ed. amplificata*. Leiden: Brill.

Strohm, H. (1987). "Ps. Aristoteles De mundo und Theilers Poseidonios", *Wiener Studien*, 100, 69–84.

Struck, P.T. (2016). *Divination and human nature: a cognitive history of intuition in classical antiquity*. Princeton: Princeton University Press.

Stückelberger, A., & G. Grasshoff, ed. (2006). *Klaudios Ptolemaios Handbuch der Geographie, griechisch – deutsch. 1. Teil*. Basel: Schwabe.

Sutton, O.G. (1960). *Understanding weather*. Harmondsworth: Penguin.

Swerdlow, N. (1969). "Hipparchus on the distance of the sun", *Centaurus*, 14, 287–305.

Taisbak, C.M. (1974). "Posidonius vindicated at all costs? Modern scholarship versus the Stoic earth measurer", *Centaurus*, 18, 253–269.

Taub, L. (2003). *Ancient meteorology*. London: Routledge.

Taub, L. (2008). *Aetna and the moon: explaining nature in ancient Greece and Rome*. Corvallis: Oregon State University Press.

Taub, L., ed. (2020). *Cambridge companion to Greek and Roman science*. Cambridge: Cambridge University Press.

Theiler, W. (1982). *Poseidonios, Die Fragmente*. 2 vols. Leipzig: Teubner.

Thom, J.C., ed. (2014). *Cosmic order and divine power: Pseudo-Aristotle On the cosmos*. Tübingen: Mohr Siebeck.

Thomson, J.O. (1948). *History of ancient geography*. Cambridge: Cambridge University Press.

Times comprehensive atlas of the world (2014). 14th ed. London: Times Books.

Todd, R., ed. (1990). *Cleomedis Caelestia*. Leipzig: Teubner.

Tredennick, H., ed. and tr. (1933-5). *Aristotle Metaphysics*. 2 vols. (Loeb classical library.) Cambridge, MA: Harvard University Press.

Tuplin, C.J., and T.E. Rihll, eds. (2002). *Science and mathematics in ancient Greek culture*. Oxford: Oxford University Press.

Uden, J. (2021). "Egnatius the Epicurean: the banalization of philosophy in Catullus", *Antichthon*, 55, 94–115.

Van Raalte and Van Straaten. *See* Raalte *and* Straaten.

Verde, F. (2016). "Posidonius against Epicurus' method of multiple explanations?", *Apeiron*, 49(4), 437–449.

Vimercati, E., ed. & tr. (2004). *Posidonio, testimonianze e frammenti*. Milano: Bompiani.

Wardle, D., tr. (2006). *Cicero on divination, De divinatione, book 1, translated with introduction and historical commentary*. Oxford: Oxford University Press.

Webster, E.W., tr. (1931). "Meteorologica", in *The works of Aristotle translated into English*, vol. 3. Oxford: Oxford University Press.

Wehrli, F. (1969a). *Die Schule des Aristoteles, 2. Aufl. Vol. 5: Straton von Lampsakos*. Basel: Schwabe.

Wehrli, F. (1969b). *Die Schule des Aristoteles, 2. Aufl. Vol. 7: Herakleides Pontikos.* Basel: Schwabe.

West, M.L. (1971). *Early Greek philosophy and the Orient.* Oxford: Oxford University Press.

von Wilamowitz-Möllendorff, U. (1906). "Der Physiker Arrian", *Hermes*, 41, 157–158.

Wilson, J.A. (1949). "Egypt", in Frankfort, H., et al. (1949), 39–135.

Wilson, M. (2013). *Structure and method in Aristotle's Meteorologica.* Cambridge: Cambridge University Press.

Index

Achilles (commentator on Aratus) 54, 63, 83, 84
Acrocorinth 47, 48
aēr see air
Aeschylus (pupil of Hippocrates of Chios) 22, 78, 80
Aëtius: range of "meteorological" topics 2; value of testimony for Posidonius 11, 133, 136, 138
Aetna (poem) 122, 214
Africa, climate of 55, 60–61
Agatharchides 156–157, 197
air-pressure, discovery of 217
air, region of, in ancient theories 41–43, 45–46
aitnēr, Stoic view of 43
Alexander of Aphrodisias 46, 164–165, 167
Alps, ancient knowledge of height of 46, 48–49, 213, 216
Ammianus Marcellinus: on earthquakes 119, 122
Anaxagoras 1; on air affecting the sun 44; on comets 77; on earthquakes 113, 116; on hail 145, 146; his meteorology 22; on Milky Way 82; on mock suns 167; on Nile floods 152; Posidonius perhaps citing Anaxagoras 80, 197; on rainbows 163; on thunder and lightning 66, 71, 72, 195
Anaximander 1; his meteorology 21; on rain 143; on thunder and lightning 65–66, 71, 185; on wind 99, 100
Anaximenes: on cloud 147; on earthquakes 113, 114, 116; on hail 145; his meteorology 21; on rainbows 163; on snow 144; on thunder and lightning 66, 71
Antiphon 23, 145

Apollodorus of Athens 47
Apollodorus the Corcyrean 137
Apollonius Myndius 81, 215
Aratus 85, 171–172, 175
Archelaus 22, 114, 115
Archimedes 43, 44
Aristarchus of Samos 43, 44
Aristides, Aelius 157
Aristotle 1, 6, 7, 11, 134, 186, 189, 216–217; analogical arguments, view and use of 199–200; his astronomy 37, 42–43; on climate and climatic zones 55–57, 61–62; on cloud 147–148; on comets 42, 76–78, 80, 172; on earthquakes 114–122; elements, theory of 37, 42, 199; on exhalations 24, 67–68, 78, 90–92, 100, 116–117, 199; final cause and meteorology 24, 29, 181; on hail 145–146; on haloes 164; mathematics and meteorology 193; *meteōrologia*, definition 1–3; meteorological theories recognised as uncertain 206–207; his meteorology, character of 24, 36, 198; meteors 42, 76–77; on Milky Way 42, 82–83; on mock suns 167; on monsoons 109; on Nile floods 152–153, 155–156; his *Organon* and its relation to his meteorology 200, 204; Posidonius' view and use of Aristotle 7, 36–37, 42, 48, 55–57, 71, 80–81, 84, 93–94, 98, 102–103, 105, 107, 109, 118–121, 128–129, 131, 137, 142–143, 146, 162–165, 167, 172, 181, 189, 193, 197, 212–213; on pre-Socratic meteorology 15, 21, 23, 66, 77, 82, 99, 113–114; on rain 142; on rainbows 161–163; on reflection 161–162; on the sea and its salt 128–129; on snow 143, 147; on

Stony Plain 123; on thunder and lightning 66–68, 71–72; on tides 131, 137; on volcanoes 122; weather, height of 45–46, 49; on weather-signs 98, 172–173; wind-rose 102–105; on winds and theory of wind 96–97, 100–101, 107–108

[Aristotle], *De inundatione Nili* 15, 21, 152–153, 155

[Aristotle], *De mundo see De mundo.*

[Aristotle], *Problemata, Book XXVI* 15, 25–26, 46, 107–108

Arrian 14, 214; on comets 81; his relation to Posidonius' meteorology 14, 15, 81; on snow 143; weather, height of 46, 48–49, 216

Artemidorus of Parium 215

Asclepiodotus 5, 71, 119, 214

astronomy: Aristotelian astronomy, the heavens consisting of a fifth element, affecting sublunary matter by their motions 37, 42, 45–46, 78, 100–101; different approaches to astronomy of philosopher and mathematician 202; divinity of heavenly bodies 41–42; size and distance of heavenly bodies 43–44; Stoic astronomy, the heavenly bodies being fire, nourished by exhalation from earth 37, 42–43, 91, 93–94, 197, 199; *see also* comets; Milky Way; Moon; Sun

Atabyrius (mountain) 47

Athenodorus 5, 130, 214

Athos (mountain) 47

Aurora borealis 2, 42, 75–76

Bakker, F. A. 16, 186–187

Betegh, G. 13, 18n29

Bion 'the astronomer' 102, 104, 197, 213

Boëthus 30, 86n30, 176

Cadiz 5, 108, 130, 132–133, 194, 207

Callippus 41

Callisthenes 24, 75–76, 153–156

Capelle, W. 46

Chrysippus 77; on astronomy 93–94; on divination 174; on exhalation 93–94; on hail 28, 145; his meteorology 16, 26–29; philosophy, parts of 191; on rain 28, 142; on seasons 27; on sight 162; on snow 28, 143–144; on thunder and lightning 28–29, 68–69

Cicero, Marcus Tullius: on biography and reputation of Posidonius 4–7; source for Posidonius' views 10; on Stoic and Posidonius' theology 29–30, 181–183; on Stoic and Posidonius' views on divination 173–178; on Stoic (and Posidonius'?) views on wind, thunder and lightning 69–71, 97

Cleanthes 26–27, 29, 30, 93, 188

Cleidemus 22, 72n6

Cleomedes 6, 214; Cleomedes' view and use of Posidonius 6, 14, 15; on exhalations 94; mountains, greatest height of 48–49, 216; on Posidonius' astronomical calculations 44, 56, 202; on Posidonius' view of the equatorial zone 58–62, 152–153, 156; on refraction 165–166; on sight 162

climatic zones: ideas before Aristotle 54–55; Aristotle's account, criticised by Posidonius and Strabo 55–57; tropical zones as described by Posidonius and other successors of Aristotle 57–61

clouds: atomists' clouds 25, 66–68, 148, 185; clouds freezing directly to snow 143–144; clouds and hail 144–146; clouds intermediate between air and water 147–148; clouds as mirrors (causing rainbows etc.) 160–168, 199; clouds source of rain 89–90, 142–143; clouds source of wind 97–100; maximum height of clouds 43, 45, 48–49, 216; properties of solid bodies attributed to clouds to explain thunder, etc. 65–72, 148, 184–185, 203–204, 206

comets 77–81

Compatangelo-Soussignan, R. 133

Crates 137

Cyllene (mountain) 46–49

Daiber, H. 16, 17

De mundo 198, 214; Aristotle, relation to 12–13; on cloud 148; on comets 81; date 13–14, 26; on earthquakes 119, 121; on exhalation 91; on hail 145; its meteorology (range of topics included) 2; Milky Way not mentioned 85; Posidonius, relation to 12–15; on snow 143; on thunder 70–71; on volcanoes 121–122; wind-rose 103–104, 106

Democritus 1, 186; air in his physical
system 41; on comets 77, 80–81; on
divination 177, 179; on earthquakes
113–114, 116; his meteorology 22, 23;
on Milky Way 82, 85; on Nile floods
156; Posidonius' possible use of
Democritus 80–81, 196; on rain 142;
on thunder and lightning 66–67, 68,
71–72, 185
Dicaearchus 7, 37, 46–48, 129, 137
Diodorus of Alexandria 84, 214
Diodorus Siculus 156–157
Diogenes of Apollonia 22, 23, 29,
78, 99, 100
Diogenes the Babylonian 78
Diogenes Laertius: on earthquakes,
Posidonius' theory of 117, 119; on
philosophy, divisions of, in Stoicism
192–193; on rainbows, Posidonius'
theory of 160, 162; range of
"meteorological" topics in account of
Stoicism 2–3; value as source for
Posidonius 6, 11, 12, 15; on wind,
Stoic theory of 97, 98
divination 147, 171, 173–179, 184, 214

earthquakes: Aristotle's account 115–117,
120–121, 123, 172–173; Callisthenes
on earthquakes 25; earthquakes and
Providence 182–183; earthquakes part
of meteorology 2; Epicurus' account
25, 115–116; in Hesiod 21; Posidonius'
account 117–122, 196–197, 214;
pre-Socratics on earthquakes 21–22,
113–114, 116; Stoics on earthquakes
before Posidonius 116; Strato's
theory 25, 115; Theophrastus' theory
24, 115; types of earthquake 12, 13,
118–119, 121; wind cause of
earthquakes 21, 114–118, 120–121,
199, 203
Edelstein, L. 10
Empedocles 22, 41, 66, 71, 72
Epicurus and Epicureans 1; on air
affecting the sun 45–46; atoms in
meteorology 25, 148, 186; on cloud 25,
148; on comets 79; on earthquakes
115–116; on exhalations 91, 99; gods
not cause of weather 26, 184, 188–189;
on hail 145; on haloes 165; their
meteorology 25, 26, 184–189;
meteorology studied to obtain peace
of mind 187–188; Milky Way not
explained 85; multiple explanations of

phenomena 16, 186–187, 212; on Nile
floods 157, 187; Posidonius' possible use
of Epicurean meteorology 15, 35–36,
185, 196; on rain 142; on rainbows 163;
on snow 144; Stoic and Epicurean
meteorology compared 188–189; on
thunder and lightning 68, 71, 184–185;
on weather-signs 184; on wind 99
Epigenes 81, 215
equatorial region, habitability of 14, 30,
56–62, 136, 157, 196–197,
213, 215–216
Eratosthenes: on circumference of earth
56; equatorial region, habitability of
59, 62; on the greatest mountain
heights 48, 49, 216; on monsoons 109;
Posidonius' use of 37; on summer
rains in Ethiopia 156; on tides 130
Etna (Mount) 120–122
Eudoxus of Cnidus 30, 31, 41, 152
Eudoxus of Cyzicus 59, 197
evaporation *see* exhalations
exhalations: concept originates from
water evaporating 89–90; exhalation
and comets 75, 78, 80, 81, 91, 172;
exhalation and earthquakes 91,
115–117; exhalation and the Milky
Way 83; exhalation and rain 89, 90,
142, 198; exhalation and the sea's salt
128–129; exhalation and thunder and
lightning 67–71, 91, 203; exhalation
and wind 91, 97–100, 106, 109, 198;
exhalation nourishing heavenly bodies
91, 93, 94, 197, 199; exhalations (post-
Aristotelian) from multiple materials
91, 92, 94, 97, 100; two exhalations,
bright and dark, from water 22, 90,
91; two exhalations, dry from earth,
wet from water 24, 90–92, 198–199

Gadeira *see* Cadiz
Galen (and pseudo-Galen) 6, 7, 83, 193,
197, 201
Galileo 216
Geminus 46–48, 59–62, 83–84
Gregoric, P. 13, 18n29
Guadalquivir (River) 132–133

Haemus (mountain) 46
hail 12, 23–24, 28, 144–145, 147–148;
divination of hail 146–147, 174;
Posidonius' theory of hail and
Seneca's account of it 144–146,
148–149, 196, 204

haloes 78, 164–166, 168, 172–173, 193
heavenly bodies *see* astronomy
Heraclides Ponticus 75, 137,
   173, 177–178
Heraclitus 22, 35, 90–91, 93, 142
Heraclitus Homericus 119, 121
Herodotus: awareness of air 41; on
   climate 55, 62; on Nile floods 30,
   152–153, 155; on rain 142; sun
   affected by wind 44
Hesiod 20, 66, 71, 89, 98–99, 113, 142
Hine, H. M. 113
Hipparchus 37, 43, 44
[Hippocrates], *Airs waters places* 90,
   104, 142
[Hippocrates], *De flatibus* 96
[Hippocrates], *De natura pueri* 98
[Hippocrates], *De victu* 98
Hippocrates of Chios 22, 77, 80
Homer, *Iliad* and *Odyssey*: gods
   controlling weather 22–23, 99; the
   Nile in Homer 153–154, 156;
   Olympus, climate of, as gods' home
   46; Posidonius' interpretation of
   Homer on meteorology 102–103,
   106–107, 109, 138, 153–154, 156, 197,
   207; tides supposedly mentioned by
   Homer 138; wind descriptions in
   Homer interpreted 102–103, 106–108;
   wind gods 99, 104

Iliad see Homer
India 61–62, 105, 108–109, 213
instruments useful for meteorology, lack
   of 216–217

La Crau (France) 123
Levanter (wind) 108, 109, 213
lightning *see* thunder and lightning
Lloyd, G. E. R. 199–200
Lucan 157
Lucretius 1; his *De rerum natura* probably
   just later than Posidonius'
   meteorology 15, 185; features of his
   meteorology 184–185; his interest in
   meteorology unusual 26; meteorology
   studied to free men from fear of the
   gods 188; multiple explanations of
   each phenomenon 185–186; on Nile
   floods 157, 187; the single cause of a
   meteorological phenomenon, if only
   one, not discoverable 187
Lydus, Johannes 119, 121, 155–156

Mansfeld, J. 83
Martianus Capella 47
meteorology, ancient: a branch of
   philosophy, mainly studied by
   philosophers 21–30, 191–193;
   difficulty of progress 216–218;
   meteorological theories dependent on
   arguments from analogy 199, 200,
   203–205; range of topics covered 1–3;
   uncertainty of their theories admitted
   by ancient meteorologists 205–207
meteors 75–77
Metrodorus of Chios 22, 82, 114, 116,
   142, 163
Milky Way 2, 22, 42, 82–85, 181, 197
mock suns 166–168, 197
Moon: Aristotle's view of 37, 42; its
   distance from earth 37, 43, 44, 213;
   lunar haloes 164–166; lunar rainbows
   160; mathematicians' study of
   192–194; the Moon a mixture of air
   and fire 37, 45; nourishment of the
   Moon by exhalations from the earth
   93, 94; Providence and the Moon 84,
   181–182; tides and the Moon 13, 101,
   129–136, 138, 199, 216–217; wind, the
   Moon a subsidiary cause of 101
mountains, greatest height of 37–38,
   46–49, 213, 216

Nile floods 21–23, 30–31, 38, 152–157,
   187, 213, 215

Odyssey see Homer
Oeta (mountain) 46
Olympiodorus 46
Olympus (mountain) 46–48
Oxyrhynchus papyrus 4458 153–155

Panaetius: his approval of Plato and
   Aristotle 7, 36; on comets 30, 86n30;
   doubted destruction and rebirth of
   cosmos 7; doubted truth of divination
   7; on equatorial region, habitability of
   30, 58–59, 62, 196; on the parts of
   philosophy 191; teacher of
   Posidonius 4, 7
parapēgmata 22, 30–31, 41, 171
parhelia *see* mock suns
Parmenides 55, 82, 196
Pelion (mountain) 47, 48
Philip of Opus 162–163
Philo of Byzantium 216

Philoponus, Johannes 46, 48–49, 216
Plato: on Atlantis 119; on hail 23, 145; on heavenly bodies 41–42; meteorology not a serious interest 7, 23; Platonism revived in late 2nd century B.C. 7; Posidonius' debt to (especially to *Timaeus*) 7, 182, 193, 201, 204; the sequence of seasons shows intelligence governs the world 29; on sight 162; on snow 23–24, 144; tides, supposed theory of 137; wind, definition of 96

Pliny the Elder: on earthquakes 121; on greatest heights of weather phenomena 43, 45, 48–49; on a meteor 76; on mountain heights 46–47; on Nile floods 157; Posidonius a source for his meteorology 214; Posidonius' views reported 43–45, 48, 62, 108–109, 213; on tides 130, 206; on torrid zone 62; on volcanoes 121

Plutarch 46–48, 92, 94

Polybius 6, 37, 48, 58–60, 62, 156

Posidonius 1–3, 6–7, 15, 27, 30, 189; on Africa, climate of 61–62; on Aristotle *see below* Posidonius' view and use of Aristotle; [Aristotle] *Problemata XXVI*, not used 107–108; Arrian's relation to Posidonius 14, 81; astronomy 37, 42–43, 91, 93–94, 202; authors he used for his meteorology 35–38, 196–197; biography 4–6; causes, his desire to find 7, 11, 35–36, 133, 179, 189, 192, 201, 214; on Cimbri and their migration 119; on circumference of earth 43; Cleomedes' relation to Posidonius 6, 14; on climatic zones 54–59, 62; on clouds 69–70, 147–148, 160–161, 164, 168; on comets 75, 79, 80, 214; *De mundo*, relation of, to Posidonius 12–14, 103, 121; his determinism 173, 175, 183; on divination 146–147, 171–179, 214; on earthquakes 117–122, 213; Epicurus and Epicureans, attitude to 35–36, 71, 184–185, 196; equatorial region, habitability of 58–62, 213; evidence about his meteorology 10–15; exhalations, theory of 91–94; on hail 144–147, 174; on haloes 164–166; Homer, attitude to 106–107, 109, 138, 156, 197; on India, climate of 61–62, 213; influence on later thinkers 6, 7, 81, 84–85, 133, 214–216; mathematics and

meteorology 161–162, 192–194, 214; meteorological theories, arguments for, and logical reasoning 198–205; meteorological theories, sources of 196–199; meteorological theories recognised as uncertain 205–207; meteorology, place of, among the parts of philosophy 191–193; meteorology, range of topics included 2–3; on meteors 75, 77; on Milky Way 83–84; on mock suns 166–168; on monsoons 108–109; moon, distance from earth 43–44; on Nile floods 152–157, 213; non-philosophers used in his meteorological studies 31, 37–33, 196–197; numerical data in his meteorology 129, 132–133, 214; oral evidence, use of 38, 108, 123, 130, 132–133, 197; own observations, use of 38, 108, 123, 130, 132–133, 197; Plato, view and use of 7, 119, 182, 193, 201; portents, belief in 76, 174; pre-Socratics, view of 54, 71, 80–81, 196; on Providence and meteorology 84, 181–183, 212; pupils 5, 214; on rain 142–143; on rainbows 160–164, 168; on reflection 162, 164–166; reputation 5, 6; sea, depth of 129; on sea's salt 128–129; on seasons of the year 27–29, 181–182; Seleucus, view and use of 37, 131–132, 136–138; Seneca's relation to Posidonius 6, 11, 69–71, 76, 97–98, 100–103, 106, 109, 117–121, 143–147, 149, 160–163, 174, 182; on sight 162; on snow 143–144; Stoic predecessors, view and use of in meteorology 14–16, 26, 35, 42–43, 68, 71, 83–84, 143, 146, 160, 181–182, 191–193, 196, 212; on 'Stony Plain' 123, 213; Strabo's relation to Posidonius 6, 10–11, 27, 54, 56–57, 102–103, 106–107, 117, 130–133, 153–156; sun, distance from earth 43–44; Theophrastus, view and use of 36, 107, 196; on thunder and lightning 68–71, 183, 185; tides, explanation of 133–138; tides, observation and description of 130–133, 136–137, 213; on volcanoes 119–122, 213; weather, greatest height of 43, 45, 48–49, 213; on weather-signs 80, 171–173; wind-rose 102–106, 108, 213; wind, theory of 96–98, 100–102; winds, individual, features of 106–108, 213

Posidonius' view and use of Aristotle 2–3, 7, 36–37, 196, 206–207, 212; on climatic zones 36–37, 55–57, 62; on comets 80, 172; on earthquakes 118–121, 203; on exhalations 91–93; on hail 145–146; on haloes 164; on Nile floods 152–154; on rainbows 160–162; on the sea's salt 128–129; on snow 143–144; on the Stony Plain 123, 213; on thunder and lightning 71; on tides 131, 137; on west Mediterranean winds 108; on wind-rose 36, 102–105, 109, 213

pre-Socratic philosophers: on comets 77–78; divine causation of weather rejected 22–23, 181; on earthquakes 113–114; evidence for their meteorology 15–16; history of their meteorology summarised 21–23; Posidonius' view and use of them 35, 80–81, 196; on rain 142; on rainbows 163; on thunder and lightning 65–67, 71, 185; volcanoes, no account of 122; on wind 100, 104; *see also the names of individual pre-Socratics*

Priscianus Lydus 6–7, 128–130, 133–136, 138

Proclus 24

Ptolemy 44, 62

Pythagoreans 22–23, 55, 78, 80–82, 163, 196

Pytheas 44, 62

rain: caused by condensation of vapour 20, 89–92, 142–143, 147–148, 199; comets disappearing a supposed sign of rain 80, 172; earthquakes, supposed connection of rain with 113–114, 116; Nile floods and rain 152–157, 213, 215; purpose seen in rain 23–24, 29, 182

rainbows 160–164, 167–168, 198–199, 203; in Aristotle's *Analytica posteriora* 205–206; rainbows and mathematics 161–162, 193; in Xenophanes, a cloud not a goddess 23

Rhodes 4–5

Runia, D. T. 83

Sandbach, F. H. 30, 95n20

Scholia to Aratus 79, 80, 165–167, 196

Scholia to Hesiod 27

Sea: greatest depth 129, 194, 214; provides fuel for the sun 93–94; its salt 128–129; *see also* tides

seasons of the year 25–29, 121, 144, 146, 181–182, 215

Seleucus 37, 129–133, 136–138, 213

Seneca 1, 214–215, 217; on comets 75–76, 78–79, 81; divine action and meteorology 182–183; on earthquakes 113–121, 199; on exhalations 97–98, 100; on hail 144–149; on haloes 165; logical arguments in meteorology 203–204; meteorological theories, uncertainty of 207; on Milky Way 85; on mock suns 167; *Naturales quaestiones*, range of topics covered 2; on portents and divination 76, 146–147, 174; Posidonius' meteorology, his evidence for 11, 69–71, 75–76, 80, 92, 118, 142, 144–145, 160–161, 198, 201–204, 214; pre-Socratics, his evidence for 21, 113–114, 116; on rainbows 161–163, 193, 199; on snow 143; on thunder and lightning 70–72; on tides 130–131, 215; on volcanoes 121; wind, cause of 97–98, 100–102, 109; wind-rose 103–106; winds, local 106

Simplicius: on the astronomy of Posidonius, studied by philosophers and mathematicians 6–7, 192, 201–202; on the element theory of Aristotle and Posidonius 92; on the greatest mountain height 48; on the Stoic view of chance 173, 178

snow 23–24, 28, 143–148

Socrates 23

Sophists 23

stade, length of 47

Stoics before Posidonius: astronomy 41, 83–84, 90–91, 93–94; causes, reluctance to discuss 11, 27–29, 192; on comets 79; on divination 7, 76, 173; on earthquakes 116; Epicurean meteorology, comparison with 188–189; on equatorial region 58; on exhalation 90–91, 93–94; induction, attitude to 205; knowledge, theory of 200–201; their meteorology 16, 26–30; on Milky Way 83–84; myths interpreted as allegories 26–27, 197; "physics" as part of philosophy 191; Posidonius' use of, in meteorology 71, 93, 143, 146, 196–197, 213; Providence and meteorology 26–27, 29, 30, 181; on rain and other precipitation 28, 142–143, 145; on rainbows 160; on seasons 27–28; on thunder and lightning 28–29, 68, 71, 185; on wind 97; *see also the names of individual Stoics*

Stony Plain, The 38, 123, 213

Strabo 5–7, 10–11, 27, 109, 201, 207; evidence Posidonius knew Aristotle's *Meteorologica* 36; on mountain heights 47; on Posidonius on the Cimbri 119, 125n46; on Posidonius and climatic zones 54–62; on Posidonius on earthquakes 117; on Posidonius and the Nile floods 152–156; on Posidonius and the Stony Plain 123; on Posidonius and tides 130–138; on Posidonius and winds 102–103, 106, 108; on the sea, and Posidonius on its greatest depth 129

Strato 7, 25, 68, 79, 91, 115–116

Struck, P.T. 177–178

Sun: affected by air and wind 44–45, 99; comets hidden by 80; controls climate 54–55, 58–61; distance and size 37, 43–45, 94, 202, 213; and exhalation, in pre-Stoic thought 90–91; fed by exhalation, in Stoic thought 92–94, 142; and haloes 164, 166; and mock suns 166–167; and rainbows 160–163; and seasons 27–29, 131–182; sunlight, refraction of 165–166; and tides 134, 136–137; and wind 97, 99–102, 109, 137

sun dogs *see* mock suns

Syriac Meteorology, The 16–17, 24–25; on earthquakes 115; on lunar haloes 165; on snow 143; as source about Theophrastus, not reliable 17, 36, 186; on thunder and lightning 68, 71–72

Thales 21, 114, 152

Theon of Alexandria 48

Theon of Smyrna 48

Theophrastus 1, 15; *De mundo*, possible source for 13; on earthquakes 115; on exhalation 91; final causes in meteorology, doubts about 25, 29; his meteorology 24, 36, 186; on Milky Way 83; Posidonius' view and use of 36, 107, 197; the Syriac Meteorology not a reliable source for 16–17; on thunder and lightning 68; on wind 101, 105–107

[Theophrastus], *De signis* 171

Thom, J. C. 14

Thrasyalces 22, 102, 153, 155–156, 196

thunder and lightning: allegorical interpretation of thunderbolt myth 26–27; cloud theory and theories of thunder incompatible 148; definition of thunder 206; development of philosophers' theories 71–72; Epicurean multiple explanations 184–187; exhalations and thunder 91, 93; final cause/purpose, if any 23, 25, 30, 182–183; logical arguments for theories of thunder and lightning 203; naturalistic theories before Posidonius 20–21, 28–29, 65–68; objections, ancient, to proposed theories 72; Posidonius' theory 35, 37, 69–71; true explanation undiscoverable without knowledge of electricity 72, 218

tides: cause, Greek theories of, before Posidonius 37–38, 129–131, 137–138; cause, Posidonius' theory of 11, 101, 133–136, 138, 197, 199, 213, 216–217; conjectures admitted in Posidonius' account 130, 207; first Greek knowledge of tides 13, 129; Homeric knowledge of tides claimed 138, 197; numerical data on tides recorded 132–133, 194, 214; observations and descriptions of tides by ancient authors 130–133, 136–137, 207, 213; Providence, tides as evidence of 181–182; tides, facts about, remembered in later antiquity 215

Timaeus (historian) 137

Timosthenes 102–105, 197, 213

Torricelli, Evangelista 217

'Troposphere' in relation to ancient ideas 49

Virgil 62

Volcanoes 116, 119–122, 213–214

weather phenomena, height of, in ancient belief 45, 48–49, 194, 213–214, 216

weather-signs 171, 193; in Aratus 31; in Aristotle 172–173; in Boëthus 30, 176; comets as 80, 172, 196; in Democritus 22; and divination 174–176, 179; in Epicurus 184, in Posidonius 171–173

wind 20, 24–25; Aristotle on cause of wind 198; cause of wind's motion 100–102; clouds as origin of wind 97–98; definition 96; difficulty of investigating true cause 216–218; earthquakes caused by wind 113–118, 120–122, 199, 203; exhalation and wind 89–91, 97–100; local winds 105–109; main winds, fixed origins of, at edge of *oikoumenē* 55, 100, 105; Nile floods and wind 152, 155–157; Posidonius' contribution to understanding wind 109; Providence and wind 182; Seneca's contribution to understanding wind 109; thunder and wind 65–71, 184–185, 203; tides and wind 130, 133, 137–138; wind-names 97, 102–106; wind-rose 97, 102–105, 108, 213; winds, particular, features of 62, 106–109, 213

Xenagoras 47

Xenophanes: on comets 77; on exhalation 89, 98, 142; his meteorology, character of 21–22; on rain 89, 142; on rainbows 23, 163; on wind 89, 98

Zeno of Citium 16, 26–27, 29, 68, 78–79, 191, 200

For Product Safety Concerns and Information please contact our EU
representative GPSR@taylorandfrancis.com
Taylor & Francis Verlag GmbH, Kaufingerstraße 24, 80331 München, Germany

* 9 7 8 1 0 3 2 5 3 0 3 0 7 *